Microbial Natural Products as a Source of Novel Antimicrobials

Microbial Natural Products as a Source of Novel Antimicrobials

Guest Editors

Patrícia Branco
Marc Maresca
Elisabete Muchagato Maurício
Luisa Brito

MDPI

Basel • Beijing • Wuhan • Barcelona • Belgrade • Novi Sad • Cluj • Manchester

Guest Editors
Patrícia Branco
Faculty of Engineering
Lusofona University
Lisbon
Portugal

Marc Maresca
ISM2
Aix-Marseille University
Marseille
France

Elisabete Muchagato Maurício
Faculty of Engineering
Lusofona University
Lisbon
Portugal

Luisa Brito
Instituto Superior de
Agronomia
Universidade de Lisboa
Lisbon
Portugal

Editorial Office
MDPI AG
Grosspeteranlage 5
4052 Basel, Switzerland

This is a reprint of the Special Issue, published open access by the journal *Antibiotics* (ISSN 2079-6382), freely accessible at: https://www.mdpi.com/journal/antibiotics/special_issues/Microsource_antibiotics.

For citation purposes, cite each article independently as indicated on the article page online and as indicated below:

Lastname, A.A.; Lastname, B.B. Article Title. *Journal Name* **Year**, *Volume Number*, Page Range.

ISBN 978-3-7258-8219-9 (Hbk)
ISBN 978-3-7258-8220-5 (PDF)
https://doi.org/10.3390/books978-3-7258-8220-5

Contents

About the Editors

Patricia Branco

Patricia Branco obtained a European Doctorate in Food Engineering from the Instituto Superior de Agronomia, University of Lisbon, in collaboration with the University of Copenhagen, Denmark. She holds a master's degree in molecular biology from the Egas Moniz Higher School of Health and a bachelor's degree in biotechnology engineering from the Faculty of Engineering, Lusófona. Patricia Branco has made significant contributions to the fields of molecular biology, food science, and foodborne pathogens and the development of eco-friendly biocides. Her research focuses on the antimicrobial properties of peptides produced by microorganisms, including their biochemical and molecular characterization, and their applications in food preservation and fermentation processes. She has published 19 articles in Scopus-indexed journals, presented her work at numerous conferences, organized four scientific events, and supervised 20 theses for biotechnology and biology students. She has also served on the thesis committee for seven biotechnology degree theses, edited four e-books, and received a scientific award for the best research paper from the International Union of Microbiological Societies. She had coordinated two projects, the ART-BIOCIDES at the HERCULES laboratory, Évora University, and the BIO-PHYTOFABs at the Linking Landscape, Environment, Agriculture, and Food (LEAF), School of Agriculture, University of Lisbon. Currently, Patricia Branco is an Assistant Professor at the Lusófona University, Lisbon, Portugal, the coordinator of the Bioengineering and Sustainability Research Group (BioRG), and the Co-director of the master's degree program in Food and Molecular Biotechnology, Lusófona University. She is also a researcher at LEAF, School of Agriculture, University of Lisbon. Patricia Branco's expertise lies at the intersection of biotechnology and microbiology, with a special focus on developing sustainable biopreservatives and biocontrol methods.

Marc Maresca

Marc Maresca is a researcher at the Institut des Sciences Moléculaires de Marseille (ISM2, UMR 7313) at Aix-Marseille University, France, where he has been working since 2011. With a strong interdisciplinary background in chemistry and biology and a PhD from Université Paul Cézanne (IMRN) in Aix-en-Provence, France, his work focuses on the development and application of antimicrobial molecules—particularly antimicrobial peptides—for use in human and animal health. In parallel, Dr. Maresca also conducts research in food safety, with a specific focus on mycotoxins and their impact on human health. Beyond academic research, he serves as a scientific expert and consultant for private pharmaceutical companies.

Elisabete Muchagato Maurício

Elisabete Muchagato Maurício has a PhD in Biomedicine: Biotechnology and Food Industries from the University of Alcalá de Henares and a master's degree in Food Technology and Quality from the Universidade Nova de Lisboa. She is currently an Assistant Professor and Researcher (career) at Universidade Lusófona. She carries out her research mainly in the area(s) of engineering sciences and technologies in the specialty of microbiology, food technology, valorization of food by-products, and detection of pathogens in food and cosmetic products. She is currently, and has been since 1995, the Technical Director at the dermocosmetics and nutraceuticals company Elisa Câmara, Lda, where she is responsible for the research and formulation laboratory for cosmetics and nutraceuticals. For more than 20 years, it has been developing dermocosmetic and nutraceutical formulas for the national and international market. Since 2020, she has been the coordinator of the Eco-Schools project, promoting sustainable practices and environmental education on campus. As a researcher at BioRG-Bioengineering & Sustainability Research Group, she is a coordinator of the Sustainability Research Area, driving projects in sustainability and environmental innovation. She has been, since 2023, a collaborator in the IPQ Technical Committee, ISO/TC 217-Cosmetics, and the European Committee for Standardization. In 2024, she developed and obtained approval for a master's program in Food and Molecular Biotechnology, of which she is the director, reinforcing her commitment to training and innovation. She has published 20 articles in specialized journals and participated in several international conferences with oral and poster presentations. She is a researcher with participation in several national and international projects and supervisor of several academic works, both for companies and master's theses, for over 20 years.

Luísa Brito

Luísa Brito is an Agro-Industrial Engineer from the Instituto Superior de Agronomia (ISA) of the Technical University of Lisbon (1983), with a master's degree in Biotechnology (1989) from the Instituto Superior Técnico of the same university and a PhD (1997) and habilitation (2012) from ISA. Her research activity is perfectly within the strategic lines of the LEAF Research Unit (Group 3—Food and Feed), of which she is a part. Her professional career has been developed in the fields of biological sciences (microbiology) and food engineering. Regarding her scientific performance, she has 56 articles referenced in WoS. She has participated in 20 funded research projects. To date, Luisa Brito has supervised or co-supervised more than 100 students of different degrees from national and international universities, including eight PhD students (four national and four international) and fifty master's students.

Preface

Microbial natural products have long been recognized as rich sources of antimicrobial compounds, offering potential solutions to combat the growing global threat of antimicrobial resistance. This Reprint, entitled "*Microbial Natural Products as a Source of Novel Antimicrobials*", compiles a curated collection of research and review articles originally published in the journal *Antibiotics*. The aim of this Reprint is to provide readers with an up-to-date overview of the discovery, characterization, and application of antimicrobial peptides, bacteriocins, and mycocins, with particular focus on their potential as safer and more sustainable alternatives to conventional antibiotics, biocides, and preservatives.

The motivation behind organizing this Reprint stems from the urgent need to identify effective and eco-friendly microbial agents to ensure food safety, improve agricultural productivity, and protect human and animal health. Although the scientific community has identified many promising microbial metabolites, few have reached commercial implementation. This Reprint seeks to bridge that gap by highlighting advances in the understanding of mechanisms of action, application contexts, and the future directions of research in this field.

The contributions included here are the result of a collaborative effort by experts working at the intersection of microbiology, biotechnology, and food engineering. The selected chapters cover a broad scope—from antimicrobial compounds derived from yeasts and lactic acid bacteria to their integration in food preservation systems agricultural practices, and pharmaceutical products.

This Reprint is intended for researchers, students, practitioners, and industry professionals in the fields of microbiology, food technology, biotechnology, and pharmacology who are interested in the development and application of microbial natural products for antimicrobial purposes. It serves as both a scientific resource and a reference point for future research and innovation in the field of microbial-based antimicrobial strategies.

We hope this Reprint will inspire continued research and interdisciplinary collaboration aimed at advancing the safe and sustainable use of microbial natural products in diverse sectors.

Patrícia Branco, Marc Maresca, Elisabete Muchagato Maurício, and Luisa Brito
Guest Editors

antibiotics

MDPI

Article

Evaluation of Biotechnological Active Peptides Secreted by *Saccharomyces cerevisiae* with Potential Skin Benefits

Elisabete Muchagato Maurício [1,2,3,*], Patrícia Branco [1,4,5,*], Ana Luiza Barros Araújo [1], Catarina Roma-Rodrigues [6,7], Katelene Lima [8], Maria Paula Duarte [9], Alexandra R. Fernandes [6,7,*] and Helena Albergaria [5]

1 BIORG—Bioengineering and Sustainability Research Group, Faculdade de Engenharia, Universidade Lusófona, Av. Campo Grande 376, 1749-024 Lisbon, Portugal; analuiza.araujo.pt@gmail.com
2 CBIOS—Research Center for Biosciences & Health Technologies, Universidade Lusófona, Campo Grande 376, 1749-024 Lisbon, Portugal
3 Elisa Câmara, Lda, Dermocosmética, Centro Empresarial de Talaíde, n°7 e 8, 2785-723 Lisbon, Portugal
4 Linking Landscape, Environment, Agriculture and Food (LEAF), Associated Laboratory TERRA, Instituto Superior de Agronomia, University of Lisbon, Tapada da Ajuda, 1349-017 Lisbon, Portugal
5 Unit of Bioenergy and Biorefinary, Laboratório Nacional de Energia e Geologia (LNEG), Estrada do Paço do Lumiar, 22, 1649-038 Lisbon, Portugal; helena.albergaria@lneg.pt
6 UCIBIO—Applied Molecular Biosciences Unit, Department Ciências da Vida, NOVA School of Science and Technology, 2829-516 Caparica, Portugal; catarina.rodrigues@ulusofona.pt
7 i4HB, Associate Laboratory—Institute for Health and Bioeconomy, Faculdade de Ciências e Tecnologia, Universidade NOVA de Lisboa, 2829-516 Caparica, Portugal
8 Research Institute for Medicines (iMed.ULisboa), Faculty of Pharmacy, Universidade de Lisboa, 1649-003 Lisbon, Portugal; k.lima@campus.ul.pt
9 The Mechanical Engineering and Resource Sustainability Center (MEtRICs), Chemistry Department, NOVA School of Science and Technology, Universidade NOVA de Lisboa, 2829-516 Caparica, Portugal; mpcd@fct.unl.pt
* Correspondence: elisabete.mauricio@ulusofona.pt (E.M.M.); patricia.branco@ulusofona.pt (P.B.); ma.fernandes@fct.unl.pt (A.R.F.)

Abstract: Biotechnological active peptides are gaining interest in the cosmetics industry due to their antimicrobial, anti-inflammatory, antioxidant, and anti-collagenase (ACE) effects, as well as wound healing properties, making them suitable for cosmetic formulations. The antimicrobial activity of peptides (2–10 kDa) secreted by *Saccharomyces cerevisiae* Ethanol-Red was evaluated against dermal pathogens using broth microdilution and challenge tests. ACE was assessed using a collagenase activity colorimetric assay, antioxidant activity via spectrophotometric monitoring of nitrotetrazolium blue chloride (NBT) reduction, and anti-inflammatory effects by quantifying TNF-α mRNA in lipopolysaccharides (LPS)-exposed dermal fibroblasts. Wound healing assays involved human fibroblasts, endothelial cells, and dermal keratinocytes. The peptides (2–10 kDa) exhibited antimicrobial activity against 10 dermal pathogens, with the Minimum Inhibitory Concentrations (MICs) ranging from 125 μg/mL for *Staphylococcus aureus* to 1000 μg/mL for *Candida albicans* and *Streptococcus pyogenes*. In the challenge test, peptides at their MICs reduced microbial counts significantly, fulfilling ISO 11930:2019 standards, except against *Aspergillus brasiliensis*. The peptides combined with Microcare® SB showed synergy, particularly against *C. albicans* and *A. brasilensis*. In vitro, the peptides inhibited collagenase activity by 41.8% and 94.5% at 250 and 1000 μg/mL, respectively, and demonstrated antioxidant capacity. Pre-incubation with peptides decreased TNF-α expression in fibroblasts, indicating anti-inflammatory effects. The peptides do not show to promote or inhibit the angiogenesis of endothelial cells, but are able to attenuate fibrosis, scar formation, and chronic inflammation during the final phases of the wound healing process. The peptides showed antimicrobial, antioxidant, ACE, and anti-inflammatory properties, highlighting their potential as multifunctional bioactive ingredients in skincare, warranting further optimization and exploration in cosmetic applications.

Keywords: *Saccharomyces cerevisiae*; antimicrobial peptides; biotechnological active peptides; biopreservatives; anti-collagenase; anti-inflammatory; anti-ageing

1. Introduction

The cosmetic market has witnessed a substantial surge in consumer preference for products derived from natural sources. As a result, the cosmetics industry is proactively seeking biological alternatives to enhance product qualities and replace synthetic chemical components [1,2].

Synthetic preservatives are widely used in the cosmetic industry to extend the shelf life of products and prevent microbial growth. However, their use has raised significant concerns due to potential adverse effects on human health and the environment. One major disadvantage is the risk of skin irritation and allergic reactions. For example, parabens, a common class of synthetic preservatives, have been linked to skin sensitivities and dermatitis (Darbre, 2006) [3]. Additionally, some synthetic preservatives, such as formaldehyde releasers, are known carcinogens and can cause long-term health issues upon prolonged exposure [4].

Many compounds sourced from bacteria, fungi, and algae, such as bio-surfactants, vitamins, antioxidants, pigments, enzymes, and biopeptides, present valuable properties in this regard [5–10].

Bioactive peptides have gained considerable attention within the cosmetics industry due to their potential to enhance skin health. Peptides exhibit a wide array of biological functions, encompassing antioxidant, anti-collagenase, anti-aging, anti-inflammatory, and antimicrobial properties [11–15], rendering them excellent candidates for inclusion in cosmetic formulations.

Natural bioactive peptides with promising cosmetic properties come from different sources, such as plants, bacteria, and yeasts [16]. Yeast peptides have garnered significant interest due to their nutraceutical and functional activities. These peptides, derived from yeast proteins, possess diverse bioactive properties. Several authors demonstrated that yeast peptides exhibit strong superoxide anion scavenging, α-glucosidase, and angiotensin-converting enzyme (ACE) inhibitory activities [17,18], indicating their potential in antioxidative, antidiabetic, and antihypertensive applications. Additionally, Bhattacharyya et al. (2003) [19] explored the functional characteristics of a yeast alcohol dehydrogenase peptide, revealing its chaperone-like activity in preventing protein aggregation, thus highlighting its potential therapeutic applications.

The search for effective and safe antimicrobial agents has led to significant interest in antimicrobial peptides (AMPs) due to their potent activity and low propensity for inducing resistance [20,21]. In cosmetic and dermatological applications, AMPs offer promising benefits for skin health [16].

AMPs, naturally occurring components of the innate immune system, play a crucial role in the defense against pathogens and in wound healing processes. Several AMPs have been identified in human skin cells, providing essential protective functions against microbial contamination [22].

Recent studies have highlighted the beneficial activities of AMPs derived from *Saccharomyces cerevisiae*, a species of yeast commonly used in fermentation processes. These AMPs were first identified in the supernatants of the wine strain *S. cerevisiae* CCMI 885, where they were found to be derived from the glycolytic enzyme glyceraldehyde 3-phosphate dehydrogenase (GAPDH) [13,23]. Afterward, these GAPDH-derived AMPs were found to be secreted by several wine *S. cerevisiae* strains [14]. In subsequent research, peptides ranging from 2–10 kDa were isolated from the supernatant of an industrial starter strain, Ethanol-Red [24]. Those peptides were separated through gel filtration chromatography, and four distinct peaks were detected and tested for their antimicrobial activity. One peak with an estimated molecular weight of 8 kDa, showed antimicrobial activity against various microbial contaminants in alcoholic fermentations, and it was determined by mass spectrometry analyses that the Ethanol-Red strain secreted the same GAPDH-derived AMPs [13,24].

Moreover, the accumulation of these GAPDH-derived AMPs on the surface of *S. cerevisiae* cells was linked to their antagonistic activity against *Lachancea thermotolerans* through a cell-cell contact death-mediated mechanism [25]. This discovery underscores the potential of GAPDH-derived AMPs in broader antimicrobial applications. Furthermore, a peptide derived from GAPDH has been implicated in tissue protection through diverse antifungal strategies and epithelial immunomodulation [26], indicating potential applications in skin care.

Microbiological contamination in topical formulations is one of the most concerning issues, especially for industry, as it can cause product degradation or, in the case of pathogenic microorganisms, can pose a risk to consumer safety. Thus, chemical preservatives are added to cosmetics and pharmaceutical products to protect them against microbial contamination that can arise due to contamination of raw materials, production water, inadequate hygiene standards during manufacture, or inappropriate use by the consumer [27]. So far, preservatives have been used in the concentrations indicated and studied to protect the product from contamination, protecting the consumers from adverse reactions. However, many questions have been raised about the safety of traditional preservation with synthetic preservatives. Approximately 6% of the population shows contact allergies or sensitization to various preservatives and ingredients in formulations, depending mainly on their concentration [28,29]. In fact, synthetic chemical preservatives are often used without specific tests on the effectiveness of the preservative system, essentially because these tests are expensive and very time-consuming [29]. In this situation, manufacturers choose to avoid the risk of contamination by introducing a greater amount of preservative, usually the maximum amount allowed in European Union regulation (EC) No 1223/2009), Annex V [30], with the consequent increase in the probability of cases of sensitivity or even skin allergies. The growing dissemination of scientific studies indicating some toxic effects associated with the safety of synthetic preservatives has encouraged their replacement by antimicrobials of natural and biotechnological origin, leading the pharmaceutical and cosmetic industries to search for alternative methods of preserving topical products [31–33]. In fact, diffusion and microdilution plate methods allow us to characterize the antimicrobial activity associated with the extract or compound under analysis but do not allow us to evaluate its preservative efficacy in relation to a real matrix, such as a cosmetic product.

The effectiveness of the preservative system in a real matrix is usually investigated by performing a challenge test. According to some authors, this method is quite rigorous and realistic and makes it possible to predict the interaction of antimicrobial agents with all the other factors and compounds present in a final formulation. In the challenge test, the product is artificially inoculated with the microorganisms most likely to contaminate cosmetics, such as bacteria, yeasts, and molds, where the ability of antimicrobial agents to remove the most contamination is assessed [34,35]. In this test, a cosmetic product should achieve specific log reductions in microbial counts at various time points and without any increase in microbial growth by day 28 (ISO 11930:2019; Alshehrei, 2024) [36,37]. This method is properly described in the European and American pharmacopeia and, more recently, in ISO 11930:2019 [36], and allows the producers to ensure the microbiological safety of the topical product during its useful life [38].

The exploration of *S. cerevisiae*-derived AMPs could pave the way for novel cosmetic formulations that harness their antimicrobial and healing properties, offering a natural and effective alternative to synthetic preservatives. As consumer demand for healthy and sustainable cosmetic products grows, the incorporation of AMPs could represent a significant advancement in the field.

Thus, taking all this together, in the present study we evaluated the antimicrobial activity of these *S. cerevisiae* AMPs against selected dermal pathogens (i.e., *Staphylococcus aureus*, *Streptococcus mutans*, *Streptococcus mitis*, *Steptococcus pyogens*, *Bacillus cereus*, *Entero-*

coccus faecalis, *Escherichia coli*, *Pseudomonas aeruginosa*, *Aspergillus brasiliensis*, and *Candida albicans*) and investigated their potential skin benefits such as anti-collagenase effect, anti-inflammatory effect, wound healing effect, and antioxidant effect (e.g., superoxide anion radical scavenging activity), as well as their cytotoxicity on dermal cells.

2. Results

2.1. Antimicrobial Activity of the S. cerevisiae Peptides (2–10 kDa) against Dermal Pathogens

2.1.1. Determination of Minimum Inhibitory Concentration (MIC)

In order to quantify and confirm the antimicrobial activity, the minimum inhibitory concentration (MIC) was determined for 10 microorganisms that have the potential to cause harm to the health of consumers. The results of the MIC tests for the peptides (2–10 kDa) against various dermal pathogens are presented in Table 1. The MIC values, expressed in μg/mL, were compared with the positive controls, vancomycin, norfloxacin, and ketoconazole, where appropriate, and ethanol (4% (*v*/*v*) (negative control). The analysis of the peptides by high-performance liquid chromatography (HPLC) demonstrated the presence of ethanol (8% (*v*/*v*)). Therefore, the maximum concentration of ethanol in the MIC assay is 4% (*v*/*v*), given that the peptides are combined with the growth medium in a 1:1 ratio, and serial dilutions of them were made in the microplate in accordance with the assay methodology.

Table 1. Minimum inhibitory concentration (MIC) of peptides (2–10 kDa) and controls against several dermal pathogens.

Strains Tested	Minimum Inhibitory Concentration (MIC) (μg/mL)				
	Peptides (2–10 kDa)	Ethanol *	Vancomycin **	Norfloxac **	Ketaconazol **
Staphylococcus aureus	125	n.d	1.56	-	-
Methicillin-resistant *S. aureus*	500	n.d	1.56	-	-
Staphylococcus epidermidis	250	n.d	1.56	-	-
Enterococcus faecalis	500	n.d	3.12	-	-
Streptococcus pyogenes	1000	n.d	3.12	-	-
Streptococcus mitis	500	n.d	3.12	-	-
Bacillus cereus	250	n.d	0.78	-	-
Candida albicans	1000	n.d	-	-	0.039
Escherichia coli	250	n.d	-	0.48	-
Pseudomonas aeruginosa	250	n.d	-	1.90	-
Aspergillus brasiliensis	>1000	n.d	-	-	-

* Negative control; ** Positive control; - not tested; n.d not detected.

The findings (Table 1) showed that all the microorganisms tested grew normally without the peptides and with ethanol as the negative control. Additionally, the microorganisms were also tested with appropriate antibiotics (positive controls) to verify the sensitivity of each strain. The antibiotics used as positive controls had MIC values that met the standards for the different microorganisms as defined in the Performance Standards for Antimicrobial Susceptibility Testing (2021) and Performance Standards for Antifungal Susceptibility Testing of Filamentous Fungi (2020) [39,40].

The peptides (2–10 kDa) demonstrated a variable degree of antimicrobial activity against several dermal pathogens (Table 1). The MIC values highlight the efficacy of the peptides, particularly against *S. aureus* with a MIC of 125 μg/mL. The MIC for *B. cereus*, *E. coli*, *P. aeruginosa*, and *S. epidermidis* was found to be 250 μg/mL, showcasing its broad-spectrum potential. The MIC for Methicillin-resistant *S. aureus* (MRSA), *E. faecalis* and *S. mitis* was 500 μg/mL. However, the peptides were less effective against *S. pyogenes* and *C. albicans*, with an MIC of 1000 μg/mL for both.

2.1.2. Peptides (2–10 kDa) Potential Preservative Effect

Physicochemical and Microbiological Analysis

After incorporation of the preservative systems, the formulations were divided into sterile 20 g containers and initial quality control was performed for organoleptic characterization, namely color, general appearance, and odor, as well as physical stability, pH, and viscosity measurements.

The results showed (Table S1) that after the incorporation of all the preservative systems (t = 0), the formulations remained stable (no physical separation was observed after centrifugation, with an initial pH of 5.2 and a viscosity of 18,000 cps, and there were no changes in color, odor, texture or pH and viscosity, after 28 days of study. Microbiological analysis of the base formulation also showed that there was no contamination, and it was eligible for further challenge testing.

Evaluation of the Antimicrobial Protection of a Cosmetic Product-Preservation Efficacy Test-Challenge Test (ISO 11930:2019)

The challenge test, according to ISO 11930:2019 [36] standards, evaluates the antimicrobial preservation efficacy of cosmetic formulations. This test is critical in ensuring that cosmetic products remain free from microbial contamination, thus safeguarding consumer health. The test involved inoculating body milk formulations with a calibrated inocula of key microorganisms: *E. coli*, *P. aeruginosa*, *S. aureus*, *C. albicans*, and *A. brasiliensis*. These microorganisms were selected due to their prevalence and potential to contaminate cosmetic products, according to European Union regulation (EC) No 1223/2009) [30]. The test monitored microbial counts over 28 days at specified intervals (0, 1, 7, 14, 21, and 28 days) [33].

To test the efficacy of peptides as a cosmetic preservative the formulations were prepared under different conditions, where the peptide (2–10 kDa) were tested alone at MIC concentration (Table 1) or in association with synthetic preservatives at lower concentrations (0.3% (v/v) Microcare® SB+2–10 kDa). This reduction in the concentration of the classic preservative was carried out for all the microorganisms to check for possible synergistic relationships with the peptide being tested. The positive control (0.6% (v/v) Microcare® BNA + 1% (v/v) Microcare SB®), was used in the proportions recommended by the suppliers and which are described as classic synthetic preservatives, normally used for effective cosmetic preservation, and the negative control (formulation without any preservative). In addition, two further controls were performed and tested in the formulations: the solvent present in the 2–10 kDa peptide (ethanol control) and a synthetic preservative control at a lower concentration (0.3% (v/v) Microcare® SB) to validate the results.

In order to analyze the efficacy of the peptides as a cosmetic preservative, a total of 35 formulations were prepared (7 combinations of preservatives × 5 different microorganisms) and the death curves were used to obtain the results shown in Figure 1.

Results showed that for all the assayed microorganisms, the formulation alone without preservatives is not self-preserving and, therefore, the addition of preservatives is required to meet criteria A or B of ISO 11930:2019 [36].

It was observed that the majority of microbial growth in the two negative controls (ethanol control and formulation without any preservative) showed a slight growth reduction in all formulations inoculated with the standard microorganisms. This indicates that the microbial population did not experience substantial mortality due to either the composition of the formulation or the ethanol at the concentration used (4% v/v). It is normal to observe a decrease in microorganisms during the 28 days of the study, as the base formulation used can cause the microorganisms to die due to the base ingredients; this has also been observed by other authors in similar studies [41–44].

As far as the positive control is concerned, the classical combination of the two synthetic preservatives proved to be effective in the reduction in all the microorganisms tested.

The results showed that peptides (2–10 kDa), when used alone at its MIC, demonstrated significant microbial reduction (Figure 1). However, its efficacy varies among different microorganisms and must be analyzed in order to comply with conservation criteria A and B as defined in the ISO 11930:2019 [36] standards (as shown in the methodology).

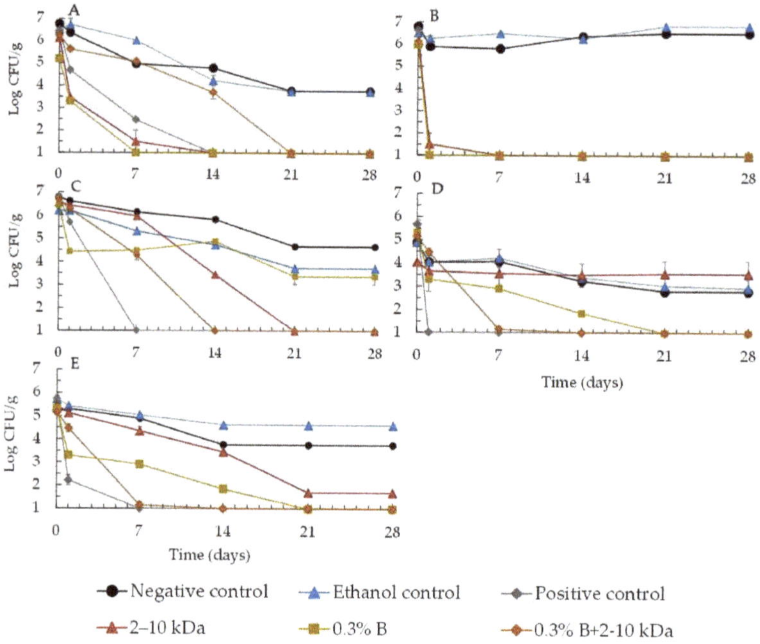

Figure 1. Growth profiles of bacteria, i.e., *E. coli* (**A**), *P. aeruginosa* (**B**), *S. aureus* (**C**) and fungi, i.e., *A. brasiliensis* (**D**) and *C. albicans* (**E**), in body milk formulation in the presence of the peptides (2–10 kDa) at the MIC value and 0.3% (*v*/*v*) of Microcare® BNA (0.3% B), and the association of both (0.3% B+2–10 kDa). A positive control with 0.6% (*v*/*v*) of Microcare® BNA and 1% (*v*/*v*) of Microcare® SB, an ethanol control, and the formulation without any preservative (negative control) were also performed.

In Figure 1A, the peptide (2–10 kDa) was shown to be effective against *E. coli* growth, with a greater reduction in growth at 7 days and 14 days, proven to be more effective than synthetic preservatives (positive control), thus fulfilling criteria A and B of the standard. In this case, the combined addition of a synthetic preservative with peptides (0.3% (*v*/*v*) Microcare® BNA+2–10 kDa) was neither effective nor synergistic, showing an effective reduction only after 14 days.

On the other hand, the peptide (2–10 kDa) proved to be effective in reducing the microbial growth of *P. aeruginosa* (Figure 1B), fulfilling both criteria A and B. In this case, the peptides behaved similarly to the positive control, effectively reducing the microorganism in question. When the peptide was added together with the preservative at a lower concentration (0.3% (*v*/*v*) Microcare® BNA+2–10 kDa), no synergy or advantage was observed in this procedure, with the best results shown for the peptides alone. This result is extremely important, as this bacteria is often associated with the degradation of preservatives and their resistance in cosmetics, which has already been described by several authors. In fact, this microorganism is considered to be one of the most difficult to eliminate from cosmetics, and its resistance to numerous preservation systems is well documented. This high resistance may result from the possibility of this microorganism being able to breakdown some preservatives and surfactants, using them as carbon sources for its development [34,41–43].

In Figure 1C, for *S. aureus* the peptides (2–10 kDa) showed a 2-log reduction after 14 days, meeting criterion B. Also, in association with synthetic preservatives at lower concentrations (0.3% (v/v) Microcare® BNA+2–10 kDa), no synergy or advantage was observed in this procedure, with the best results showing a reduction in the growth of 2 log after 7 days and more than 3 log after 14 days, meeting criterion B of the standard with no increase in growth during the test. So, it can be seen that in this case, the additional addition of 0.3% (v/v) Microcare® BNA to peptides (2–10 kDa) did not show any advantage or synergy and continued to meet criterion B of the standard. As observed in Figure 1D, the peptides (2–10 kDa) did not show an effective reduction in growth of *A. brasilensis*, with similar values to those observed for the two controls (negative control and ethanol control), thus failing to fulfill criteria A and B of the standard. On the other hand, the peptides, when combined with the synthetic preservative at a lower concentration (0.3% (v/v) Microcare® BNA+2–10 kDa), showed synergy, thus intensifying the action of the preservative alone and making it possible to observe compliance with criteria A and B of the standard.

As observed in Figure 1E the peptides (2–10 kDa) showed an effective reduction in *C. albicans* growth of more than 1 log after 7 days and 3 log after 14 days, meeting criteria A and B of the standard. In addition, after 7 days of testing, a synergy was observed when the peptide was mixed with a synthetic preservative (0.3% (v/v) Microcare® BNA+2–10 kDa) reinforcing the effect of the isolated peptide.

It is therefore concluded that the peptides are effective against the tested bacteria and *C. albicans*, showing that they meet the safety criteria required by legislation A or B. It can be seen that the peptides are able to act alone, meeting the A criteria, against, *P. aeruginosa*, *E. coli*, and *C. albicans* and criteria B against *S. aureus* but not against mold *A. brasilensis*. However, when the peptides (2–10 kDa) are used in association with a preservative at a lower concentration (0.3% (v/v) Microcare® BNA+2–10-kDa), synergy can be observed in both *C. albicans* and *A. brasiliensis*. Therefore, effective preservation of the formulation can be achieved by adding the peptides (2–10 kDa) at MIC concentration to a synthetic preservative in order to comply with the criteria of ISO 11930:2019 [36]. It was also observed that the introduction of the different preservation systems with the peptides 2–10 kDa did not affect the physicochemical and physical stability of the emulsions produced and that at the end of the test, 28 days later, there were no changes in the initial conditions established; they were stable and showed no changes in color, odor, pH, or viscosity.

2.2. *Potential Skin Benefits of the Peptides (2–10 kDa)*

In Vitro Collagenase Inhibition Assay

The peptides (2–10 kDa) were evaluated for their ability to inhibit collagenase. The peptides when tested at a concentration of 250 μg/mL, a reduction of 41.8% in collagenase activity, was observed. Furthermore, the peptides (2–10 kDa) exhibited significant anti-collagenase effects at higher concentrations, with inhibition rates of 81.9% at 500 μg/mL and 94.5% at 1000 μg/mL (Figure 2).

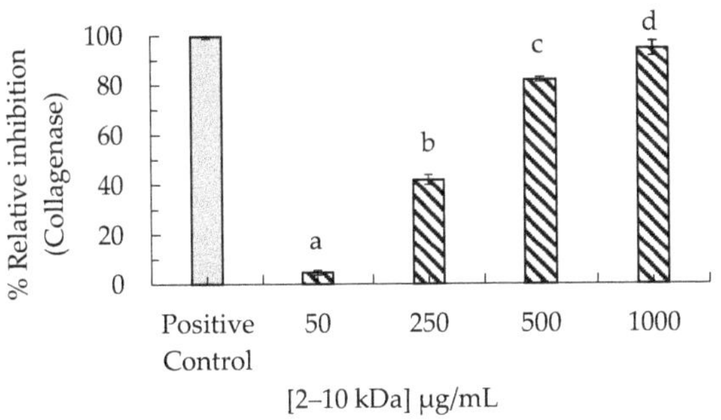

Figure 2. Inhibition of in vitro collagenase activity by 1,10-phenanthroline (positive control) and by peptides (2–10 kDa) at final concentrations of 50, 250, 500, and 1000 μg/mL. Data are presented as means ± SD (error bars) from three independent measurements. Different letters (a–d) indicate significant differences ($p < 0.05$) between the peptide concentrations tested (50–1000 μg/mL).

2.3. The Peptides (2–10 kDa) Cytotoxicity in Dermal Cells

The cytotoxicity of the peptides (2–10 kDa) in cells from the skin was evaluated on a melanoma cell line, MNT-1, and in normal dermal fibroblasts, keratinocytes, and melanocytes. Cells were exposed for 48h with crescent concentrations of the peptides and their viability was examined using the 3-(4,5-dimethylthiazol-2-yl)-5-(3-carboxymethoxyphe nyl)-2-(4-sulfophenyl)-2H tetrazolium, inner salt (MTS) colorimetric assay.

Results showed that the peptides (2–10 kDa) present no significant alteration in cell viability until at least 500 µg/mL (Figure 3).

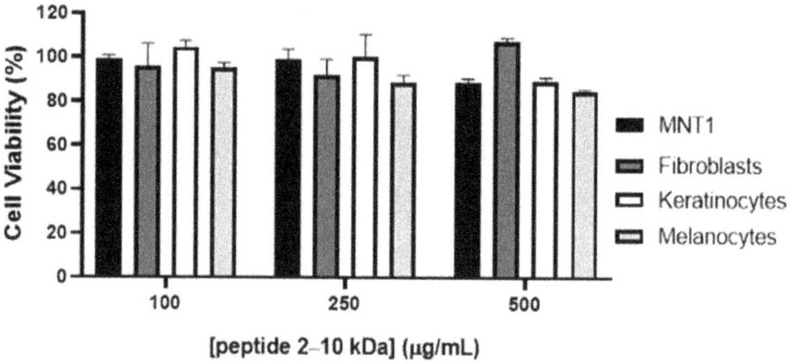

Figure 3. Cytotoxicity of the peptides (2–10 kDa) in melanoma cell line MNT1 and normal dermal fibroblasts, keratinocytes, and melanocytes. Cells were exposed to increasing concentrations of the peptide for 48 h and viability was evaluated with the MTS assay. Bars represent the average ± standard deviation.

2.4. Analysis of Anti-Inflammatory Effect of Peptide (2–10 kDa)

To evaluate the effect of the peptides (2–10 kDa) on the dermal inflammatory process, normal dermal fibroblasts were first incubated with 250 µg/mL of the peptides, with 1% (*v*/*v*) ethanol (vehicle control) or with the same volume of medium (untreated control). After 2 h, they were also exposed to lipopolysaccharides (LPS) and incubated for another 2 h. The level of inflammation was measured by quantification of tumor necrosis factor-alpha (TNF-α) mRNA via RT-qPCR, through comparison with respective untreated control cells. Results showed that incubation with LPS after 2 h exposure to vehicle control resulted in an increased expression of TNF-α (Figure 4). On the other hand, pre-incubation with the peptides resulted in a protective effect against inflammation, with a decrease of TNF-α of 0.15 ± 0.02 in cells incubated first with the peptides and then with LPS to untreated cells incubated with LPS (Figure 4, orange bars), while it was only observed a decrease of 0.39 ± 0.27 in peptides (2–10 kDa) treated cells compared to untreated cells (Figure 4, blue bars).

2.5. Analysis of the Effect of Peptides (2–10 kDa) on Angiogenesis

To further understand the effect of the peptides (2–10 kDa) in the proliferative phase of the wound healing process, it was examined the in vitro development of capillary-like tubes by endothelial cells [45]. After HUVEC seeding on top of a matrigel, images were acquired every 15 min, originating a video that allows to evaluate the effect of 250 µg/mL of the peptides (Movie S1 (Supplementary Materials) and representative images in Figure 5) in comparison to the effect of 1% (*v*/*v*) ethanol (vector control, Movie S2 (Supplementary Materials) and representative images in Figure 5). Movie S1 (Supplementary Materials) shows that the formation of tubes in the presence of the peptides starts at 1 h 30 min after seeding and increases until maximum after 6 h (representative images in Figure 5). A similar behavior was observed when cells were exposed to the vector control (Movie S2 (Supplementary Materials), representative images in Figure 5).

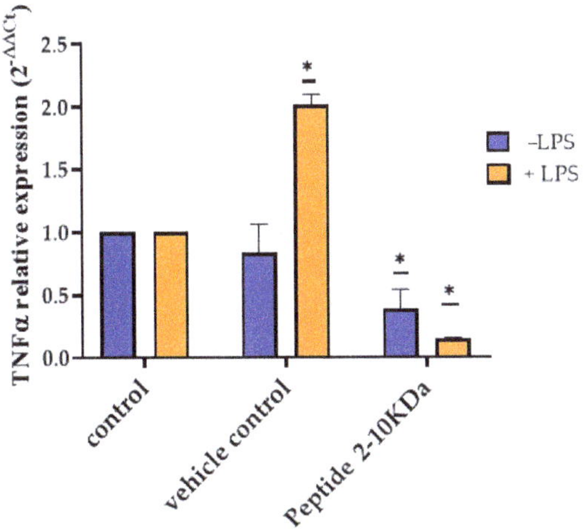

Figure 4. Expression levels of TNF-α in normal dermal fibroblasts after incubation with peptides (2–10 kDa). TNF-α expression after 2 h incubation of fibroblasts with 250 μg/mL of the peptides (2–10 kDa), 1% (*v*/*v*) ethanol (vehicle) or medium (control), followed by further 2 h incubation with (+LPS, orange bars) or without (−LPS, blue bars). Bars represent the average and standard deviation of at least three experiments. * p-value < 0.05.

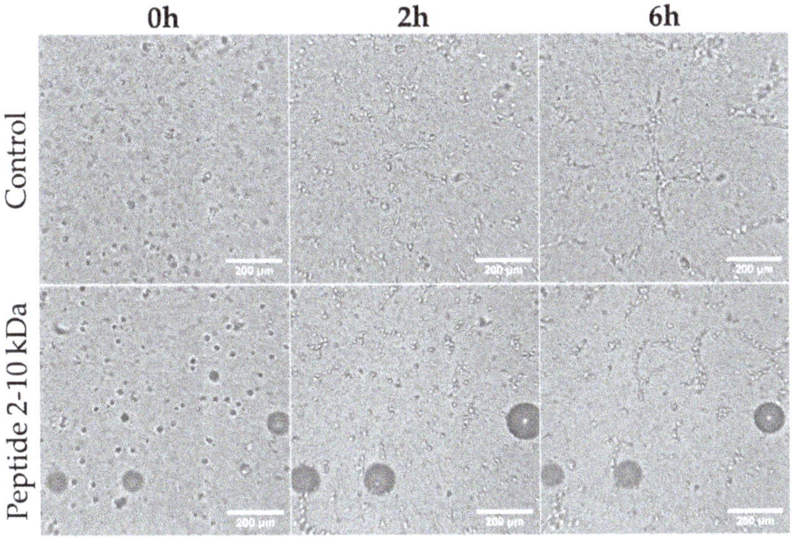

Figure 5. Capillary-like tube formation by HUVEC. Representative images of cells after 0 h, 2 h and 6 h of cells seeding on top of Matrigel in F12-K medium supplemented with 250 μg/mL peptides (2–10 kDa), or 1% (*v*/*v*) ethanol (Control). Scale bar corresponds to 200 μm.

2.6. Scratch Test

To further understand the peptides (2–10 kDa) effect on the cell migration ability that occurs during the proliferative phase of the wound healing process, a scratch assay was performed in fibroblasts, keratinocytes, and endothelial cells (HUVEC) [46,47]. Results showed that after 24 h, the percentage of regeneration of the scratch in fibroblasts was 93.7 ± 3.0% in the control sample, while it registered a remission of only 17.7 ± 2.7% when cells were incubated with 250 μg/mL of the peptides (Figure 6A,D), suggesting a delay in the migration of fibroblasts in the presence of the peptides (2–10 kDa). On the other hand, there were no significant differences observed in the regeneration of the scratch in HUVEC (Figure 6B,E) and keratinocytes (Figure 6C,F) in the presence of the peptides (2–10 kDa) or in the presence of the vector control.

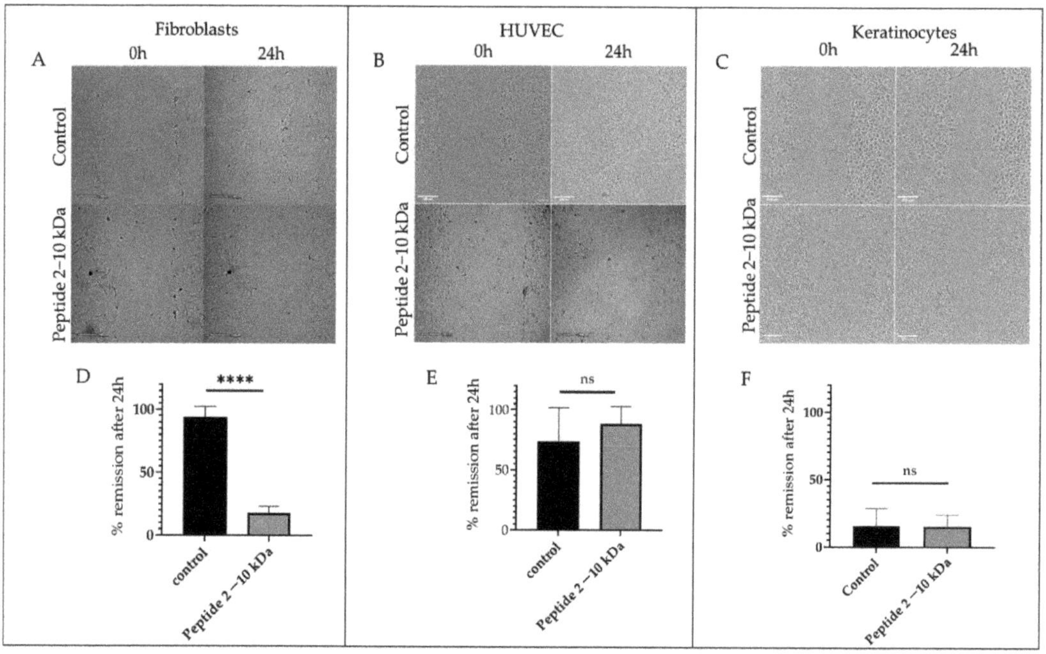

Figure 6. Wound healing assay. Representative images of the wound healing assay at 0 h and 24 h after incubation with 1% (*v*/*v*) ethanol (control) or 250 μg/mL of peptides (2–10 kDa) in (**A**) normal dermal fibroblasts, (**B**) HUVEC, and (**C**) keratinocytes. Scale bars correspond to 200 μm. Percentage of wound scratch closure (% remission) after 24 h incubation with 1% (*v*/*v*) ethanol (control) or 250 μg/mL of peptides (2–10 kDa) in (**D**) normal dermal fibroblasts, (**E**) HUVEC and (**F**) keratinocytes. Bars represent the mean ± SEM of at least two independent experiments. **** p value < 0.0001, ns—statistically not significant.

2.7. Superoxide Anion Radical Scavenging Capacity of the Peptides (2–10 kDa)

The superoxide scavenging capacity was assayed by the inhibition of the nitrotetrazolium blue chloride (NBT) reduction. The antioxidant capacity of the peptides was indicated by a dose-dependent decrease in absorbance at 560 nm, which results from the inhibition of the NBT reduction by the PMS/NADH-generated superoxide radical (Figure 7).

The peptides (2–10 kDa) under study presented a superoxide radical scavenging capacity (IC50 value of 30.6 ± 1.4 μg/mL) close to that of the positive control, gallic acid, (IC50 value of 20.7 ± 3.7 μg/mL) (Table 2).

Table 2. Superoxide anion radical scavenging capacity. Values are expressed as mean standard deviation.

Samples	IC_{50} (μg/mL)
Peptides (2–10 kDa)	30.6 ± 1.4
Gallic acid (positive control)	20.7 ± 3.7

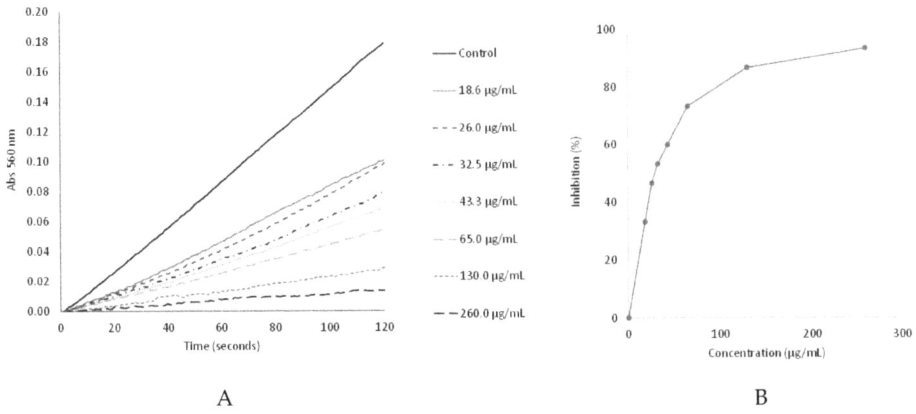

Figure 7. Superoxide scavenging capacity: (**A**) Plot of absorbance (Abs 560 nm) as a function of time for the different peptides concentrations; (**B**) Effect of the peptides on the inhibition of the NBT reduction by the PMS/NADH generated superoxide radical.

3. Discussion

Both traditional medicine and contemporary scientific research recognize the utility of bioactive peptides in the development of skincare products [48]. This investigation represents the first assessment of peptides secreted by the industrial *S. cerevisiae* strain Ethanol-Red for various bioactivities relevant to skincare, including antimicrobial activity against dermal pathogens, efficacy as a cosmetic preservative (both alone and in combination with synthetic preservatives), collagenase inhibition, antioxidant activity, anti-inflammatory effects, and wound healing properties.

The tested peptides exhibited variable antimicrobial activity against several dermal pathogens (Table 1). In fact, according to the same authors, the antimicrobial activity of natural substances can be defined as strong inhibitors (MIC up to 500 µg/mL), moderate inhibitors (MIC between 600 and 1500 µg/mL), and weak inhibitors (MIC above 1600 µg/mL) [49–51]. Using this classification, the peptides demonstrated significant efficacy against both Gram-positive and Gram-negative bacteria, with strong inhibition against *S. aureus* (MIC 125 µg/mL); *S. epidermidis, E. coli,* and *P. aeruginosa* (MIC 250 µg/mL); Methicillin-resistant *S. aureus* (MRSA), *E. faecalis, S. mitis* (MIC 500 µg/mL), while the peptides showed moderate activity against *S. pyogenes* and *C. albicans*, with a MIC of 1000 µg/mL.

These findings underscore the potential of the peptides secreted by *S. cerevisiae* as natural antimicrobial agents and a potential active ingredient to be incorporated in topical products.

The differential efficacy observed across various pathogens suggests that while the peptide exhibits broad-spectrum activity, its application may be more suitable to inhibit specific microorganisms. Overall, this study highlights the promising role of *S. cerevisiae*-derived peptides in developing effective and natural skincare solutions.

The challenge test results (Figure 1) underscore the importance of selecting appropriate preservatives for cosmetic formulations. The results obtained showed that the peptide can also be used in the preservation of cosmetic formulations. It can be seen that the peptide is able to act alone against *P. aeruginosa, E. coli, C. albicans,* and *S. aureus* complying with the criteria A and B of the f ISO 11930:2019 [36] but not against mold *A. brasiliensis*. However, when the peptides (2–10 kDa) were added in association with a preservative at a lower concentration (0.3% (*v*/*v*) Microcare® BNA+2–10 kDa), synergy can be observed in both *C. albicans* and *A. brasiliensis* (Figure 1). The co-preservation system (0.3% (*v*/*v*) Microcare® BNA+2–10 kDa) appears to be the most suitable for preserving the emulsion studied, allowing the concentration of the conventional preservative to be reduced. This result may be mainly due to the possibility of having established some synergistic relationships between the peptides and the Microcare® BNA preservative, which proved to be efficient in accordance with the efficacy criteria A and B. Similar results were also described by Kunicka-Styczyńska, et al. (2011) [52] when they compared the antimicrobial efficacy of a conventional synthetic preservation system with a plant-based preservation system, these authors also found a synergistic effect between the two systems. In fact, the use of natural substances as potential preservatives in cosmetics has been mentioned by some authors [33,51]. However, they have found that it is often only when combined with other preservatives or substances with antimicrobial properties that they can ensure the strongest and longest-lasting preservation of cosmetics [32,53]. Thus, based on the results of this study, the possibility of using the peptides in combination with a reduced preservative system for the correct preservation of the emulsions makes it possible to reduce the dose of conventional preservatives recommended by the suppliers. This also makes it possible to reduce the toxic effects often associated with the use of these chemical synthetic preservatives.

It should also be emphasized that any change to the formulation will necessarily require a new challenge test to be performed due to the complexity and diversity of topical formulations and the possibility of interaction between the different ingredients of the formulation.

The *S. cerevisiae* peptides (2–10 kDa), particularly when combined with Microcare® BNA, demonstrate promising preservation efficacy, potentially allowing for reduced concentrations of synthetic preservatives. These findings highlight a pathway towards safer, more effective cosmetic products with optimized preservative systems, which align with the growing consumer demand for natural and less harmful ingredients in skincare formulations.

Collagen is the most abundant protein in the extracellular matrix (ECM) of skin and plays a crucial role in maintaining tissue structure and function [54,55]. However, excessive collagen degradation by collagenases can lead to tissue damage and contribute to the progression of several diseases, including arthritis, chronic wounds, and skin aging [55–57]. Therefore, the search for natural inhibitors of collagenases has gained significant attention. Natural agents with anti-aging properties, such as plant extracts and natural peptides from fruits, spirulina, and yeasts, have been reported [58–62]. Peptides and proteins exhibit notable anti-collagenase activities, which are of significant interest for various therapeutic and cosmetic applications. For instance, Ennaas et al. (2016) [63] identified collagencin, an antimicrobial peptide from fish collagen, which has potential antibacterial properties and interactions with membrane structures, indicating its relevance in skin health and food safety. Also, Ryu et al. (2010) [64] isolated a novel peptide (SPH-1) from a seahorse, that inhibits of the NF-κB/p38 signaling pathway, blocking the phosphorylation and activation of NF-κB and p38 kinase, which are critical signaling molecules in the expression of collagenases 1 and 3 [64]. By inhibiting this pathway, this peptide reduces the production of these collagenases, thus preventing collagen degradation.

The present study found that the peptides secreted by *S. cerevisiae* were able to inhibit collagenase activity in a dose-dependent manner, with inhibition percentages of 41.8% and 81.9% for concentrations of 250 and 500 μg/mL, respectively (Figure 2), highlighting their potential as a natural inhibitor of collagenase and their applicability in anti-aging skincare products.

Wound healing is a complex process triggered after tissue injury, involving a complex network of different types of cells, and molecular mediators including cytokines and the endothelial system [47,65]. The process is divided into four major steps: (i) hemostasis, which occur immediately after tissue damage, and consists of blood vessel constriction, formation of a platelet plug, and fibrin production to seal the wound and prevent excessive bleeding; (ii) inflammatory phase, occurs after 2–3 days, and consist in the attraction of neutrophiles and macrophages to remove debris and bacteria; (iii) the proliferative phase, which happens 2 to 10 days after the tissue injury, and is associated with fibroblasts migration with consequent production of collagen and other extracellular matrix components, and release of cytokines that triggers angiogenesis and epithelialization with migration and proliferation of epithelial cells to cover the wound surface; and lastly, the remodeling phase, which begins 2 to 3 weeks after the tissue damage, and is associated with the rearrangement and realignment of the collagen fibbers in the wound to enhance tissue strength and scar formation [47,65]. The four stages are tightly regulated through growth factors, cytokines, and other signaling molecules, resulting in a dynamic process [47,66]. Depending on the physiological condition of the organism, wound healing may result in wounds that heal with minimal scars, fibrosis, or in chronic wounds [47,67]. Fibrosis is a process that may arise during the remodeling phase and result in hypertrophic scars and keloids, which have detrimental effects since it impairs tissue function and has an aesthetic impact on the patient [67]. It is characterized by aberrant extracellular matrix deposition, fibroblast accumulation, and chronic inflammation [67]. Chronic wounds are generally characterized by a nonmigratory and hyperproliferative epidermis, the presence of infection with biofilm formation and unresolved inflammation, exhibiting elevated levels of pro-inflammatory cytokines, like TNF-a [66]. Taken together, results showed that fibroblast incubation with the peptides under study resulted in a decreased cell migration (Figure 6) and a protective effect against inflammation induction (Figure 4). Moreover, incubation of HUVEC with 250 µg/mL of the peptides (2–10 kDa) showed no significant alterations in the endothelial cells' movement (Figure 6) or angiogenic ability (Figure 5). The keratinocyte's movement was also not affected by the presence of the peptides (Figure 6F). Although the reduced fibroblast migration in the presence of the peptides makes it less suitable for its application during the initial proliferative phase of the wound healing process, its anti-inflammatory activity, together with antibacterial and antioxidant activities, and no cytotoxicity at concentrations up to 500 µg/mL (Figure 3), makes the peptides suitable for application during the late proliferative phase or in the beginning of the remodeling phase of the wound healing process. In fact, by attenuating fibroblast migration, the application of the peptides may reduce fibrosis, and consequently minimize the aesthetic impact of the scar and loss of tissue characteristics and counteract the mechanisms that are impaired in chronic wounds, including chronic inflammation or bacterial growth [65,66].

The superoxide anion radical can arise when a single electron is supplied to molecular oxygen. The most important source of superoxide radical in vivo is probably the mitochondrial electron-transport chain [68]. However, this radical can arise also from other enzymatic and non-enzymatic reactions that use molecular oxygen as an electron acceptor [69]. The oxidizing power of superoxide itself is moderate, but other radicals and reactive intermediates, like hydrogen peroxide, hydroxyl radical, or peroxynitrite radicals, are formed in reaction sequences starting with superoxide anion radical [68,70]. Thus, the production of superoxide can give rise to more powerful and reactive species and eventually trigger the state of oxidative stress, with the occurrence of oxidative damage to important cellular constituents such as DNA, proteins, and lipids. Oxidative stress plays an important role in the development of several diseases and in the aging process, including the skin aging process [70,71].

The peptides secreted by *S. cerevisiae* under study presented a superoxide radical scavenging capacity close to that of the positive control (gallic acid) (Table 2) and higher than those reported to other peptides, namely to peptides derived from chicken skin gelatin hydrolysate [72] or to peptides from collagen hydrolysate of redlip croaker (*Pseudosciaena polyactis*) scales [73]. Factors influencing peptides' antioxidant capacity include their amino acid composition, sequence, and molecular weight [74]. Hydroxyl groups present in aromatic amino acids, like tryptophan, tyrosine, and phenylalanine, play a critical role in free radical scavenging [72]. Additionally, histidine, proline, valine, leucine, methionine, glycine, and alanine have been reported to contribute to antioxidant capacity [71,72,74], The superoxide radical scavenging capacity of the peptides suggests its potential to reduce oxidative skin damage, thereby potentially slowing the skin aging process.

These findings highlight the potential of the peptides secreted by *S. cerevisiae* under study as natural antimicrobial agents in skincare products. These peptides have been demonstrated to be effective against both Gram-positive and Gram-negative bacteria, and their synergistic effects when combined with Microcare® BNA suggest a viable pathway for reducing the concentration of conventional synthetic preservatives in cosmetic formulations. This approach not only fulfills consumer expectations regarding the use of natural and safer ingredients, as well as prevents skin allergic reactions and avoids synthetic preservatives that might provoke health problems [3,4]. Further research and optimization are required in order to fully harness the capabilities of these peptides, particularly their applicability across diverse cosmetic formulations. These developments indicate a significant advancement in the field of skincare products, facilitating the creation of preservative systems that are both effective and compliant with safety standards. This represents an important step towards integrating innovative preservative solutions into the cosmetics industry.

4. Materials and Methods

4.1. Production of the Peptides (2–10 kDa)

The peptides (2–10 kDa) of *S. cerevisiae* fermentation supernatants were obtained as described in Branco et al. (2014) [13]. Briefly, an alcoholic fermentation was performed in 2 L shake-flasks containing 1 L of a synthetic medium. The synthetic medium consisted of 110 g/L glucose, 110 g/L fructose, an acid solution containing 6 g/L tartaric acid, 3 g/L malic acid, and 0.5 g/L citric acid; an amino acid (aa) solution composed of 1.7 g/L yeast nitrogen base without aa, 2 g/L casamino acids, 0.2 g/L calcium chloride, 0.8 g/L arginine, 1 g/L proline, 0.1 g/L tryptophan, and 2.5 g/L yeast extract, with a pH of 4.5. All solutions were autoclaved, except for the amino acid solution, which was sterilized by filtration (0.22 µm). The medium was inoculated with *S. cerevisiae* Ethanol-Red strain at an initial cell density of 1.0×10^5 cells/mL and then incubated at 25 °C, without agitation, for 7 days. After 7 days, cells were removed by centrifugation and the supernatant was then sterilized by filtration through 0.45 µm Millipore membrane and then by 0.22 µm Millipore membrane (Burlington, MA, USA). Afterward, this cell-free supernatant was first ultrafiltrated through centrifugal filter units (Vivaspin 15R, Sartorius, Göttingen, Germany) equipped with 10 kDa membranes and then the permeate (<10 kDa) was concentrated (40-fold) in similar centrifugal units equipped with 2 kDa membranes.

4.2. High-Performance Liquid Chromatography Analysis

The 2–10 kDa chromatography profile (Figure S1) was obtained by gel filtration chromatography, using a Superdex-Peptide column (10/300 GL, GE Healthcare, London, UK) coupled to a High-Performance Liquid Chromatography (HPLC) system (Merck Hitachi, Darmstadt, Germany) equipped with a UV detector (Merck Hitachi, Darmstadt, Germany). Two hundred microlitres of fraction was eluted with ammonium acetate 0.1 M at a flow rate of 0.7 mL/min.

The ethanol concentration in the 2–10 kDa peptidic fraction was analyzed in duplicate using a High-Performance Liquid Chromatography (HPLC) system from Merck Hitachi (Darmstadt, Germany), which was equipped with a refractive index detector (L-7490, Merck Hitachi, Darmstadt, Germany). Initially, the 2–10 kDa fraction was filtered through 0.22 µm Millipore membranes (Burlington, MA, USA) before being injected into a Sugar-Pak column, 10 µm, 6.5 mm × 300 mm (Waters Hitachi, Milford, CT, USA). The elution was carried out with a degassed aqueous mobile phase containing CaEDTA (50 mg/L) at a temperature of 90 °C, with a flow rate of 0.5 mL/min.

4.3. Microorganisms and Preparation of Inoculum Cultures

In this work, the following microorganisms from Culture Collection of Elisa Câmara Laboratory were used: Gram-positive bacteria: *Bacillus cereus* (ATCC® 11778); Methicillin-resistant *Staphylococcus aureus* (MRSA ATCC® 33591), *Staphylococcus aureus* (ATCC® 6538), *Streptococcus mutans* (ATCC® 25175), *Streptococcus mitis* (NCIMB® 13770), Gram-negative bacteria: *Escherichia coli* (ATCC® 8739) and *Pseudomonas aeruginosa* (ATCC® 9027); the mold *Aspergillus brasiliensis* (ATCC® 16404) and yeasts: *Candida albicans* (ATCC® 10231) and

Saccharomyces cerevisiae Ethanol-red, obtained from the Lesaffre Advanced Fermentations company (Marcq-en-Baroeul, France).

A subculture was prepared to create the ATCC working culture of the microorganisms, using the stock culture on plates with the recommended media for each microorganism, following the suppliers' guidelines and the Clinical and Laboratory Standards Institute [39]. *S. cerevisiae* Ethanol-red was maintained on YEPD-agar slants (containing 20 g/L glucose, 20 g/L peptone, 10 g/L yeast extract, and 20 g/L agar, pH 6) and stored at 4 °C. *S. cerevisiae* Ethanol-red inoculums were obtained by transferring one YEPD-agar slant of each strain (pre-grown at 30 °C for 48–72 h) into 100 mL of YEPD medium (composed of 10 g/L yeast extract, 20 g/L peptone, and 20 g/L glucose) and incubating the cultures at 30 °C with agitation at 150 rpm for 16 h.

4.4. Antimicrobial Activity

The antimicrobial activity of the peptides (2–10 kDa) was assessed against all the microorganisms which were mentioned in point 4.3 and determined using the broth microdilution method, according to Clinical and Laboratory Standards Institute guidelines [39]. All assays were performed in triplicate, and positive controls were used for bacteria (vancomycin and norfloxacin at 1 mg/ ml, Merck KGaA, Darmstadt, Germany) or for yeasts (ketoconazole at 0.1 mg/mL, Merck KGaA, Darmstadt, Germany). The negative controls (ethanol at 8% (*v*/*v*)) were also included in all the assays and all the controls were used in the same conditions and dilutions.

Minimum Inhibitory Concentration (MIC)

The MIC was assessed using the broth microdilution method in 96-well microtiter plates, as described by Pereira et al. (2023) [50]. Peptides (2–10 kDa) were added at a final concentration of 2000 μg/mL, with the first row undergoing a 1:2 dilution, therefore the maximal concentration tested was 1000 μg/mL for all microorganisms, lowed by a series of 2-fold dilutions in different broths: Brain Heart Infusion (BHI) broth (Biokar Diagnostics, Allonne, France) for *Streptococcus* strains, Sabouraud Dextrose broth (Biokar Diagnostics, Allonne, France) for yeasts and molds, and Mueller-Hinton broth (Biokar Diagnostics, Allonne, France) for the remaining microorganisms. A standardized saline suspension (NaCl, 0.85% *w*/*v*) was prepared to inoculate the plates, with turbidity adjusted to match the 0.5 McFarland standard (DEN-1, McFarland Densitometer, Biosan, Riga, Latvia), equivalent to 1–2 × 10^8 CFU/mL for bacteria, and 1–5 × 10^6 CFU/mL for yeasts and molds. The microplates were incubated at 35 °C for 24 h for all microorganisms, except yeasts, which were incubated at 30 °C for 24 h, and molds, which required 5–7 days of incubation at 25 °C. After incubation, the MIC was identified as the lowest concentration of peptides at which no visible microbial growth was observed.

4.5. Peptides (2–10 kDa) Potential Preservative Effect

4.5.1. Tested Product

The cosmetic base formulation tested was a nutritive royal jelly cream® (O/W emulsion) without preservation, provided by Elisa Câmara, Lda. (Lisbon, Portugal) and Ingredients in compliance with the International Nomenclature of Cosmetics Ingredients (INCI) are: Aqua, Caprylic/Capric Trygliceride (Trigliceridos MCT; Quimidroga; Barcelona, Spain); Cetearyl Alcohol, Sodium Cetearyl Sulfate (Lanette N®; BASF, Germany); Polyglyceryl-3 Methylglucose Distearate (Tego Care® 450; Evonik nutrition and care GmbH, Essen, Germany); Glycerin (Glycerine USP, PH. Eur.; Interfat-natural oils, Barcelona, Spain); Panax Ginseng extract (Ginseng Extract; Provital, Barcelona, Spain); Royal Jelly (Lipoplastidine® Pappa regalis, Vevy Europe s.p.a., Genova, Italy). The formulations were prepared in the Elisa Câmara laboratory in accordance with good manufacturing practices (ISO 22716:2007) [75].

4.5.2. Physicochemical and Microbiological Analysis

To determine the physicochemical parameters of the royal jelly formulations, pH (ISO 4316:1977) [76] (Metrohm; 692pH/Ion Meter 09187, Herisau, Sweden), viscosity (ISO 2555:2000) [77] (Brookfield (sp5/20rpm); HADV-II+RT609008, Worcester, MA, USA) and physical stability were by centrifuging during 1 h at 5000 rpm (P. Selecta; Centronic 3ZZ102, Barcelone, Spain). Also, microbiological analyses were also carried out, testing for enumeration and detection of mesophilic bacteria (ISO 21149:2017) [78], *S. aureus* (ISO 22718:2015) [79], *E. coli* (ISO 21150:2015) [80], *P. aeruginosa* (ISO 22717:2015) [81], *C. albicans* (ISO 18416:2015) [82] and total yeast and molds count (ISO 16212:2017) [83] to ensure that the formulation was not contaminated according standard methods.

4.5.3. Evaluation of the Antimicrobial Protection of a Cosmetic Product-Preservation Efficacy Test-Challenge Test (ISO 11930:2019)

The evaluation of the preservation of a cosmetic formulation is based on inoculation of the formulation with a calibrated inocula. The microorganisms' standards recommended and used for this test were *P. aeruginosa*, *S. aureus*, *E. coli*, *C. albicans*, and *A. brasilliensis*, which are considered the most important microorganisms that can easily contaminate cosmetics products [36].

This standard method is employed during the product development phase to prevent microbial growth in cosmetic formulations. It determines the necessary preservative concentration for effective preservation by assessing the reduction in microorganisms over a 28-day period. All culture media, diluents, and inocula were prepared according to the manufacturer's guidelines and in compliance with ISO 11190:2019 standards [36].

Several cosmetic milk emulsions were prepared using the previously described base formulation (see Section 4.5.1), with different concentrations/combinations of preservatives: 1-peptides (2–10 kDa) at MIC concentration; 2-Microcare® BNA (benzyl alcohol at 0.3% (*v*/*v*); 3-Microcare® BNA (benzyl alcohol (0.3% (*v*/*v*)) + peptides (2–10-kDa) at MIC concentration; 4-Microcare® BNA (benzyl alcohol 0.6% (*v*/*v*)) + Microcare SB® (sodium benzoate, potassium sorbate ((1% (*v*/*v*)); 5-Ethanol at 8% (*v*/*v*); 6-formulation without preservative (negative control). Then, the different formulations prepared (samples of 20 g) were introduced in sterile containers, separately and were inoculated with each one of the five standard strains considered to obtain 1×10^6 CFU/g for bacteria and 1×10^5 CFU/g for fungus. All the inoculated formulations were stored and observed for 28 days at 25 °C. Colony counts were performed at contact times of 0, 1, 7, 14, 21, and 28 days, using the plate count method of ISO 11930:2019 [36]. The results were expressed as log CFU/g. All determinations were performed in triplicate.

Thus, the efficacy of the peptides (2–10 kDa) as a cosmetic preservative was further assessed alone at MIC concentration (Table 1) or in association with synthetic preservatives at a lower concentration (0.3% (*v*/*v*) Microcare® SB+2–10-kDa). This reduction in the concentration of the classic preservative was carried out to check for possible synergistic relationships with the peptides being tested. The positive control (0.6% (*v*/*v*) Microcare® SB+ and 1% (*v*/*v*) Microcare® SB was used in the proportions recommended by the suppliers, which are described as classic synthetic preservatives, normally used for effective cosmetic preservation, and the negative control (formulation without any preservative). Additionally, two more controls were performed and tested in formulations: the solvent present in the peptides (2–10-kDa) (ethanol control) and a control for synthetic preservative at a lower concentration (0.3% (*v*/*v*) Microcare® SB) to validate the results.

The formulation's preservation is deemed effective if it meets either Criterion A or B:

Criterion A-For bacteria, the formulation is effective if the total bacterial count decreases by two to three logarithmic units between the second and seventh days of the challenge test, with no subsequent increase in population until the test concludes on the 28th day. For fungi, the criterion is met if the fungal count is reduced by two logarithmic units by the 14th day, with no further growth until the end of the test. If these conditions are satisfied for all recommended microorganisms, the cosmetic product is considered protected against microorganisms that could pose a health risk to consumers.

Criterion B-For bacteria, the formulation is effective if the bacterial count decreases by three logarithmic units by the 14th day, with no growth observed until the 28th day. For fungi, the criterion is satisfied if the fungal count is reduced by one logarithmic unit by the 14th day, with no growth observed until the test concludes. While meeting this criterion indicates the formulation is up to standard, additional risk analysis is advised to identify control factors unrelated to the formulation, such as specialized packaging. If the formulation fails to meet either criterion A or B, it faces a high risk of contamination. In such cases, a thorough risk analysis is necessary to prevent potential contaminants during the product's lifespan, or the formulation should be revised with new preservatives to ensure its safety.

4.6. Potential Skin Benefits of the S. cerevisiae Peptides (2–10 kDa)

In Vitro Collagenase Inhibition Assay

To evaluate the peptide's collagenase inhibitory activity, the Collagenase Activity Colorimetric Assay Kit (Merck KGaA, Darmstadt, Germany) was used. The kit measured collagenase activity using a synthetic peptide (FALGPA) that mimics collagen's structure. The assay was performed according to the manufacturer's instructions. Briefly, aliquots of the peptides were prepared in the solvent, ethanol 8% (*v*/*v*) at a final concentration ranging from 50 to 1000 µg/mL. The peptides (2 µL aliquots) were then mixed with 10 µL of collagenase (0.35 U/mL), and 88 µL of assay buffer. An enzyme control containing only the enzyme and the buffer and 2 µL of solvent (ethanol 8% (*v*/*v*)) was also prepared. A collagenase inhibitor (positive control) provided by the manufacturer was used, i.e., 1,10 phenanthroline. Additionally, a blank containing only the buffer was also prepared. The reaction was started by adding 40 µL of FALGPA and 60 µL of buffer, the absorbance was then measured at 345 nm every 2 min for at least 40 min to plot a kinetic curve. Readings were performed using a BioTek SYNERGY Microplate Reader (Agilent Technologies, Santa Clara, CA, USA). Collagenase activity (enzyme control) was calculated by the following equation:

$$\text{Collagenase activity (U/mL)} = \frac{\left(\frac{-\Delta\ A345nm}{\Delta T} enzyme\ control - \frac{-\Delta\ A345nm}{\Delta T}\ blank\right) \times RV \times DF}{EC \times V}$$

Δ A345nm corresponds to the absorbance difference between the beginning and the end of the acquisition readings, ΔT corresponds to the time difference between the beginning and the end of the acquisition readings, RV is the reaction volume (0.2 mL); DF is the dilution factor; EC corresponds to the extinction coefficient of collagenase substrate (0.53 mM); V is the enzyme volume (mL).

To determine the inhibitor, i.e., 1,10-Phenanthroline or the peptides activity, the same equation to calculate the collagenase activity was applied.

The percent relative inhibition of 1,10-Phenanthroline or peptides with respect to enzyme control was calculated using this equation:

$$\%\ \text{Inhibition} = [(\text{Activity (enzyme control)} - \text{Activity (inhibitor)})/\text{Activity (enzyme control)}] \times 100$$

4.7. Cytotoxicity of Peptides (2–10 kDa) in Dermal Cells

The cytotoxicity of the peptides under study in dermal cells was evaluated in the melanoma cell line (MNT-1, CRL-3450, ATCC, Manassas, VA, USA), primary dermal fibroblasts (PCS-201-012, ATCC), primary epidermal keratinocytes (PCS-200-010, ATCC) and primary epidermal melanocytes (PCS-200-012, ATCC). Cultures were maintained and experiments were performed according to the cell culture. For MNT-1 and fibroblasts, it was used Dulbecco's modified eagle medium (DMEM, ThermoFisher Scientific, Waltham, MA, USA) supplemented with a mixture of 100 µg/mL Streptomycin and 100 U/mL Penicillin (ThermoFisher Scientific) and 10% (*v*/*v*) fetal bovine serum (FBS, ThermoFisher Scientific). The medium for melanocytes and keratinocytes was Dermal cells basal media

(PCS-200-030, ATCC) supplemented with melanocytes growth kit (PCS-200-041, ATCC) or keratinocytes growth kit (PCS-200-040, ATCC), respectively. In a 96-well plate, cells were seeded at a density of 7500 cells per well, and after 24 h, either crescent concentrations of the peptides or equivalent crescent concentrations of ethanol (vector control) were applied.

After a 48-h incubation at 37 °C, with 5% CO_2 and saturated humidity, the medium was replaced with a mixture of fresh medium and the Cell Titer 96® Aqueous One solution cell proliferation assay (Promega, Madison, WI, USA), following the manufacturer's guidelines. The viability of cells exposed to each peptide concentration was then compared to the viability of cells exposed to the corresponding concentration of the vector (ethanol), with the results standardized as a percentage.

4.8. Wound-Healing Assay

For the wound healing assay, 4×10^5 cells of primary human fibroblasts were seeded in 24 well plates in DMEM supplemented with 10% (v/v) FBS and antibiotic mixture and incubated at 37 °C, 5% (v/v) CO_2 and 99% (v/v) humidity. Primary umbilical vein endothelial cells (HUVEC, CRL-1730, ATCC) were seeded in 24 well plates with a cell density of 4×10^5 cells/mL in F12-K medium supplemented with 10% (v/v) FBS (ATCC), 1000 µg/mL Heparin (Sigma Aldrich, Merck, Darmstadt, Germany), 30 µg/mL endothelial cell growth factor (ECGF, Sigma Aldrich, Merck) and 1% (v/v) of antibiotic/antimycotic mixture and incubated at 37 °C, 5% (v/v) CO_2 and 99% (v/v) humidity. Human dermal keratinocytes were seeded with a 2×10^5 cell/mL density in 24-well plates in dermal cells basal medium (ATCC) supplemented with keratinocytes growth kit (ATCC). After 24 h, a pipette tip was used to perform a scratch in the cell monolayer, and the medium was replaced by fresh medium supplemented with 250 µg/mL of the peptides or 1% (v/v) of ethanol (vector control). The scratch was imaged for 24 h with a Cytosmart Lux2 (Axion Biosystems, Atlanta, GA, USA) and the percentage of remission was calculated by measuring the size of the wound scratch at 0 h and after 24 h with ImageJ software.

4.9. Inflammation Assay

To analyze the peptides (2–10 kDa) effect in the inflammation mediated by fibroblasts, normal dermal fibroblasts were seeded in T25 flasks with a cell density of 4×10^5 cells/flask and after 24 h, they were incubated for 2 h with 250 µg/mL of the peptides, 1% (v/v) ethanol (vector control), or the same volume of medium (untreated control) and then for 2 h with 7 µg/mL lipopolysaccharide (LPS, Merck). Samples without LPS were prepared in parallel for control purposes. The following procedures were similar to those described in Branco et al. (2023) [15] and the expression of TNF-α in samples was determined with the 2-Ct method [84] using 18S mRNA as internal control and respective untreated control samples.

4.10. Endothelial Cell Tube Formation

The procedure to evaluate the in vitro angiogenesis followed the guidelines of the protocol published by Arnaoutova and Kleinman (2010) [45]. In the first step, the wells from a 96-well culture plate were coated with 50 µL Matrigel basement membrane high concentration (Corning, Merck, Corning, NY, USA) and incubated for 30 min at 37 °C. Next, HUVEC at 80% confluence in a T25 flask were detached with Tryple express (ThermoFisher Scientific) and a mixture was prepared with 1.5×10^5 cells/mL and 250 µg/mL of peptides or 1% (v/v) ethanol, or prepared in F12-K medium supplemented with 5% (v/v) FBS (ATCC), 0.5 mg/mL Heparin (Sigma Aldrich, Merck), 15 µg/mL endothelial cell growth factor (ECGF, Sigma Aldrich, Merck) and 0.5% (v/v) of antibiotic/antimycotic mixture (ThermoFisher Scientific). The mixture was gently poured on top of the matrix and images were acquired each 15 min for 24 h using Cytosmart Lux2 microscope and respective software (Axion Biosystems, Atlanta, GA, USA).

4.11. Superoxide Anion Radical Scavenging Assay

The Superoxide anion radical scavenging activity was determined spectrophotometrically at 560 nm by monitoring the effect of the peptides (2–10 kDa) on the reduction in NBT to the blue chromogen formazan by superoxide anion, generated by the NADH/PMS system according to the procedure described by Valentão et al. (2001) [85]. Briefly, the reaction mixtures were prepared in 96 well microplates and consisted of 20 µL of solutions with different concentrations of the peptides (0 to 3.9 µg/µL), 30 µL of NADH (Alfa Aesar, Karlsruhe, Germany) (1.66 mM in phosphate buffer 19 mM, pH 7.4), 30 µL of NBT (Sigma-Aldrich, St. Louis, MO, USA) (430 µM in phosphate buffer 19 mM, pH 7.4) and phosphate buffer (19 mM, pH 7.4) up to 295 µL. The reaction was initiated by the addition of 5 µL of PMS (Sigma-Aldrich, St. Louis, MO, USA) (162 µM in phosphate buffer 19 mM, pH 7.4) and the extent of NBT reduction was followed by measuring the increase in the absorbance at 560 nm, for 2 min, at room temperature, in a microplate reader in kinetic mode (FLUOstar® Omega Plate Reader, BMG Labtech, Ortenberg, Germany). Gallic acid (Alfa Aesar, Karlsruhe, Germany) was used as a positive control. Each sample (peptides or gallic acid) concentration generated a time-dependent curve and the slope (ΔAbs/sec) of the linear domain of these curves was used to determine the inhibition, in percentage, of the NBT reduction to blue chromogen formazan, according to the formula:

$$\text{Inhibition (\%)} = [(\text{Slope control} - \text{Slope sample})/(\text{Slope control})] \times 100$$

Values were assessed in triplicate and the results were expressed as the final concentration (µg/mL) in the reaction mixture, which inhibited 50% of the NBT reduction (IC_{50}).

4.12. Statistical Analysis

Data are expressed as the average and standard deviation of at least three independent experiments. Statistical analysis was performed using GraphPad Prism v8.2.1, and data were considered statistically significant when the p-value < 0.05, calculated by the unpaired t-test.

5. Conclusions

In conclusion, this study underscores the multifaceted potential of the peptides secreted by *S. cerevisiae* Ethanol-red, as a promising bioactive ingredient for skincare applications. The antimicrobial activity of these peptides against a range of dermal pathogens, including *S. aureus* and *P. aeruginosa*, highlights their potential as a natural preservative. Their efficacy against Methicillin-resistant *S. aureus* (MRSA) and other pathogens such as *E. coli* and *S. epidermidis* suggests a broad-spectrum of antimicrobial capability. Moreover, the challenge test results demonstrate the efficacy of these peptides, particularly when combined with Microcare® BNA at 0.3% (v/v), showing promising preservation efficacy and potentially allowing for reduced concentrations of synthetic preservatives. These findings highlight a pathway towards safer, more effective cosmetic products with optimized preservative systems, aligning with the growing consumer demand for natural and less harmful ingredients in skincare formulations.

In addition, the collagenase inhibitory properties of the peptides studied highlight their utility in anti-aging formulations, while their anti-inflammatory effects and suitability for the remodeling phase of wound healing suggest therapeutic benefits.

The capacity of the peptides (2–10 kDa) to scavenge superoxide radicals and reduce oxidative stress further supports their role in protecting the skin from oxidative damage and slowing the aging process. Taken together, these findings advocate for the continued exploration of peptides secreted by *S. cerevisiae* in skincare, leveraging their natural bioactivity to meet consumer demand for safer and more effective cosmetic products.

Supplementary Materials: The following supporting information can be downloaded at: https://www.mdpi.com/article/10.3390/antibiotics13090881/s1, Movie S1. HUVEC tube formation in the presence of 250 μg/mL Peptide fraction 2–10 KDa.mp4; Movie S2. HUVEC tube formation in the presence of 1% (*v*/*v*) ethanol.mp4. Figure S1. Gel filtration chromatography profile of the 2–10 kDa peptide fraction extracted from the 7-day-old supernatant of *S. cerevisiae* Ethanol-red. Table S1. Quality control of the base formulation at t = 0 and t = 28 days for organoleptic, physico-chemical, and microbiological analysis. References [43,86] are cited in the Supplementary Materials.

Author Contributions: E.M.M. and P.B. conceived and designed the research. A.L.B.A., C.R.-R., E.M.M., K.L., M.P.D. and P.B. conducted experiments. A.R.F., E.M.M., H.A., M.P.D. and P.B. analyzed data. A.R.F. made a special contribution to the conception of the anti-inflammatory and wound healing studies in human cells, C.R.-R. conducted these experiments, analyzed the data, and wrote, revised, and edited this part of the manuscript. E.M.M., M.P.D., H.A. and P.B. wrote and edited the manuscript. E.M.M., H.A., M.P.D. and P.B. revised the manuscript. Funding acquisition: A.R.F., M.P.D. and H.A. All authors have read and agreed to the published version of the manuscript.

Funding: The funder is the Foundation for Science and Technology (FCT, Portugal), the numbers refer to the projects: PTDC/BII-BIO/31761/2017, 10.54499/UIDP/04378/2020 and 10.54499/UIDB/04378/2020 of the Research Unit on Applied Molecular Biosciences-UCIBIO and the project LA/P/0140/2020, of the Associate Laboratory Institute for Health and Bioeconomy-i4HB. CRR was funded by FCT/MCTES, grant number SFRH/BPD/124612/2016.

Institutional Review Board Statement: This article does not contain any studies with human participants or animals performed by any of the authors.

Informed Consent Statement: Not applicable.

Data Availability Statement: All data generated and analyzed in this study is included in this manuscript.

Acknowledgments: The authors thank FCT for the financial support of this work.

Conflicts of Interest: The authors declare they have no conflict of interest.

References

1. Alves, A.; Sousa, E.; Kijjoa, A.; Pinto, M. Marine-derived compounds with potential uses as cosmeceutical and nutricosmetics. *Molecules* **2020**, *25*, 2536. [CrossRef] [PubMed]
2. Liu, J.K. Natural products in cosmetics. *Nat. Prod. Bioprospect.* **2022**, *12*, 40. [CrossRef] [PubMed]
3. Darbre, P.D. Environmental oestrogens, cosmetics and breast cancer. *Best Pract. Res. Clin. Endocrinol. Metab.* **2006**, *20*, 121–143. [CrossRef] [PubMed]
4. de Groot, A.C.; Veenstra, M. Formaldehyde-releasers in cosmetics: Relationship to formaldehyde contact allergy. *J. Am. Acad. Dermatol.* **2010**, *62*, 60–68. [CrossRef] [PubMed]
5. Wu, Y.; Choi, M.-H.; Li, J.; Yang, H.; Shin, H.-J. Mushroom Cosmetics: The Present and Future. *Cosmetics* **2016**, *3*, 22. [CrossRef]
6. Gupta, P.L.; Rajput, M.; Oza, T.; Trivedi, U.; Sanghvi, G. Eminence of Microbial Products in Cosmetic Industry. *Nat. Prod. Bioprospect.* **2019**, *9*, 267–278. [CrossRef]
7. Bianchet, R.T.; Vieira Cubas, A.L.; Machado, M.M.; Siegel Moecke, E.H. Applicability of bacterial cellulose in cosmetics—Bibliometric review. *Biotechnol. Rep.* **2020**, *27*, e00502. [CrossRef]
8. Kiki, M.J. Biopigments of Microbial Origin and Their Application in the Cosmetic Industry. *Cosmetics* **2023**, *10*, 47. [CrossRef]
9. Dini, I. The Potential of Algae in the Nutricosmetic Sector. *Molecules* **2023**, *28*, 4032. [CrossRef]
10. Rybczyńska-Tkaczyk, K.; Grenda, A.; Jakubczyk, A.; Kiersnowska, K.; Bik-Małodzińska, M. Natural Compounds with Antimicrobial Properties in Cosmetics. *Pathogens* **2023**, *12*, 320. [CrossRef]
11. Min, M.W.; Kim, C.E.; Chauhan, S.; Park, H.J.; Park, C.S.; Yoo, T.H.; Kang, T.J. Identification of peptide inhibitors of matrix metalloproteinase 1 using an in-house assay system for the enzyme. *Enzyme Microb. Technol.* **2019**, *127*, 65–69. [CrossRef] [PubMed]
12. Ferreira, M.S.; Magalhães, M.C.; Sousa-Lobo, J.M.; Almeida, I.F. Trending Anti-Aging Peptides. *Cosmetics* **2020**, *7*, 91. [CrossRef]
13. Branco, P.; Francisco, D.; Chambon, C.; Hébraud, M.; Arneborg, N.; Almeida, M.G.; Caldeira, J.; Albergaria, H. Identification of novel GAPDH-derived antimicrobial peptides secreted by *Saccharomyces cerevisiae* and involved in wine microbial interactions. *Appl. Microbiol. Biotechnol.* **2014**, *98*, 843–853. [CrossRef] [PubMed]
14. Branco, P.; Francisco, D.; Monteiro, M.; Almeida, M.G.; Caldeira, J.; Arneborg, N.; Prista, C.; Albergaria, H. Antimicrobial properties and death-inducing mechanisms of saccharomycin, a biocide secreted by *Saccharomyces cerevisiae*. *Appl. Microbiol. Biotechnol.* **2017**, *101*, 159–171. [CrossRef] [PubMed]

Branco, P.; Maurício, E.M.; Costa, A.; Ventura, D.; Roma-Rodrigues, C.; Duarte, M.P.; Fernandes, A.R.; Prista, C. Exploring the Multifaceted Potential of a Peptide Fraction Derived from *Saccharomyces cerevisiae* Metabolism: Antimicrobial, Antioxidant, Antidiabetic, and Anti-Inflammatory Properties. *Antibiotics* **2023**, *12*, 1332. [CrossRef]

Ngoc, L.T.N.; Moon, J.-Y.; Lee, Y.-C. Insights into Bioactive Peptides in Cosmetics. *Cosmetics* **2023**, *10*, 111. [CrossRef]

Mirzaei, M.; Mirdamadi, S.; Ehsani, M.; Aminlari, M.; Hosseini, E. Purification and Identification of Antioxidant and ACE-Inhibitory Peptide from *Saccharomyces cerevisiae* Protein Hydrolysate. *J. Funct. Foods* **2015**, *19*, 259–268. [CrossRef]

Ni, H.; Li, L.; Liu, G.; Hu, S. Inhibition Mechanism and Model of an Angiotensin I-Converting Enzyme (ACE)-Inhibitory Hexapeptide from Yeast (*Saccharomyces cerevisiae*). *PLoS ONE* **2012**, *7*, e37077. [CrossRef]

Bhattacharyya, J.; Santhoshkumar, P.; Sharma, K. A peptide sequence-YSGVCHTDLHAWHGDWPLPVK [40–60]-in yeast alcohol dehydrogenase prevents the aggregation of denatured substrate proteins. *Biochem. Biophys. Res. Commun.* **2003**, *307*, 1–7. [CrossRef]

Hancock, R.E.W.; Sahl, H.-G. Antimicrobial and host-defense peptides as new anti-infective therapeutic strategies. *Nat. Biotechnol.* **2006**, *24*, 1551–1557. [CrossRef]

Mahlapuu, M.; Håkansson, J.; Ringstad, L.; Björn, C. Antimicrobial Peptides: An Emerging Category of Therapeutic Agents. *Front. Cell. Infect. Microbiol.* **2016**, *6*, 194. [CrossRef] [PubMed]

Wang, G. Human Antimicrobial Peptides and Proteins. *Pharmaceuticals* **2014**, *7*, 545–594. [CrossRef] [PubMed]

Albergaria, H.; Francisco, D.; Gori, K.; Arneborg, N.; Gírio, F. *Saccharomyces cerevisiae* CCMI 885 secretes peptides that inhibit the growth of some non-Saccharomyces wine-related strains. *Appl. Microbiol. Biotechnol.* **2010**, *86*, 965–972. [CrossRef] [PubMed]

Branco, P.; Diniz, M.; Albergaria, H. Evaluation of the Biocontrol Potential of a Commercial Yeast Starter against Fuel-Ethanol Fermentation Contaminants. *Fermentation* **2022**, *8*, 233. [CrossRef]

Branco, P.; Kemsawasd, V.; Santos, L.; Diniz, M.; Caldeira, J.; Almeida, M.G.; Arneborg, N.; Albergaria, H. *Saccharomyces cerevisiae* accumulates GAPDH-derived peptides on its cell surface that induce death of non-*Saccharomyces* yeasts by cell-to-cell contact. *FEMS Microbiol. Ecol.* **2017**, *93*, fix055. [CrossRef]

Wagener, J.; Schneider, J.J.; Baxmann, S.; Kalbacher, H.; Borelli, C.; Nuding, S.; Küchler, R.; Wehkamp, J.; Kaeser, M.D.; Mailänder-Sanchez, D.; et al. A Peptide Derived from the Highly Conserved Protein GAPDH Is Involved in Tissue Protection by Different Antifungal Strategies and Epithelial Immunomodulation. *J. Investig. Dermatol.* **2013**, *133*, 144–153. [CrossRef]

Varvaresou, A.; Papageorgiou, S.; Tsirivas, E.; Protopapa, E.; Kintziou, H.; Kefala, V.; Demetzos, C. Self-preserving cosmetics. *Int. J. Cosmet. Sci.* **2009**, *31*, 163–175. [CrossRef]

Nielsen, N.H.; Linneberg, A.; Menne, T.; Madsen, F.; Frolund, L.; Dirksen, A.; Jorgensen, T. Allergic contact sensitization in an adult Danish population: Two cross-sectional surveys eight years apart (the Copenhagen Allergy Study). *Acta Derm.-Venereol.* **2001**, *81*, 31–34. [CrossRef]

Lundov, M.D.; Moesby, L.; Zachariae, C.; Johansen, J.D. Contamination versus preservation of cosmetics: A review on legislation, usage, infections, and contact allergy. *Contact Dermat.* **2009**, *60*, 70–78. [CrossRef]

European Parliament and Council. Regulation (EC) No 1223/2009 of the European Parliament and of the Council of 30 November 2009 on Cosmetic Products. *Off. J. Eur. Union* **2009**, *342*, 59–209.

Epstein, H. Cosmetics preservation: Sense and nonsense. *Clin. Dermatol.* **2006**, *24*, 551–552. [CrossRef] [PubMed]

Papageorgiou, S.; Varvaresou, A.; Tsirivas, E.; Demetzos, C. New alternatives to cosmetics preservation. *J. Cosmet. Sci.* **2010**, *61*, 107. [PubMed]

Maurício, E.; Rosado, C.; Duarte, M.P.; Verissimo, J.; Bom, S.; Vasconcelos, L. Efficiency of Nisin as Preservative in Cosmetics and Topical Products. *Cosmetics* **2017**, *4*, 41. [CrossRef]

Manou, I.; Bouillard, L.; Devleeschouwer, M.J.; Barel, A.O. Evaluation of the preservative properties of *Thymus vulgaris* essential oil in topically applied formulations under a challenge test. *J. Appl. Microbiol.* **1998**, *84*, 368–376. [CrossRef] [PubMed]

Moser, C.L.; Meyer, B.K. Comparison of Compendial Antimicrobial Effectiveness Tests: A Review. *AAPS PharmSciTech* **2011**, *12*, 222–226. [CrossRef]

ISO 11930:2019; Cosmetics—Microbiology—Evaluation of the Antimicrobial Protection of a Cosmetic Product. International Organization for Standardization: Geneva, Switzerland, 2019.

Alshehrei, F. Microbiological Quality Assessment of Skin and Body care Cosmetics by using Challenge test. *Saudi J. Biol. Sci.* **2024**, *31*, 103965. [CrossRef]

Dao, H.; Lakhani, P.; Police, A.; Kallakunta, V.; Ajjarapu, S.S.; Wu, K.W.; Ponkshe, P.; Repka, M.A.; Narasimha Murthy, S. Microbial Stability of Pharmaceutical and Cosmetic Products. *AAPS PharmSciTech* **2018**, *19*, 60–78. [CrossRef]

CLSI. *Performance Standards for Antimicrobial Susceptibility Testing*; Seventeenth Information Supplement; CLSI Document M100-S17 (M2-A7 and M7-A7); Clinical and Laboratory Standards Institute: Wayne, PA, USA, 2007; Volume 27.

CLSI. *Performance Standards for Antifungal Susceptibility Testing of Filamentous Fungi*, 3rd ed.; Clinical and Laboratory Standards Institute: Wayne, PA, USA, 2017.

Wong, S.; Street, D.; Delgado, S.I. Recalls of foods and cosmetics due to microbial contamination reported to the US Food and Drug Administration. *J. Food Prot.* **2000**, *63*, 1113–1116. [CrossRef]

Ferrarese, L.; Paglia, R.; Ghirardini, A. Bacterial resistance in cosmetics industrial plant: Connected problems and their solution. *Ann. Microbiol.* **2003**, *53*, 477–490.

43. Halla, N.; Fernandes, I.P.; Heleno, S.A.; Costa, P.; Boucherit-Otmani, Z.; Boucherit, K.; Rodrigues, A.E.; Ferreira, I.C.F.R.; Barreiro, M.F. Cosmetics preservation: A review on present strategies. *Molecules* **2018**, *23*, 1571. [CrossRef]
44. Kočevar Glavač, N.; Lunder, M. Preservative efficacy of selected antimicrobials of natural origin in a cosmetic emulsion. *Int. J. Cosmet. Sci.* **2018**, *40*, 276–284. [CrossRef] [PubMed]
45. Arnaoutova, I.; Kleinman, H. In vitro angiogenesis: Endothelial cell tube formation on gelled basement extract. *Nat. Protoc.* **2010**, *5*, 628–635. [CrossRef] [PubMed]
46. Kaplani, K.; Koutsi, S.; Armenis, V.; Skondra, F.G.; Karantzelis, N.; Tsaniras, S.C.; Taraviras, S. Wound healing related agents: Ongoing research and perspectives. *Adv. Drug Deliv. Rev.* **2018**, *129*, 242–253. [CrossRef] [PubMed]
47. Gurtner, G.; Werner, S.; Barrandon, Y.; Longaker, M.T. Wound repair and regeneration. *Nature* **2008**, *453*, 314–321. [CrossRef]
48. Wen, C.; Zhang, J.; Zhang, H.; Duan, Y.; Ma, H. Plant protein-derived antioxidant peptides: Isolation, identification, mechanism of action and application in food systems: A review. *Trends Food Sci. Technol.* **2020**, *105*, 308–322. [CrossRef]
49. Duarte, M.C.T.; Leme, E.E.; Delarmelina, C.; Soares, A.A.; Figueira, G.M.; Sartoratto, A. Activity of essential oils from Brazilian medicinal plants on *Escherichia coli*. *J. Ethnopharmacol.* **2007**, *111*, 197–201. [CrossRef]
50. Vieitez, I.; Maceiras, L.; Jachmanián, I.; Alborés, S. Antioxidant and antibacterial activity of different extracts from herbs obtained by maceration or supercritical technology. *J. Supercrit. Fluids* **2018**, *133*, 58–64. [CrossRef]
51. Pereira, P.; Mauricio, E.M.; Duarte, M.P.; Lima, K.; Fernandes, A.S.; Bernardo-Gil, G.; Cebola, M.J. Potential of supercritical fluid myrtle extracts as an active ingredient and co-preservative for cosmetic and topical pharmaceutical applications. *Sustain. Chem. Pharm.* **2022**, *28*, 100739. [CrossRef]
52. Kunicka-Styczyńska, A.; Sikora, M.; Kalemba, D. Lavender, tea tree and lemon oils as antimicrobials in washing liquids and soft body balms. *Int. J. Cosmet. Sci.* **2011**, *33*, 53–61. [CrossRef]
53. Patrone, V.; Campana, R.; Vittoria, E.; Baffone, W. In vitro synergistic activities of essential oils and surfactants in combination with cosmetic preservatives against *Pseudomonas aeruginosa* and *Staphylococcus aureus*. *Curr. Microbiol.* **2010**, *60*, 237–241. [CrossRef]
54. Bandzerewicz, A.; Gadomska-Gajadhur, A. Into the Tissues: Extracellular Matrix and Its Artificial Substitutes: Cell Signalling Mechanisms. *Cells* **2022**, *11*, 914. [CrossRef]
55. Zhou, H.; Li, W.; Pan, L.; Zhu, T.; Zhou, T.; Xiao, E.; Wei, Q. Human extracellular matrix (ECM)-like collagen and its bioactivity. *Regenerative Biomaterials* **2024**, *11*, rbae008. [CrossRef]
56. Thudium, C.S.; Frederiksen, P.; Karsdal, M.A. Changes in Type VI Collagen Degradation Reflect Clinical Response to Treatment in Rheumatoid Arthritis Patients Treated with Tocilizumab. *Arthritis Res. Ther.* **2024**, *26*, 3. [CrossRef]
57. Brett, D. A Review of Collagen and Collagen-Based Wound Dressings. *Wounds* **2008**, *20*, 347–356.
58. He, T.; Fisher, G.J.; Kim, A.J.; Quan, T. Age-Related Changes in Dermal Collagen Physical Properties in Human Skin. *PLoS ONE* **2023**, *18*, e0292791. [CrossRef]
59. Thring, T.S.; Hili, P.; Naughton, D.P. Anti-collagenase, anti-elastase and anti-oxidant activities of extracts from 21 plants. *BMC Complement. Altern. Med.* **2009**, *9*, 27. [CrossRef]
60. Gruber, J.V.; Smith, D.; Bouldin, L. Personal Care Composition Containing Yeast Extract and Hexapeptide. U.S. Patent Application No. 13/247,689, 28 September 2014.
61. Vijayakumar, R.; Abd Gani, S.S.; Mohd Mokhtar, N.F. Anti-elastase, anti-collagenase and antimicrobial activities of the underutilized red pitaya peel: An in vitro study for anti-aging applications. *Asian J. Pharm. Clin. Res.* **2017**, *10*, 251–255. [CrossRef]
62. Kang, Y.-A.; Kim, Y.-J.; Jin, S.-K.; Choi, H.-J. Antioxidant, Collagenase Inhibitory, and Antibacterial Effects of Bioactive Peptides Derived from Enzymatic Hydrolysate of Ulva australis. *Mar. Drugs* **2023**, *21*, 469. [CrossRef]
63. Ennaas, N.; Hammami, R.; Gomaa, A.; Bédard, F.; Biron, É.; Subirade, M.; Beaulieu, L.; Fliss, I. Collagencin, an antibacterial peptide from fish collagen: Activity, structure and interaction dynamics with membrane. *Biochem. Biophys. Res. Commun.* **2016**, *473*, 642–647. [CrossRef]
64. Ryu, B.; Qian, Z.J.; Kim, S.K. SHP-1, a novel peptide isolated from seahorse inhibits collagen release through the suppression of collagenases 1 and 3, nitric oxide products regulated by NF-kappaB/p38 kinase. *Peptides* **2010**, *31*, 79–87. [CrossRef]
65. Tottoli, E.M.; Dorati, R.; Genta, I.; Chiesa, E.; Pisani, S.; Conti, B. Skin wound healing process and new emerging technologies for skin wound care and regeneration. *Pharmaceutics* **2020**, *12*, 735. [CrossRef]
66. Eming, S.A.; Martin, P.; Tomic-Canic, M. Wound repair and regeneration: Mechanisms, signaling, and translation. *Sci. Transl. Med.* **2014**, *6*, 265sr6. [CrossRef]
67. Criollo-Mendoza, M.S.; Contreras-Angulo, L.A.; Leyva-López, N.; Gutiérrez-Grijalva, E.P.; Jiménez-Ortega, L.A.; Heredia, J.B. Wound healing properties of natural products: Mechanisms of action. *Molecules* **2023**, *28*, 598. [CrossRef]
68. Halliwell, B. Reactive species and antioxidants. Redox biology is a fundamental theme of aerobic life. *Plant Physiol.* **2006**, *141*, 312–322. [CrossRef]
69. Fujii, J.; Homma, T.; Osaki, T. Superoxide radicals in the execution of cell death. *Antioxidants* **2022**, *11*, 501. [CrossRef]
70. Gulcin, İ. Antioxidants and antioxidant methods: An updated overview. *Arch. Toxicol.* **2020**, *94*, 651–715. [CrossRef]
71. Akbarian, M.; Khani, A.; Eghbalpour, S.; Uversky, V.N. Bioactive peptides: Synthesis, sources, applications, and proposed mechanisms of action. *Int. J. Mol. Sci.* **2022**, *23*, 1445. [CrossRef]
72. Sarbon, N.M.; Badii, F.; Howell, N.K. Purification and characterization of antioxidative peptides derived from chicken skin gelatin hydrolysate. *Food Hydrocoll.* **2018**, *85*, 311–320. [CrossRef]

- Wang, W.Y.; Zhao, Y.Q.; Zhao, G.X.; Chi, C.F.; Wang, B. Antioxidant peptides from collagen hydrolysate of redlip croaker (*Pseudosciaena polyactis*) scales: Preparation, characterization, and cytoprotective effects on H_2O_2-damaged HepG2 cells. *Mar. Drugs* **2020**, *18*, 156. [CrossRef]
- Tadesse, S.A.; Emire, S.A. Production and processing of antioxidant bioactive peptides: A driving force for the functional food market. *Heliyon* **2020**, *6*, e04765. [CrossRef]
- *ISO 22716:2007*; Cosmetics—Good Manufacturing Practices (GMP)—Guidelines on Good Manufacturing Practices. International Organization for Standardization: Geneva, Switzerland, 2007.
- *ISO 4316:1977*; Surface Active Agents—Determination of pH of Aqueous Solutions—Potentiometric Method. International Organization for Standardization: Geneva, Switzerland, 1977.
- *ISO 2555:2018*; Plastics—Resins in the Liquid State or as Emulsions or Dispersions—Determination of Apparent Viscosity Using a Single Cylinder Type Rotational Viscometer Method. International Organization for Standardization: Geneva, Switzerland, 2018.
- *ISO 21149:2017*; Cosmetics—Microbiology—Enumeration and Detection of Aerobic Mesophilic Bacteria. International Organization for Standardization: Geneva, Switzerland, 2017.
- *ISO 22718:2015*; Cosmetics—Microbiology—Detection of *Staphylococcus aureus*. International Organization for Standardization: Geneva, Switzerland, 2015.
- *ISO 21150:2015*; Cosmetics—Microbiology—Detection of *Escherichia coli*. International Organization for Standardization: Geneva, Switzerland, 2015.
- *ISO 22717:2015*; Cosmetics—Microbiology—Detection of *Pseudomonas aeruginosa*. International Organization for Standardization: Geneva, Switzerland, 2015.
- *ISO 18416:2015*; Cosmetics—Microbiology—Detection of *Candida albicans*. International Organization for Standardization: Geneva, Switzerland, 2015.
- *ISO 16212:2017*; Cosmetics—Microbiology—Enumeration of Yeast and Mould. International Organization for Standardization: Geneva, Switzerland, 2017.
- Livak, K.J.; Schmittgen, T.D. Analysis of Relative Gene Expression Data Using Real-Time Quantitative PCR and the $2^{-\Delta\Delta CT}$ Method. *Methods* **2001**, *25*, 402–408. [CrossRef] [PubMed]
- Valentão, P.; Fernandes, E.; Carvalho, F.; Andrade, P.B.; Seabra, R.M.; Bastos, M.L. Antioxidant activity of *Centaurium erythraea* infusion evidenced by its superoxide radical scavenging and xanthine oxidase inhibitory activity. *J. Agric. Food Chem.* **2001**, *49*, 3476–3479. [CrossRef] [PubMed]
- Huang, J.; Hitchins, A.D.; Tran, T.T.; McCarron, J.E. Microbiological methods for cosmetics. In *Bacteriological Analytical Manual (BAM)*, 8th ed.; FDA: Silver Spring, MD, USA, 2017.

antibiotics MDPI

Article

Hybrid Caffeic Acid-Based DHFR Inhibitors as Novel Antimicrobial and Anticancer Agents

Renu Sehrawat [1], Ritu Pasrija [2], Priyanka Rathee [3], Deepika Kumari [2], Anurag Khatkar [1,*], Esra Küpeli Akkol [4,*] and Eduardo Sobarzo-Sánchez [5,6]

1 Department of Pharmaceutical Sciences, Maharshi Dayanand University, Rohtak 124001, India; rsehrawat29@gmail.com
2 Department of Biochemistry, Maharshi Dayanand University, Rohtak 124001, India; ritupasrija.biochem@mdurohtak.ac.in (R.P.); kalirawandeepika@gmail.com (D.K.)
3 Faculty of Pharmaceutical Sciences, Baba Mastnath University, Rohtak 124021, India; priyankapharmacy@bmu.ac.in
4 Department of Pharmacognosy, Faculty of Pharmacy, Gazi University, Ankara 06330, Türkiye
5 Instituto de Investigación y Postgrado, Facultad de Ciencias de la Salud, Universidad Central de Chile, Lord Cochrane 417, Santiago 8330507, Chile; eduardo.sobarzo@ucentral.cl
6 Department of Organic Chemistry, Faculty of Pharmacy, University of Santiago de Compostela, 15782 Santiago de Compostela, Spain
* Correspondence: dranurag.pharma@mdurohtak.ac.in (A.K.); esrak@gazi.edu.tr (E.K.A.)

Abstract: A novel series of 1,2,4-triazole analogues of caffeic acid was designed, synthesized, characterized, and assessed for their capacity to inhibit DHFR, as well as their anticancer and antimicrobial properties. A molecular docking analysis was conducted on DHFR, utilizing PDB IDs 1U72 and 2W9S, aiming to design anticancer and antimicrobial drugs, respectively. Among all the synthesized derivatives, compound CTh7 demonstrated the highest potency as a DHFR inhibitor, with an IC_{50} value of 0.15 μM. Additionally, it exhibited significant cytotoxic properties, with an IC_{50} value of 8.53 μM. The molecular docking analysis of the CTh7 compound revealed that it forms strong interactions with key residues of *homo sapiens* DHFR such as Glu30, Phe34, Tyr121, Ile16, Val115, and Phe31 within the target protein binding site and displayed excellent docking scores and binding energy (−9.9; −70.38 kcal/mol). Additionally, synthesized compounds were screened for antimicrobial properties, revealing significant antimicrobial potential against bacterial strains and moderate effects against fungal strains. Specifically, compound CTh3 exhibited notable antibacterial efficacy against *Staphylococcus aureus* (MIC = 5 μM). Similarly, compound CTh4 demonstrated significant antibacterial activity against both *Escherichia coli* and *Pseudomonas aeruginosa*, with MIC values of 5 μM for each. A docking analysis of the most active antimicrobial compound CTh3 revealed that it forms hydrogen bonds with Thr121 and Asn18, a π–cation bond with Phe92, and a salt bridge with the polar residue Asp27.

Keywords: caffeic acid; anticancer; antimicrobial; DHFR inhibitor; in silico design

1. Introduction

Cancer encompasses a vast array of diseases in which abnormal cells proliferate uncontrollably, originating from various areas within the body and potentially spreading to nearby or distant regions [1]. It ranks among the primary causes of mortality worldwide, and its occurrence is steadily rising. As per a World Health Organization (WHO) report, it is a significant factor contributing to mortality that led to approximately 10 million deaths in the year 2020, accounting for roughly one in every six fatalities [2]. Overall, the development of novel anticancer drugs is crucial for improving cancer treatment options, enhancing patient recovery, overcoming resistance, and addressing the unique challenges associated with different types of cancer. Antimicrobial resistance (AMR) is also a growing health concern that poses a threat to effectively treat an increasing range of microbial infections. The WHO has acknowledged AMR as one of the ten major global public health challenges, and about 10 million deaths each year by 2050 could be caused by drug-resistant diseases. In response, the WHO has categorized resistant bacteria into different groups based on their priority. Priority 1 includes critically important bacteria like *Acinetobacter baumannii* and *Pseudomonas aeruginosa*, which are associated with severe healthcare-associated infections. Priority 2 consists of highly important bacteria such as *Staphylococcus aureus* and *Helicobacter pylori*, responsible for common infections. Priority 3 includes moderately important bacteria like *Streptococcus pneumoniae* and *Haemophilus influenzae*, associated with community-acquired infections [3–5]. The emergence of multidrug-resistant (MDR) pathogens has severely compromised the efficacy of existing antimicrobial drugs. Current chemotherapeutic agents are also associated with severe side effects and resistance. Hence, *Dihydrofolate reductase* (DHFR; E.C. 1.5.1.3) emerges as a compelling target as it is a ubiquitous enzyme found in the cells of nearly all living organisms. It plays a crucial role in converting 7,8-dihydrofolate into 5,6,7,8-tetrahydrofolate by using the NADPH coenzyme in this process. Tetrahydrofolate is crucial in the synthesis of essential building blocks of DNA and important amino acids by the de novo process [6–8]. The inhibition of this enzyme leads to a deficiency of the active form of folic acid, which, in turn, disturbs nucleotide biosynthesis and, ultimately, leads to cell death. This trait is used to inhibit cancerous cells and microbial growth [9–13]. DHFR inhibitors have been utilized for many years in treating both cancer and microbial infections. However, there is still room for exploring new lead compounds for DHFR inhibition activity due to the development of resistance against existing DHFR inhibitors. Methotrexate, an established antimetabolite, is extensively used as a major chemotherapeutic agent. Nevertheless, methotrexate also faces resistance in mammalian cancer cells. It could occur through various mechanisms. These include a reduced methotrexate uptake by the reduced folate carrier (RFC), increased methotrexate efflux facilitated by over-expression ATP-binding cassette (ABC) transporters, decreased methotrexate polyglutamation by *Folypolyglutamate synthetase*, elevated expression of DHFR protein via DHFR gene amplification, reduced methotrexate affinity resulting from DHFR mutation, and combinations of these mechanisms [14–16]. Resistance to trimethoprim has indeed been documented, attributed to various mechanisms, including the activation of alternative metabolic pathways and the overproduction of resistant chromosomal DHFR enzymes [17]. However, the persistent resistance to existing DHFR inhibitors and the absence of novel DHFR inhibitors in the pipeline for decades have compelled us to conduct concentrated research in this area.

To identify a novel structural class for DHFR inhibition, caffeic acid was selected as a lead to design novel derivatives as in our previous investigation aimed at identifying novel compounds from natural sources that could inhibit DHFR; caffeic acid demonstrated significant activity and favorable binding scores [18]. Although numerous studies have been conducted to assess the antimicrobial and anticancer properties of caffeic acid, the lack of understanding regarding its mechanism of action and its interaction with DHFR has significantly impeded the utilization of caffeic acid as a lead and the discovery of promising novel derivatives based on caffeic acid [19–23]. Screening various heterocyclic compounds for the development of novel derivatives of caffeic acid, we found that compounds containing nitrogen and oxygen atoms have demonstrated significant biological efficacy [24]. Extensive studies have shown that 1,2,4-triazole analogues, depending on their substituents, possess diverse biological properties such as antimicrobial, analgesic [25–27], anti-tumor [28,29], anti-inflammatory [30], anti-hypertensive [31], and antiviral [32] activities. The 1,2,4-triazole moiety is present in numerous standard medicines, emphasizing its significance [33]. Likewise, Schiff bases have garnered interest in the medicinal field owing to their extensive range of biological activities, including antimicrobial [34–36], antitubercular [37,38], anti-HIV, anticancer [39], anti-inflammatory, analgesic [40], and anticonvulsant properties [41]. In the meantime, nitrogen and sulfur small-ring heterocycles have been regarded as appealing foundational structures for an extended period because of their ease of synthesis, versatility, and relevance in therapeutics [42]. By considering the notable significance of Schiff bases, caffeic acid, and 1,2,4-triazole in the fields of biology and medicine, we have designed and synthesized hybrid caffeic acid-based derivatives containing Schiff bases derived from 1,2,4-triazole and evaluated their potential as bioactive compounds. A library of analogues was designed by using different aldehydes with aliphatic, aromatic, and heterocyclic rings with different substitutions. This approach aimed to inspect the influence of the substituent's and ring size on the effectiveness of compounds. Through computational docking studies, we have identified the most promising compounds for further analysis.

Based on the observations outlined, we hereby present the design and synthesis of novel molecular scaffolds containing 1,2,4-triazole rings using caffeic acid as a synthon in the hope of obtaining medicinally significant DHFR inhibitors as antimicrobial and anticancer agents. The compounds were characterized using data from the FTIR, NMR, and elemental analysis. Their DHFR inhibition, antibacterial, antifungal, and anticancer capabilities were examined against specific standard strains and cell line.

2. Results and Discussion

2.1. Molecular Modeling

Structure-based drug design (SBDD) was utilized to design novel compounds and assess their affinities towards a target protein. This approach is widely accepted for discovering novel leads and eliminating incompatible derivatives [43]. A molecular docking analysis was performed for all designed derivatives of caffeic acid, targeting the three-dimensional coordinates of *homo sapiens* DHFR (PDB ID 1U72) co-crystallized with methotrexate (MTX). Additionally, for microbial infection, the analysis included the S. aureus trimethoprim-resistant variant S1DHFR (PDB ID: 2W9S) [44]. Based on our previous research, caffeic acid, a lead compound, was selected as it exhibited favorable interactions with the oxidative enzyme DHFR [18]. Consequently, modifications were made at the carboxylic acid group of caffeic acid to design ligands, and a 1,2,4-triazole ring was introduced to mimic the pyrimidine ring of trimethoprim. DHFR inhibition activity has also been reported for the 1,2,4-triazole ring [13,45]. The necessity of the free phenolic hydroxyl group in caffeic acid arises from its ability to establish hydrogen bonds with crucial amino acid residues in the target protein. The designed derivatives were subjected to virtual screening using the Schrödinger suite of molecular modeling software 13.1. Ligands exhibiting better and comparable binding energy and docking scores than caffeic acid, standard trimethoprim, and methotrexate were retained for further exploration of the mechanistic approach toward DHFR. The well-docked compounds were subsequently synthesized and evaluated for their DHFR inhibitory potential, as well as their anticancer and antimicrobial properties. The most active compounds were further investigated to understand their mechanistic interactions within the binding site.

2.1.1. Molecular Docking Interaction Analysis of *Homo Sapiens* DHFR (PDB ID 1U72)

A molecular docking analysis was performed to design caffeic acid derivatives using the structure of homo sapiens DHFR (PDB ID 1U72), co-crystallized with methotrexate as a protein target. Among all the designed derivatives, compound CTh7 exhibited a similar docking performance to methotrexate, with a binding energy of −70.38 kcal/mol and a docking score of −9.9. It is bound to the same site, interacting with key residues such as Glu30, Ser59, and Asp21 by hydrogen bonding and with Phe34 by π–π stacking, as shown in Figure 1. This finding aligns with M. Wang's 2017 results [46]. It also snugly fits into the hydrophobic pocket of the modeled protein. These findings were further corroborated by cross-docking investigations involving the target protein, ligand, and MTX. The most effective anticancer agent CTh7 with MTX may be seen in the superimposed image (Figure 2), revealing similar pharmacophores and occupying the same active site regions within the binding pocket of DHFR. Methotrexate, a standard inhibitor of DHFR, was also tested against PDB ID 1U72. It forms hydrogen bonds with Glu30, Ile7, Val115, Asn64, and Arg70 (Table 1). Redocking the co-crystallized ligands was performed to validate the docking protocol, and the RMSD value was found to be 1.19, which was satisfactory for approving the docking protocol. The molecular docking analysis revealed that nearly all the synthesized ligands engaged the same core of the target enzyme, surrounded by residues such as Leu67, Pro61, Ile60, Tyr121, Ile16, Val115, Leu22, Ile7, Val8, Ala9, Phe31, and Phe34. To validate the docking results, selected designed ligands were subsequently synthesized in the laboratory and assessed for their anticancer and DHFR inhibitory activities.

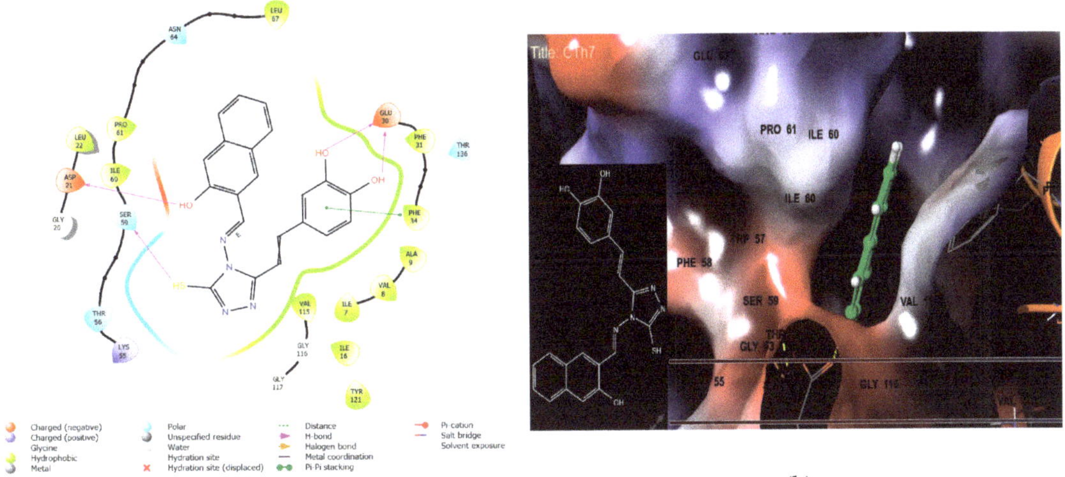

(a) (b)

Figure 1. (**a**) 2D model illustrating the interaction between compound CTh7 and active site amino acids, highlighting various types of bonds. (**b**) A three-dimensional representation of CTh7 within the active pocket of the DHFR (PDB ID 1U72) receptor, highlighting amino acid residues in close proximity to the active site.

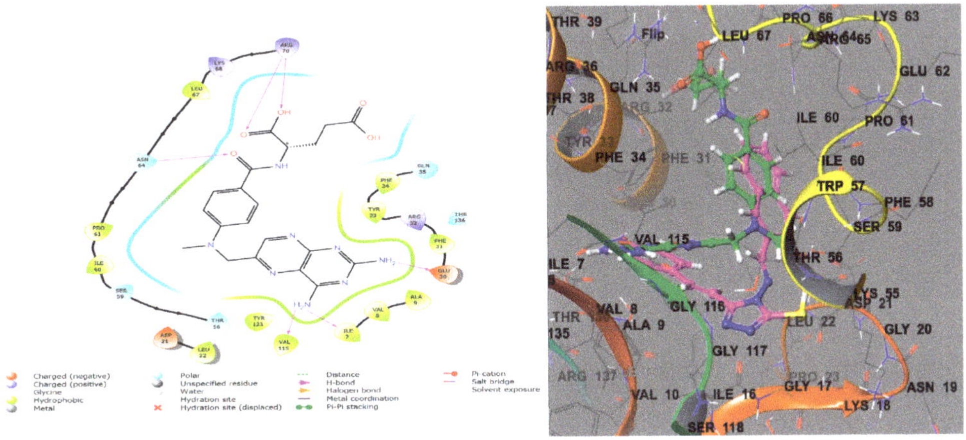

(a) (b)

Figure 2. (**a**) Depicts a 2D interaction perspective of methotrexate with PDB ID 1U72. (**b**) Superimposed image of CTh7 (magenta) and MTX (green) in the binding pocket of the DHFR (1U72) receptor with amino acid residues present in proximity of the active site.

Table 1. Molecular docking interaction analysis (PDB ID 1U72) and biological assessment of potential anticancer synthesized compounds and standard drugs.

Docking Parameter/Biological Assessment	CTh1	CTh4	CTh6	CTh7	Methotrexate
Docking score	−8.8	−8.2	−8.8	−9.9	−9.8
Binding energy [▲G (kcal/mol)]	−69.80	−60.61	−65.34	−71.38	−73.93
Hydrogen bond	2-hydrogen-bond with Glu30	2-hydrogen-bond with Glu30	2-hydrogen-bond with Glu30	Asp21, Ser59, and 2-hydrogen-bond with Glu30	Glu30, Ile7, Val115, Asn64, and 2-hydrogen-bond with Arg70
Hydrophobic interaction	Pro61, Ile60, Leu22, Tyr121, Ile16, Val115, Phe34, Phe31, Ala9, Val8, Ile7	Phe31, Phe34, Ile7, Val8, Ala9, Val115, Tyr121, Ile16, Le22, Ile60, Pro61	Leu67, Pro61, Ile60, Tyr121, Ile16, Val115, Leu22, Ile7, Val8, Ala9, Phe31, Phe34	Leu67, Pro61, Ile60, Tyr121, Ile16, Val115, Leu22, Ile7, Val8, Ala9, Phe31, Phe34	Leu67, Phe34, Phe31, Tyr33, Ala9, Val8, Ile7, Tyr121, Val115, Leu22, Pro61, Ile60
π–π stacking	Phe34	Phe34	-	Phe34	-
π–cation bond	-	-	Phe34	-	-
Salt bridge	-	-	Lys55	-	-
DHFR Inhibition IC_{50} (μM) ± SD, $n = 3$	17.03 ± 1.10	2.04 ± 0.12	10.54 ± 0.7	0.15 ± 0.02	0.088 ± 0.001
Anticancer IC_{50} (μM), $n = 3$	147.74 ± 11.4	122.17 ± 7.3	89.25 ± 4.17	8.53 ± 0.19	*

(-): No interaction; (*): Not tested.

Molecular Dynamics Simulation Studies

The stability of the docked pose of the most active DHFR inhibitor (CTh7) in the *homo sapiens*-DHFR binding site (PDB ID: 1U72) was assessed through a molecular dynamics simulation. The RMSD (root mean square deviation) of both the protein backbone and heavy atoms over the entire 100 ns MD simulation was carried out. The results indicated that the compound CTh7 exhibited a remarkable stability within the active site of the enzyme. The study further highlighted significant interactions between CTh7 and crucial amino acids such as Glu30, Trp24, Asp21, Gly20, and Phe34. Throughout the simulation, CTh7 consistently maintained interactions with the active site residues of the protein (Figure 3). The Ligand RMSD (shown on the right *Y*-axis in Figure 3a) provided insights into the stability of the ligand in the binding pocket of the protein. The 'Lig Fit Prot' shows the RMSD of a ligand when the protein–ligand complex is first aligned on the protein backbone of the reference, and then the RMSD of the ligand heavy atoms is measured. Notably, RMSD values remained below 2 Å for the first 10 ns after reaching equilibrium, confirming the conformational stability of the protein–ligand (CTh7) complex.

(a) (b)

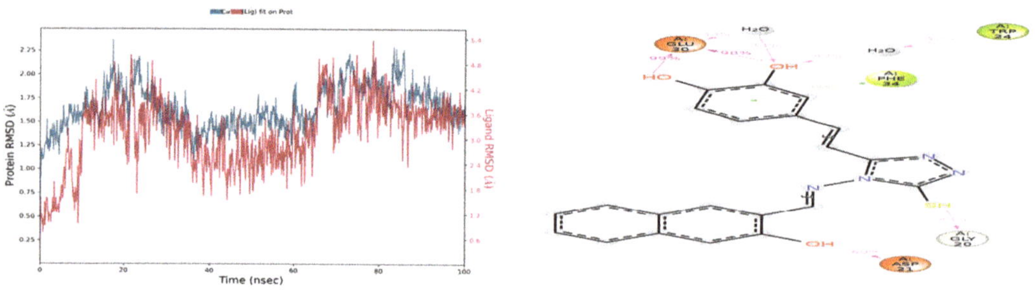

Figure 3. *Cont.*

(c)

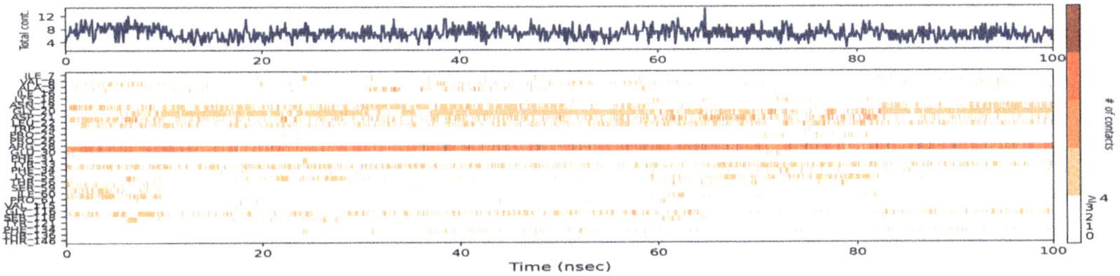

(d)

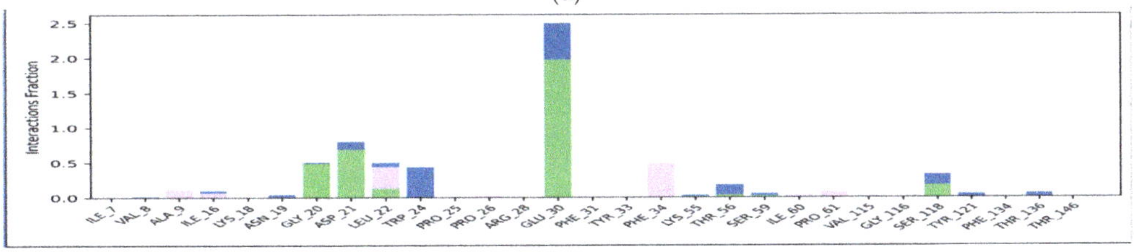

(e)

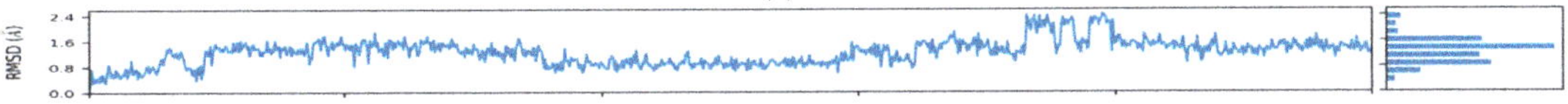

Figure 3. (**a**) RMSD relative to the initially minimized complex of the molecule during MD simulations {protein (PDB ID 1U72): left Y-axis and ligand CTh7: right Y-axis}. (**b**) A schematic of detailed ligand CTh7 atom interactions with the protein residues (interactions that occur more than **30.0%** of the simulation time in the selected trajectory). (**c**) A timeline representation of the interactions and contacts between protein and ligand CTh7. (**d**) Protein interaction with ligand (green: hydrogen bond; purple: hydrophobic; blue: water bridges). (**e**) Root mean square deviation of a ligand with respect to the reference conformation.

2.1.2. Molecular Docking Interaction Analysis of Microbe DHFR (PDB ID 2W9S)

The molecular docking technique was utilized to design caffeic acid derivatives, utilizing the *S. aureus* trimethoprim-resistant variant S1DHFR (PDB ID: 2W9S), co-crystallized with TMP (trimethoprim) as the protein target [44]. Among all the designed derivatives, the docking pose of the highly effective antibacterial compound CTh3 revealed hydrogen bonds forming between the phenolic hydroxyl groups of the compound and the Asn18 and Thr121 residues of the target protein. The nitro group of the compound formed a π–cation bond with Phe92 and a salt bridge with the polar (negatively charged) residue Asp27, as shown in Figure 4. This type of interaction resembled that of the co-crystallized ligand TMP. The compound was observed to be firmly positioned within the binding site of the enzyme and engaging in hydrophobic interactions with residues Tyr98, Ile50, Ile31, Leu28, Ala7, Val6, Ile5, Phe92, Leu20, and Ile14, resulting in excellent docking scores and binding energies (−9.8, −71.12 kcal/mol).

These desirable types of interactions might explain the superior antimicrobial activity of CTh3. Figure 5 demonstrates the various interactions between the co-crystallized ligand trimethoprim and the target protein. These interactions include hydrogen bond formation with Ile5, Phe92, and Asp27 residues and salt bridge formation with Asp27.

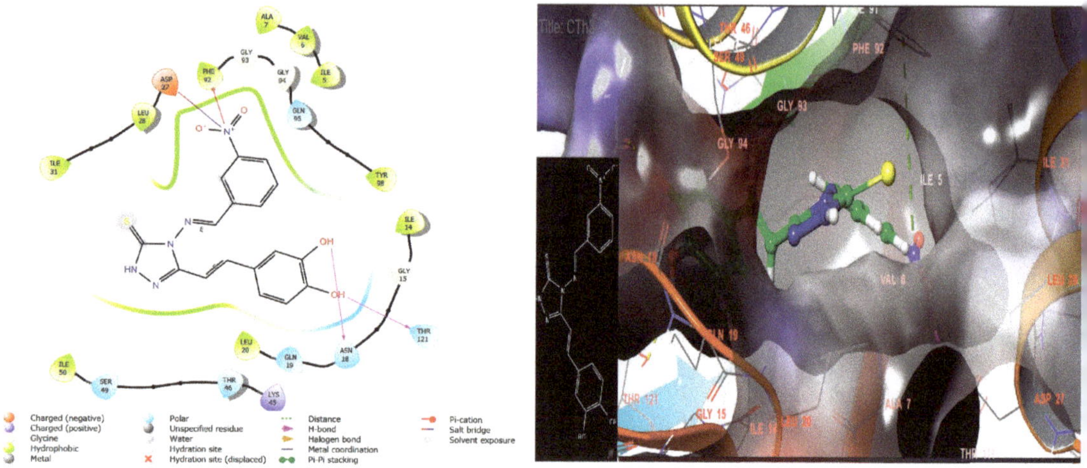

(a) (b)

Figure 4. (**a**) 2D model illustrating the interaction between compound CTh3 and active site amino acids, highlighting various types of bonds. (**b**) A three-dimensional representation of CTh3 within the active pocket of the DHFR (2W9S) receptor, highlighting amino acid residues in close proximity to the active site.

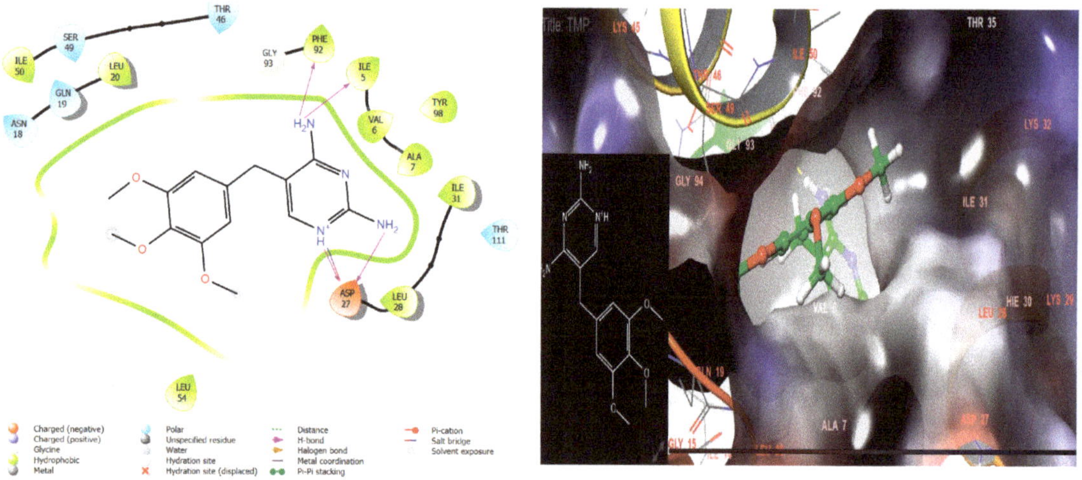

(a) (b)

Figure 5. (**a**) 2D model illustrating the interaction between trimethoprim and binding site amino acids, highlighting various types of bonds. (**b**) A three-dimensional representation of TMP within the active pocket of the DHFR (2W9S) receptor, highlighting amino acid residues in close proximity to the active site.

Another ligand, CTh4, exhibited remarkable interactions within the target protein, forming hydrogen bonds with key residues Asp27, Asn18, and Thr121, as shown in Figure 6. The phenolic group of the ligand and the triazole moiety are embedded in a hydrophobic cavity, leading to excellent docking scores and binding energies (−9.3, −72.84 kcal/mol).

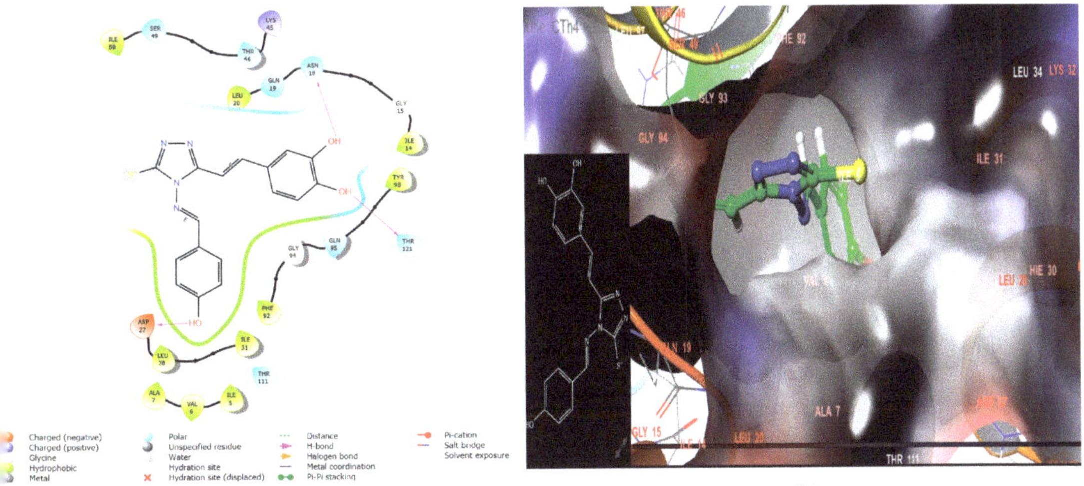

(a) (b)

Figure 6. (**a**) 2D model illustrating the interaction between compound CTh4 and active site amino acids, highlighting various types of bonds. (**b**) A three-dimensional representation of CTh4 within the active pocket of the DHFR (2W9S) receptor, highlighting amino acid residues in close proximity to the active site.

Hydrophobic interactions were also seen between the ligand and hydrophobic amino acid residues, securely holding the compound CTh4 within the compact substrate cavity (superimposed image of CTh4 and TMPs shown in Figure 7).

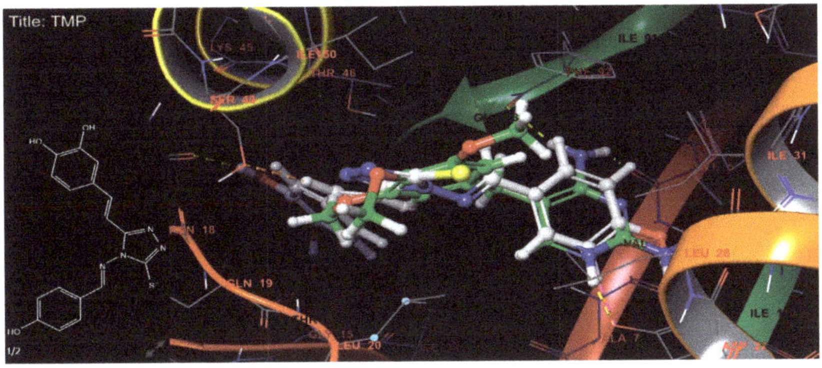

Figure 7. Superimposed image of CTh4 (green) and TMP (white) in the pocket of the DHFR (2W9S) receptor active site with amino acid residues present in proximity to the active site.

Compound CTh6 exhibited promising docking outcomes, having a docking score of −9.4 and a binding energy of −73.77 kcal/mol (Figure 8). The molecular docking analysis revealed that nearly all tested ligands occupied the same core of enzymes, surrounded by residues such as Asp 27, Leu 54, Ile 50, Phe 92, Tyr 98, Ile 14, Leu 20, Trp 22, Leu 28, Tyr 109, Ala 7, Val 6, Ile 5, Ile 31, Leu 20, Gly 93, Gly 94, Gln 19, Asn 18, Thr 121, and Thr 46 (additional information given in Table 2). These findings are further supported by the results of Ikram et al. and Adnane Aouidate [47,48]. The redocking of the co-crystallized ligand had been performed to confirm the docking protocol, resulting in a satisfactory RMSD value of 0.56 Å.

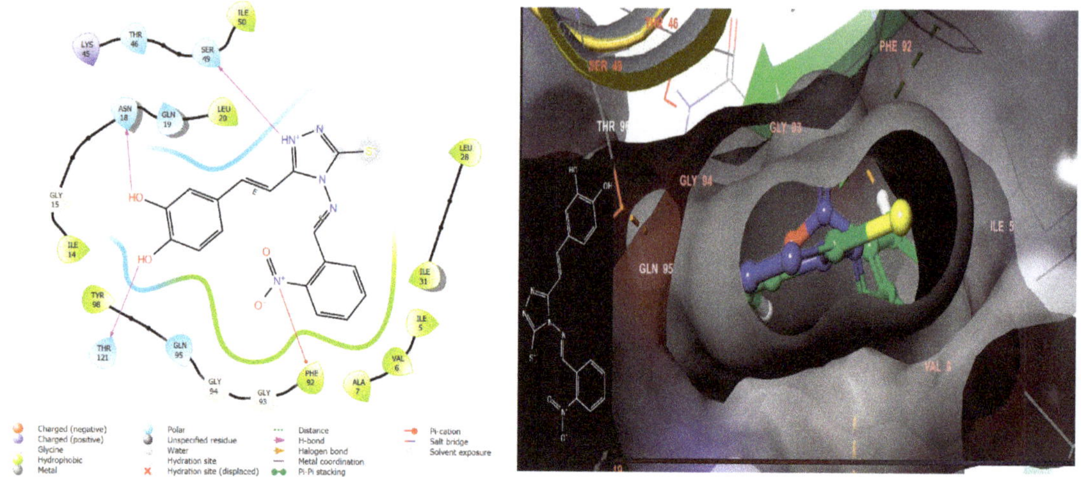

(a) (b)

Figure 8. (**a**) 2D model illustrating the interaction between compound CTh6 and active site amino acids, highlighting various types of bonds. (**b**) Three-dimensional representation of CTh6 within the active pocket of the DHFR (2W9S) receptor, highlighting amino acid residues in close proximity to the active site.

Table 2. Molecular docking interaction analysis (PDB ID 2W9S) and DHFR inhibition assay of potential antimicrobial synthesized compounds and standard drug.

Compound	Docking Score	Binding Energy [▲G (kcal/mol)]	Hydrogen Bond Interaction	Hydrophobic Interaction	Other Interaction	DHFR Inhibition [IC_{50} (μM) ± SD, n = 3]
CTh1	−8.2	−45.28	Asn18, Thr121	Ile50, Leu20, Ile14, Tyr98, Phe92, Tyr109, Ile5, Val6, Ala7, Leu28, Ile31	-	17.03 ± 1.10
CTh3	−9.8	−71.12	Asn18, Thr121	Tyr98, Ile50, Ile31, leu 28, Ala7, Val 6, Ile5, Phe 92, Leu20, Ile 14	π–cation with Phe92, salt bridge with Asp27	8.88 ± 0.4
CTh4	−9.5	−72.84	Asn18, Thr 121, Asp 27	Tyr98, Ile50, Ile31, leu 28, Ala7, Val 6, Ile5, Phe 92, Leu20, Ile 14	-	2.04 ± 0.12
CTh6	−9.4	−73.77	Asn18, Thr 121, Ser 49	Tyr98, Ile50, Ile31, Ala7, Val 6, Ile5, Phe 92, Leu20, Ile 14	π–cation Phe92	10.54 ± 0.7
CTh7	−9.6	−59.68	Asp27, Ser 49	Tyr98, Ile50, Ile31, leu 28, Ala7, Val 6, Ile5, Phe 92, Leu20, Ile 14, Trp 22	-	0.15 ± 0.02
CTh10	−9.3	−57.23	Asn 18, Ile5, Thr 121	Tyr98, Ile50, Ile31, leu 28, Ala7, Val 6, Ile5, Phe 92, Leu20, Ile 14	-	11.30 ± 0.92
CTh	−9.2	−61.45	Ala 7, Phe 92, 2-H-bond interaction with Asn18	Tyr98, Ile50, Ile14, Ile31, Ala7, Val 6, Ile5, Phe 92, Leu20, Ile 14	-	16.92 ± 1.2
TMP	−9.3	−70.12	2-H-bond with Asp 27, Phe 92, Ile5	Ala7, Ile 31, Leu 20, Leu28, Phe 92, Tyr 98, Ile 50, Leu 54, Ile 5, and Val 6	Salt bridge with Asp27	15.33 ± 1.9

(-): No interaction.

2.1.3. ADMET Evaluation

ADMET prediction plays a crucial role in drug development by enabling the rapid and preliminary testing of ADMET properties before compounds undergo an in vitro evaluation.

The evaluation of ADMET properties revealed that the new compounds exhibit excellent human intestinal absorption, as predicted by Caco-2 permeability, MDCK permeability, P-glycoprotein inhibition (Pgp), and HIA values (Table 3). This advantage is attributed to the superior lipophilicity of the compounds, which facilitates their passage through biological membranes. Human oral bioavailability predictions (F20%) indicate a good oral bioavailability for all compounds except CTh4, CTh7, and CTh. The distribution, as evaluated by PPB permeability, indicated significant plasma protein binding for all compounds except compound CTh. This finding highlights the need for tailoring compounds to improve distribution. However, the distribution assessed by VD and BBB for the novel synthesized compounds was found to be excellent. An analysis revealed that the compounds could inhibit most cytochrome P450 family isoenzymes involved in drug metabolism, particularly cytochrome 2D6 and 2C19. Excretion, evaluated in terms of clearance, highlighted that the novel derivatives exhibit excellent total clearance. Toxicity, the last parameter studied in the ADMET profile, revealed a critical drawback concerning the positive probability of AMES toxicity for the derivatives, indicating potential mutagenicity and carcinogenicity. Additionally, compounds CTh7 and CTh10 showed positive results for skin sensitivity, and compound CTh10 exhibited a respiratory toxicity probability. However, all other synthesized compounds were free from rat oral acute toxicity, respiratory toxicity, and skin sensitization. All derivatives are devoid of the probability of cardiotoxicity, indicating no inhibitory effect on hERG.

Table 3. ADMET evaluation of synthesized compounds.

Properties	**CTh1**	**CTh3**	**CTh4**	**CTh6**	**CTh7**	**CTh10**	**CTh**
Lipinski rule	Accepted	Accepted	Accepted	Accepted	Accepted	Accepted	Accepted
			Absorption				
Caco-2 permeability	−4.971	−5.046	−5.055	−5.064	−5.12	−4.947	−5.332
MDCK permeability	8.5×10^{-6}	1.28×10^{-5}	7.2×10^{-6}	9.3×10^{-6}	7.68×10^{-6}	1.9×10^{-5}	3.3×10^{-6}
Pgp-inhibitor	---	---	---	---	---	---	---
HIA	---	---	---	---	---	---	---
$F_{20\%}$	--	---	+++	---	+++	---	+++
$F_{30\%}$	+++	--	+++	-	+++	---	+++
			Distribution				
PPB	97.807%	98.39%	97.023%	98.224%	99.004	99.41%	50.853%
VD	0.455	0.22	0.404	0.273	0.51	0.924	0.365
BBB penetration	---	---	---	---	---	---	---
			Metabolism				
CYP1A2 inhibitor	++	++	++	++	++	++	-
CYP1A2 substrate	--	---	---	---	---	---	---
CYP2C19 inhibitor	-	---	--	-	-	--	---
CYP2C9 substrate	+++	---	+++	---	---	--	++
CYP2D6 inhibitor	--	--	-	---	-	--	---
CYP2D6 substrate	++	-	+	-	+	++	-
CYP3A4 inhibitor	++	++	+++	++	++	+	-
CYP3A4 substrate	--	--	--	--	--	--	---

Table 3. *Cont.*

Properties	CTh1	CTh3	CTh4	CTh6	CTh7	CTh10	CTh
			Excretion				
CL	9.676	8.233	10.334	8.609	9.194	8.044	13.595
$T_{1/2}$	0.746	0.738	0.885	0.790	0.711	0.75	0.909
			Toxicity				
hERG blockers	---	---	---	---	---	---	---
AMES toxicity	++	+++	+	++	++	+++	+++
Rat oral acute toxicity	---	---	---	---	---	---	---
Skin sensitization	-	-	-	-	++	++	--
Respiratory toxicity	---	---	---	---	---	++	---

The prediction probability values are transformed into six symbols: 0–0.1 (---), 0.1–0.3 (--), 0.3–0.5 (-), 0.5–0.7 (+), 0.7–0.9 (++), and 0.9–1.0 (+++).

2.2. Chemistry

The in silico screening revealed that the introduction of a 1,2,4-triazole ring to caffeic acid could enhance the interaction between the derivatives and the DHFR receptor. To further improve the interaction and activity, hundreds of Schiff bases of 4-[2-(4-amino-5-mercapto-4H-triazol-3-yl)vinyl]benzene-1,2-diol were designed. These derivatives were synthesized following a multistep process outlined in Scheme 1 [45].

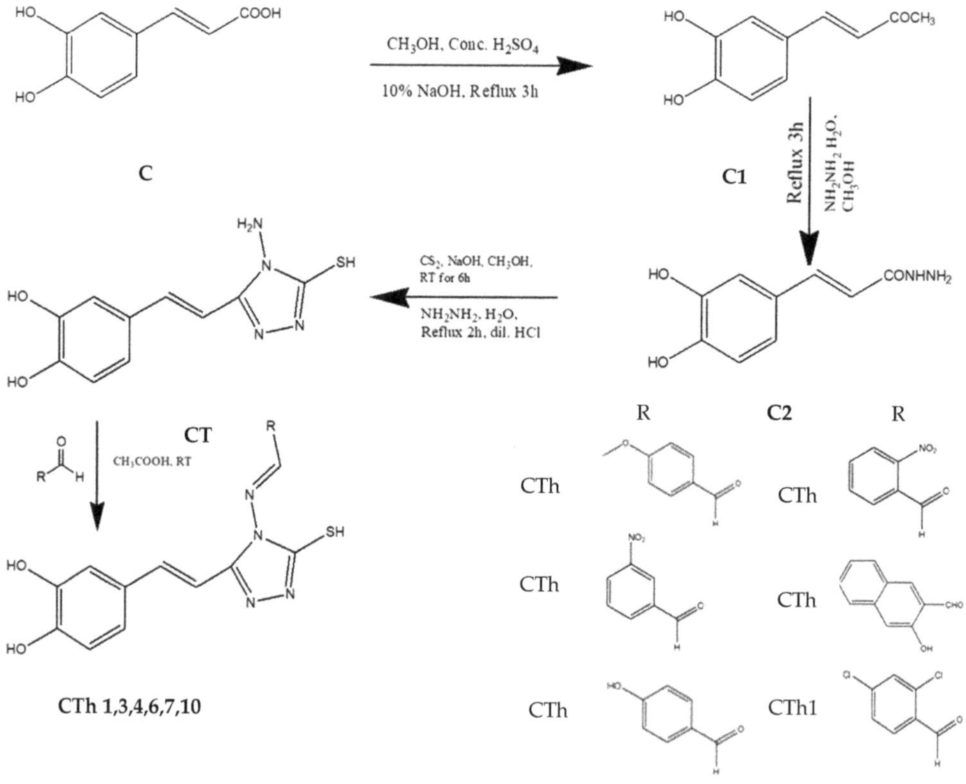

Scheme 1. General synthetic scheme to synthesize caffeic acid derivatives.

The process involved substitution, hydrazinolysis, and condensation reactions. Starting with the compound 4-(3,4-dihydroxyphenyl)but-3-en-2-one (C1), which was prepared through a Fischer–Speier esterification reaction in methanol, it was refluxed with hydrazine hydrate to obtain 3-(3,4-dihydroxyphenyl)acrylohydrazide (C2). The compound C2 underwent cyclization to form 4-[2-(4-amino-5-mercapto-4H-triazol-3-yl)vinyl]benzene-1,2-diol (CTh). Various analogues of the 1,2,4-triazole were synthesized by reacting CTh with different aldehydes and utilizing acetic acid as a catalyst. A single-spot TLC was used to confirm that the reaction had been completed successfully. The structures of the synthesized derivatives were elucidated by using ^{1}H NMR, ^{13}C NMR, FTIR, and elemental analysis. In the IR spectra of all the synthesized compounds, distinctive peaks were observed, signifying the existence of N-H stretching (3100–3300 cm^{-1}), phenolic OH stretching (3600–3400 cm^{-1}), aromatic C-H stretching (3000–2900 cm^{-1}), S-H stretching (2350–2361 cm^{-1}), C-S bending (740–760 cm^{-1}), aromatic ring vibrations (1796–1700 cm^{-1}), and -C=N stretching (1520–1590 cm^{-1}). These observations suggest the formation of the triazole ring and Schiff bases in the synthesized compounds. A band near 1600 cm^{-1} was observed in all synthesized derivatives, which highlights the formation of the imine bond C=N of the Schiff base [49]. The absence of a broad band of carboxylic acid between 3200–2500 also indicates the conversion of caffeic acid into a product. The formation of these compounds is additionally substantiated by the ^{1}H NMR signals, as evidenced by the absence of the carboxylic acid peak at around δ 12 throughout every compound that was synthesized.

Characterization of Synthesized Compounds

4-(2-(4-(4-methoxybenzylideneamino)-5-mercapto-4H-1,2,4-triazol-3-yl)vinyl)benzene-1,2-diol (**CTh1**). M.Wt.: 368.41; Yield: 72%; M.P.: 220–221; R_f: 0.67; FTIR in cm^{-1}: 3612 (O-H stretching, phenol), 3116 (C-H stretching, aromatic), 2825 (C-H stretching, aliphatic), 1197 (C-O-C stretching, methoxy), 2358 (S-H stretching), 1654 (C=N stretching, triazole), 817 (C-S bending); ^{1}H NMR (DMSO-d6, 400 MHz) in δ: 9.88 (s, 1H; Schiff base; imine), 8.66 (s, IH; OH),8.64 (s, 1H; OH)), 7.88 (d, *J* = 8.8 Hz, 1H; Ar-H), 7.82 (d, *J* = 8.7 Hz, 1H; Ar-H), 7.14 (d, *J* = 8.7 Hz, 2H; Ethylene-H), 7.06 (d, *J* = 8.7 Hz, 1H; Ar-H), 6.76–6.69 (m, 1H; Ar-H), 3.87 (s, 3H; OCH3), 3.83 (s, 1H; SH); ^{13}C NMR (DMSO-d6) in δ: 191.81(Ar-C; 2C), 164.67(imine-C; 1C), 161.00(Ar-C; 1C), 133.71(Ar-C; 3C), 131.81(Het Ar-C; 2C), 130.45(ethylene-C; 1C), 130.09(ethylene-C; 1C), 114.98(Ar-C; 3C), 114.86(Ar-C; 1C), 114.26(Ar-C; 1C), 56.16(methyl-C; 1C), 56.00(methyl-C; 1C, S), Anal calculated for $C_{18}H_{16}N_4O_3S$: C, 58.68; H, 4.38; N, 15.21; O, 13.03; S, 8.70 Found: C, 58.69; H, 4.39; N, 15.20; O, 13.04; S, 8.69.

4-(2-(4-(3-nitrobenzylideneamino)-5-mercapto-4H-1,2,4-triazol-3-yl)vinyl)benzene-1,2-diol (**CTh3**). M.Wt.: 383.38; Yield: 75%; M.P.: 232–233; R_f: 0.66; FTIR in cm^{-1}: 3617 (O-H stretching, phenol), 3114 (C-H stretching, aromatic), 2926 (C-H stretching, aliphatic), 2358 (S-H stretching), 1525 (C=N stretching, triazole), 698 (C-S bending); ^{1}H NMR (DMSO-d6, 400 MHz) in δ: 8.72 (s, 1H; Schiff base; imine), 8.47 (s, 1H; OH), 8.47 (s, 1H; OH), 7.84 (d, *J* = 8.6 Hz, 1H; Ar-H), 7.74 (d, *J* = 8.4 Hz, 3H; ethylene-H), 7.67–7.56 (m, 1H; Ar-H), 6.64 (m, 2H; Ar-H), 5.70 (s, 1H; Ar-H), 3.74 (dd, *J* = 7.1 Hz, 1H; Ar-H), 2.89 (s, 1H; SH); 13C NMR (DMSO-d6) in δ: 175.86(Ar-C; 1C), 162.77(imine-C; 1C), 161.03(Het Ar-C; 2C), 148.76–148.73(Ar-C; 1C), 134.91(Ar-C; 1C), 133.76(Ar-C; 1C), 133.64(Ar-C; 1C), 131.51(Ar-C; 1C), 131.11(ethylene-C;1C), 130.85(ethylene-C;1C), 130.64(0C, S), 126.30(Ar-C; 1C), 123.97(Ar-C; 1C), 123.11(0C, S), 121.82(Ar-C; 1C), 119.23(Ar-C; 1C), 115.92(Ar-C; 1C); Anal calculated for $C_{17}H_{13}N_5O_4S$: C, 53.12; H, 3.67; N, 18.22; O, 16.65; S, 8.34 Found: C, 53.10; H, 3.68; N, 18.23; O, 16.66; S, 8.35.

4-(2-(4-(4-hydroxylbenzylideneamino)-5-mercapto-4H-1,2,4-triazol-3-yl)vinyl)benzene-1,2-diol (**CTh4**). M.Wt.: 354.08; Yield: 77%; M.P.: 219–220; R_f: 0.59; FTIR in cm^{-1}: 3613 (O-H stretching, phenol), 3082 (C-H stretching, aromatic), 2360 (S-H stretching), 1620 (C=N stretching, triazole), 730 (C-S bending); ^{1}H NMR (DMSO-d6, 400 MHz) in δ: δ 8.91 (s, 1H; Schiff base; imine), 8.70 (s, 2H; OH), 8.57 (d, *J* = 7.7 Hz, 1H; Ar-H), 8.37 (d, *J* = 8.3 Hz, 1H; Ar-H), 8.32 (d, *J* = 7.7 Hz, 2H; ethylene-H), 7.82 (t, *J* = 8.0 Hz, 3H; Ar-H), 2.89 (s, 1H; SH); ^{13}C NMR (DMSO-d6) in δ: 161.03(imine-C; 1C), 148.76–148.73 (Het Ar-C; 2C, S), 148.63(Ar-C;

2C), 135.67 (Ar-C; 2C), 134.91 (Ar-C; 1C), 133.76–133.64 (Ar-C; 3C), 131.51(ethylene-C; 1C), 131.11 (ethylene-C; 1C), 130.74 (Ar-C; 1C), 126.30(Ar-C; 1C), 124.77(Ar-C; 1C), 124.15 (Ar-C; 1C); Anal calculated for $C_{17}H_{14}N_4O_3S$: C, 57.62; N, 15.81; H, 3.98; O, 13.54; S, 9.05 Found: C, 57.64; N, 15.82; H, 3.99; O, 13.53; S, 9.04

4-(2-(4-(2-nitrobenzylideneamino)-5-mercapto-4H-1,2,4-triazol-3-yl)vinyl)benzene-1,2-diol (**CTh6**). M.Wt.: 383.38; Yield: 72%; M.P.: 216–217; R_f: 0.6; FTIR in cm^{-1}: 3615 (O-H stretching, phenol), 3112 (C-H stretching, aromatic), 2359 (S-H stretching), 1652 (C=N stretching, triazole), 732 (C-S bending), 1525 (NO_2 stretching); ^{1}H NMR (DMSO-d6, 400 MHz) in δ: 8.92 (s, 1H; Schiff base; imine), 8.71 (s, 2H; OH), 8.41–8.29 (m, 3H; Ar-H), 7.82 (d, J = 7.6 Hz, 2H; Ar-H), 2.73 (s, 1H; SH); ^{13}C NMR (DMSO-d6) in δ: 160.95 (imine-C; 1C), 148.82–148.66 (Het Ar-C; 2C), 135.70(Ar-C; 2C), 134.90(Ar-C; 2C), 131.11(ethylene-C; 3C), 130.73(Ar-C; 1C), 126.28(Ar-C; 2C), 124.77(Ar-C; 1C), 123.11(Ar-C; 3C); Anal calculated for $C_{17}H_{13}N_5O_4S$: C, 53.12; H, 3.67; N, 18.22; O, 16.65; S, 8.34 Found: C, 53.13; H, 3.69; N, 18.23; O, 16.66; S, 8.35.

4-(2-(4-(4-naphthalen-2-ol)-5-mercapto-4H-1,2,4-triazol-3-yl)vinyl)benzene-1,2-diol (**CTh7**). M.Wt.: 404.44; Yield: 72%; M.P.: 201–202; R_f: 0.55; FTIR in cm^{-1}: 3613 (O-H stretching, phenol), 3115 (C-H stretching, aromatic), 2929 (C-H stretching, aliphatic), 2358 (S-H stretching), 1621 (C=N stretching, triazole), 784 (C-S bending); ^{1}H NMR (DMSO-d6, 400 MHz) in δ: 10.01 (s, 1H; Schiff base; imine), 8.67 (d, J = 8.6 Hz, 1H; Ar-H), 8.61 (s, 1H-Ar -H), 8.47 (s, 1H; Ar -H), 8.05 (d, J = 9.0 Hz, 1H; ethylene-H), 7.93 (dd, J = 8.0, 1.4 Hz, 1H; naphthalene-H), 7.63 (dd, J = 8.4, 6.8, 1.5 Hz, 1H; naphthalene- H), 7.57 (dd, J = 8.5, 6.8, 1.5 Hz, 1H; naphthalene- H), 5.70 (s, 3H; OH), 5.29 (d, J = 2.8 Hz, 1H; Ar -H), 4.50 (s, 1H; SH); ^{13}C NMR (DMSO-d6) in δ: 162.77(Ar-C; 1C), 161.03(imine-C; 1C), 148.76–148.63(Het Ar-C; 2C), 135.67(Ar-C; 2C), 134.91(Ar-C; 1C), 133.76(Ar-C; 1C), 133.64(Ar-C; 1C), 131.51(ethylene-C; 1C), 131.11(ethylene-C; 1C), 130.74(Ar-C; 1C), 130.64(Ar-C; 1C), 126.30(Ar-C; 1C), 124.77(Ar-C; 1C), 124.15(Ar-C; 1C), 123.97(Ar-C; 1C), 123.11(Ar-C; 3C), 121.82(Ar-C; 1C); Anal calculated for $C_{21}H_{16}N4O_3S$: C, 62.36; N, 13.85; H, 3.99; O, 11.87; S, 7.93 Found: C, 62.37; N, 13.84; H, 4.00; O, 11.87; S, 7.94.

4-(2-(4-((2,4-dichlorocyclohexa-1,5-dienyl)methyleneamino)-5-mercapto-4H-1,2,4-triazol-3-yl) vinyl)benzene-1,2-diol (**CTh10**). M.Wt.: 409.28; Yield: 72%; M.P.: 222–223; R_f: 0.62; FTIR in cm^{-1}: 3741 (O-H stretching, phenol), 3114 (C-H stretching, aromatic), 2924 (C-H stretching, aliphatic), 2361 (S-H stretching), 767 (C-S bending), 1651 (C=N stretching, triazole), 1521 (C=C stretching, aromatic), 767 (C-Cl aryl stretching); ^{1}H NMR (DMSO-d6, 400 MHz) in δ: 10.28 (s, 2H; OH), 7.89 (s, 1H; Schiff base; imine), 7.88–7.85 (m, 2H; Ar-H), 7.73 (t, J = 2.5 Hz, 1H; Ar-H), 7.65 (d, J = 1.9 Hz, 1H; Ar-H), 7.62 (d, J = 2.1 Hz, 1H; ethylene-H), 7.57 (dd, J = 2.3 Hz, 1H; Ar-H), 6.69 (d, J = 2.4 Hz, 1H; Ar-H), 4.10–4.04 (m, 1H; Ar-H), 3.77–3.67 (m, 4H; Ar-H), 2.55 (s, 1H; SH); 13C NMR (DMSO-d6) in δ: 189.37(Ar-C; 3C), 172.61 (imine-C; 1C), 147.97(Ar-C; 1C), 140.21(Het Ar-C; 1C), 137.62(Het Ar-C; 1C), 131.55(Ar-C; 1C), 131.42(ethylene-C; 2C), 130.77(Ar-C; 1C), 128.78(Ar-C; 1C), 115.83(Ar-C; 1C), 61.62(1C, S), 60.59(Ar-C; 1C), 60.45(Ar-C; 1C), 56.07–56.00(Ar-C; 2C); Anal calculated for $C_{17}H_{14}C_{12}N4O_2S$: C, 49.89; H, 3.45; N, 13.69; O, 7.82; S, 7.83 Found: C, 49.91; H, 3.46; N, 13.70; O, 7.82; S, 7.84.

4-(2-amino 1-mercapto-4H-1,2,4-triazol-3-yl)vinyl)benzene-1,2-diol (**CTh**). M.Wt.: 250.28; Yield: 89%; M.P.: 235–236; R_f: 0.5; FTIR in cm^{-1}: 2925 (C-H stretching, aromatic), 2855 (C-H stretching, aliphatic), 3198 (N-H stretching, amine), 2360 (S-H stretching), 1680 (C=N stretching, triazole), 769 (C-S bending); ^{1}H NMR (DMSO-d6, 400 MHz) in δ: 8.72 (s, 2H; OH), 8.47 (s, 1H; Ar-H), 7.84 (d, J = 8.6 Hz, 2H; ethylene-H), 7.74 (d, J = 8.4 Hz, 2H; Ar-H), 5.70 (s, 2H; NH2), 3.75 (s, 1H; SH); ^{13}C NMR (DMSO-d6) in δ: 161.03 (Het Ar-C; 1C), 148.63(Ar-C; 2C), 135.67 (Het Ar-C; 1C, S), 134.91 (Ar-C; 1C), 131.11 (ethylene-C; 2C), 126.30 (Ar-C; 1C), 124.77 (Ar-C; 1C), 123.11(Ar-C; 1C); Anal calculated for $C_{10}H_{10}N_4O_2S$: C, 47.99; N, 22.39; H, 4.03; O, 1.79; S, 12.81; Found: C, 48.01; N, 22.39; H, 4.05; O, 1.78; S, 12.82.

2.3. Biological Activity

2.3.1. DHFR Inhibition Assay

To evaluate the efficacy, mechanism, and confirmation of the results obtained from molecular docking, the in vitro inhibitory effect of the synthetic derivatives on DHFR was assessed. This assessment was carried out using the DHFR assay kit (catalog number CS0340, Sigma-Aldrich, Burlington, MA, USA) by using a spectrophotometric method. The enzymatic degradation of dihydrofolic acid to tetrahydrofolic acid was monitored at 340 nm for 180 s, following the procedure recommended by the manufacturer of the assay kit. IC_{50} values of the synthesized derivatives were calculated and presented in Table 2. The obtained in vitro activity results aligned with the computational predictions, with a few minor deviations. Methotrexate, an established standard for DHFR inhibition, was included in the screening process under identical conditions for comparison purposes and as a positive control. The results demonstrated the significant potency of the studied compounds in inhibiting the target enzyme at micromolar concentrations. Among all the synthesized derivatives, the most potent lead compound was CTh7, with an IC_{50} value of 0.15 μM. It might be due to the fact that it forms hydrogen bonding with key amino acid residues such as Glu30, Ser59, and Asp21, and π–π stacking with Phe34 against the modeled *homo sapiens* DHFR. Compound CTh4 also demonstrated a remarkable inhibition against DHFR, with an IC_{50} value of 2.04 μM. As the IC_{50} range for Schiff bases of 4-[2-(4-amino-5-mercapto-4H-triazol-3-yl)vinyl]benzene-1,2-diol fell between 0.15 and 17.03 μM, these findings suggest that DHFR may be the molecular target of these analogues, potentially exerting an antimetabolite effect. These compounds exhibit strong DHFR inhibitory activity, which is consistent with their cytotoxic and antibacterial properties.

2.3.2. Assessment of Anticancer Activity

The anticancer potential of synthesized derivatives was assessed by targeting the MCF-7 breast cancer cell line. The results, summarized in Table 1, revealed that compound CTh7 exhibited remarkable anticancer activity, with an IC_{50} value of 8.53 μM, but less potent than the standard drug doxorubicin (IC_{50} = 3.62 μM). This outcome is in line with the findings from docking studies and DHFR inhibition assays, suggesting that CTh7 might act as a potent DHFR inhibitor. These results provide evidence that caffeic derivatives can be effectively designed to function as DHFR inhibitors, which is a promising direction for anticancer drug development. However, a further evaluation on normal cells is necessary in order to ascertain whether the observed effects are truly selective towards cancer cells and not simply indicative of the general toxicity of the compounds. Overall, this research sheds an important light on the potential of this novel class of compounds as anticancer agents, opening avenues for future studies and further optimization to improve their efficacy and therapeutic potential against breast cancer.

2.3.3. Antimicrobial Activity

The minimum concentration required to halt the growth of a micro-organism within 18 to 24 h is referred to as the minimum inhibitory concentration (MIC) [50]. The serial dilution technique was utilized to evaluate the antimicrobial activity of synthesized derivatives against Gram-negative, Gram-positive, and fungal strains. Trimethoprim, ampicillin, tetracycline, fluconazole, and voriconazole were used as reference drugs. To dissolve the reference and sample derivatives, DMSO was employed as a solvent. The synthesized compounds were evaluated for their antibacterial effectiveness against *E. coli*, *P. aeruginosa*, and *S. aureus*, and their antifungal potential was tested against *Candida albicans* and *Aspergillus niger*. Based on the results of the MIC values presented in Table 4, it was evident that compound CTh3 displayed a remarkable efficacy against *S. aureus* with an MIC value of 5 μM, surpassing the performance of trimethoprim (63 μM) and ampicillin (11 μM). Compound CTh6 also demonstrated excellent activity against *S. aureus*, exhibiting an MIC value of 10 μM. Additionally, compound CTh4 exhibited notable effectiveness against *P. aeruginosa*, *E. coli*, and *C. albicans*, with MIC values of 5 μM, 5 μM, and 11 μM, respectively, outperforming the reference drugs. The synthesized compounds showed significant antibacterial activity, comparable to or even better than standard drugs such as trimethoprim and ampicillin. However, their inhibitory effects against *C. albicans* and *A. niger* were relatively lower compared to their antibacterial activity, with only CTh3, CTh4, and CTh10 showing marginal effectiveness. Compound CTh3 exhibited an MIC of 20 μM against *C. albicans*, while compound CTh10 displayed the highest efficacy against *A. niger* with an MIC of 39 μM. *E. coli* proved to be the most sensitive to all synthesized compounds, and *A. niger* was the least sensitive. Overall, the antibacterial activity was influenced by the scaffold type and the substituent in the aromatic moiety. It is concluded from this research that Schiff bases containing electron-withdrawing groups such as -NO_2 and -OH on their aromatic ring exhibit significant antimicrobial activity. These groups are found to interact through various bonds, such as hydrogen bonding, π–cation interactions, and salt bridge formations with essential amino acid residues of DHFR, which might contribute to their effectiveness against micro-organisms.

Table 4. Antimicrobial activity (MIC in μM) of synthesized derivatives.

Compound	MIC (μM)				
	Bacterial Strain			Fungal Strain	
	S. aureus	*P. aeruginosa*	*E. coli*	*C. albicans*	*A. niger*
CTh1	43	86	86	173	-
CTh3	5	10	10	20	83
CTh4	22	5	5	11	-
CTh6	10	10	20	41	83
CTh7	19	19	39	-	-
CTh10	19	39	39	39	39
CTh	63	127	255	-	127
Trimethoprim	63	31	31	255	127
Ampicillin	11	22	11	NT	NT
Tetracycline	144	36	36	NT	NT
Fluconazole	NT	NT	NT	0.81	260.12
Voriconazole	NT	NT	NT	0.34	5

NT (Not tested): Ampicillin and tetracycline are antibiotics and were, therefore, not tested against fungal strains; trimethoprim was tested against all strain as it was the co-crystallized ligand of selected PDB ID 2W9S during the design process. Similarly, fluconazole and voriconazole are antifungals and were not tested against bacterial strains. -: (No inhibition observed in the studied range).

3. Structure–Activity Relationship

To study the correlation between biological activity and the structures of newly synthesized caffeic acid derivatives, an SAR analysis was performed, and the findings are given below:

- The introduction of nitro substitution at the ortho and meta positions of an aromatic ring has been found to enhance both DHFR inhibitory activity and antimicrobial activity in derivatives. For instance, compounds CTh3 and CTh6 exhibit good DHFR inhibitory and excellent antimicrobial activity, with CTh3 showing an MIC of 5 μM against *S. aureus*.
- Incorporating a hydroxyl group, as seen in CTh4, significantly increases the DHFR inhibitory activity, likely due to the hydrogen bond formation with the essential negatively charged amino acid residue Asp27. This compound also demonstrates excellent antimicrobial activity with MICs of 5 μM, 5 μM, and 11 μM against *E. coli*, *P. aeruginosa*, and *C. albicans*, respectively.

- Furthermore, when a chloro group is substituted on the aromatic ring of a Schiff base, the antifungal activity is further enhanced, particularly against *A. niger*. For example, compound CTh10 exhibits strong activity against *A. niger* with an MIC of 39 μM.
- The introduction of a bulky aromatic group leads to a sharp increase in DHFR inhibitory activity and remarkable anticancer activity. Compound CTh7, for example, displays an IC_{50} value of 8.53 μM against the MCF-7 cell line.

Upon analyzing the biological activities, it can be concluded that the occurrence of a phenolic hydroxyl group in caffeic acid is essential for DHFR inhibitory activity. Additionally, introducing a triazole ring on the carboxylic functional group of the acid further enhances antimicrobial and anticancer activity. The Schiff base formation with the triazole ring significantly improves biological activity. Substituting electronegative groups like nitro, hydroxyl, and chloro groups enhances antimicrobial activity remarkably. The incorporation of bulky groups in Schiff bases may increase the DHFR inhibition and anticancer potential against the MCF-7 tumor cell line, possibly due to the increased lipophilic character of the compound. Overall, these findings suggest that specific substitutions and modifications can significantly impact the biological activities of the compounds under study, as summarized in Figure 9.

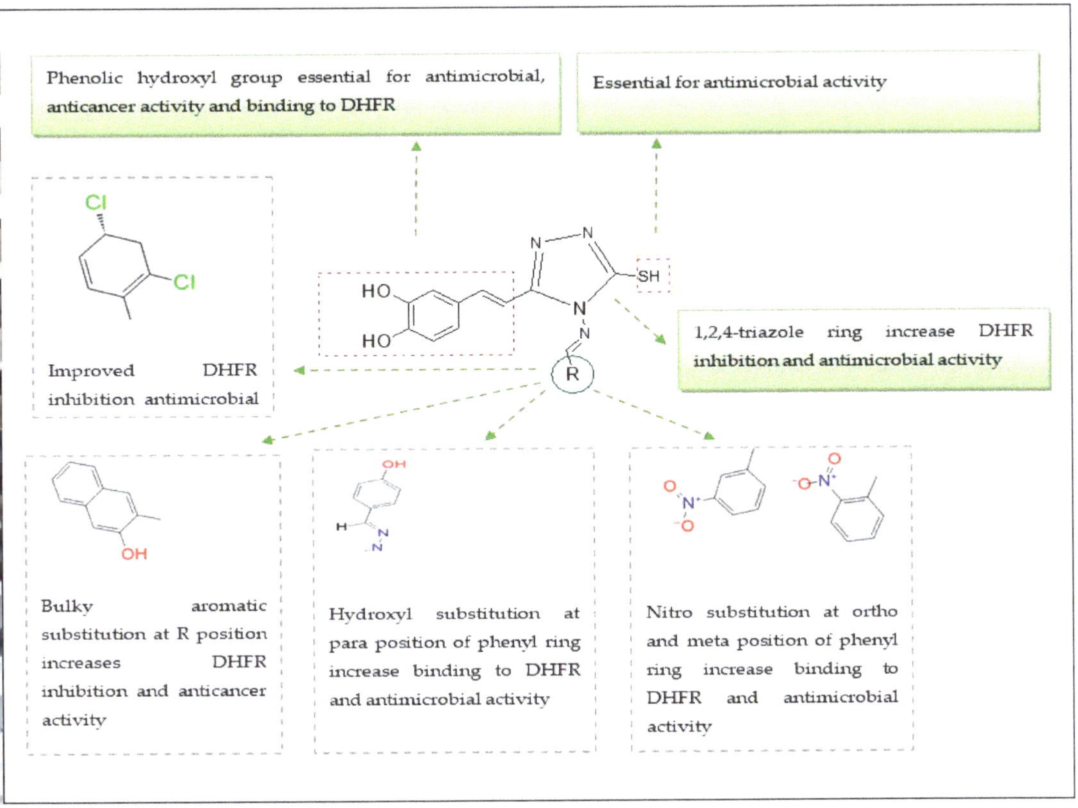

Figure 9. Explication of SAR of newly synthesized caffeic acid derivatives.

4. Materials and Methods

4.1. Instruments and Chemicals

The research employed analytical-grade chemicals and reagents, which were obtained from Merck (Darmstadt, Germany), Sigma-Aldrich (Burlington, MA, USA), Loba Chemie (Mumbai, India), and CDH (New Delhi, India). The progression of the reactions was monitored on pre-coated TLC plate (60 GF 254, silica gel 0.25 mm; Merck, Germany). Dichloromethane, methanol, and acetic acid in a ratio of 9.4:0.5:0.1 mL, respectively, served as the developing solvent system, and visualization of spots was carried out under a UV lamp. The melting point of synthesized compounds was determined utilizing the open capillary method using a sonar melting point apparatus, and the reported values were uncorrected. The infrared spectra of the synthesized compounds were recorded using an FTIR Bruker ATR instrument (Vertex 70v; Bruker Optik GmbH, Rosenheim, Germany) in cm^{-1}. The ^{1}H NMR and ^{13}C NMR spectra were acquired using a Bruker Avance II 400 NMR (Bruker, Wissembourg, France) spectrophotometer operating at frequencies of 400 MHz and 100 MHz, respectively. DMSO-d6 was used as the solvent, while tetramethylsilane was utilized as the internal standard. Chemical shifts were recorded in δ unit. Additionally, elemental analysis was conducted using the Perkin–Elmer 2400 Series II CHNS-O analyzers. For the in vitro assessment of DHFR inhibitory activity, a DHFR assay kit (Sigma-Aldrich, catalog number CS0340) was used.

4.2. Molecular Modelling

4.2.1. Molecular Docking

A molecular docking investigation was conducted to check enzyme–ligand interaction using Maestro Glide software 13.1 (New York, NY, USA) [51]. X-ray crystal structures of DHFR were retrieved from the RCSB Protein Data Bank with PDB IDs 1U72 for anticancer and 2W9S for antimicrobial drug design [52]. Selections of PDB IDs were carried out on the basis of resolution and species. The caffeic acid derivatives library was built using the Chemdraw Ultra software 12.0, and Maestro's LigPrep tool was utilized to prepare ligands for energy minimization, correct the co-ordinates, and generate tautomers to obtain appropriate conformation; 32 stereoisomers per ligand were allowed; at target, pH 7 +/− 2 was set as a default option, and the force field was OPLS3e. The protein preparation wizard of Schrödinger was used to prepare proteins to obtain optimized and energy-minimized protein configurations [53–55]. DHFR protein is directly downloaded on the Maestro workspace interface from the RCSB Protein Data Bank. The protein preparation involved executing pre-process steps, and the Epik tool was employed to ionize heteroatoms at biological pH, ensuring the maintenance of a biosimilar environment. Following the pre-process, the ultimate step involved obtaining an energy-minimized structure using OPLS3e as the force field. Grid was generated by picking molecules, and docking site (active site) validation was carried out by splitting co-crystallized ligands from the minimized prepared protein complex and then re-docking to the active site. For the 1U72 and 2W9S PDB IDs, the RMSD (root mean square deviation) values were found to be 1.19 and 0.56, respectively, which is very satisfactory for approving the docking protocol. To analyze the interaction of proteins and ligands, docking with extra precision (XP) was used. The most promising ligand was chosen for laboratory synthesis and subsequent assessment of its pharmacological activity.

4.2.2. Binding Free Energies and MM/GBSA

The MM/GBSA energy calculation prime module of Maestro was used to determine the free binding energy of designed derivatives of caffeic acid. This was accomplished using the VSGB solvation model and other default set options for calculating the binding energy. This approach is widely acknowledged and accepted for assessing the free binding energy of designed compounds to proteins. It aids in rationalizing experimental observations and virtual screening outcomes, effectively distinguishing between drugs and binders only [56,57]. The following formula was used to determine the energy differences:

$$\Delta E = Ecomplex - Eligand - Eprotein$$

4.2.3. ADMET Evaluation

In this research, the ADMETlab 2.0 online tool was used to evaluate various pharmacokinetic parameters and toxicity properties [58]. The ADMET prediction for the synthesized compounds was predicted to ensure adherence to Lipinski's rules of five, which indicate their drug-like characteristics as designed ligands. The absorption of ligands was assessed by predicting values across various factors, including intestinal absorption, membrane and skin permeability, P-glycoprotein substrate or inhibitor status, and human intestinal absorption (HIA). The drug distribution of ligands was evaluated by predicting parameters such as plasma protein binding (PPB), volume of distribution (VD), and blood–brain barrier permeability (BBB) values. Metabolism predictions were carried out by CYP (cytochrome P450) models for substrate or inhibition, while excretion was assessed using clearance (CL) and drug half-life (T1/2). Toxicity is predicted using various parameters like AMES toxicity, hERG inhibition, rat oral acute toxicity, respiratory toxicity, and skin sensitization.

4.2.4. Molecular Dynamics Simulations

MD simulations of most potential compound was conducted at the binding site of *human*-DHFR (PDB ID: 1U72) using Desmond package within Maestro. The construction of the system involved utilizing the OPLS-AA force field within an explicit solvent, employing the single-point-charge (SPC) water model, denoted as OPLS-AA/SPC, and adding Cl^- counter-ions to neutralize the overall charge. The resulting complex system consisted of around 28,215 atoms. Before running equilibration and extensive-production MD simulations, the systems were minimized and pre-equilibrated using Desmond's default relaxation protocol. The MD simulations ran for 100 ns, employing the NPT ensemble for the integration of equations of motion. The temperature was held at 300 K through the Nosé–Hoover thermostat method with a relaxation time of 1 ps. Trajectory recording occurred at 20 ns intervals throughout the entire MD run [59].

4.3. Synthesis of Caffeic Acid Derivatives

4.3.1. Synthesis of 4-(3,4-Dihydroxyphenyl)but-3-en-2-one (C1)

A solution was prepared by dissolving 0.131 mol of caffeic acid in 250 mL of methanol, and then concentrated H_2SO_4 (0.131 mol) was added dropwise to the solution (Scheme 1). The resultant solution was refluxed for duration of 3 h. The single-spot TLC technique was utilized to check the completion of the reaction. After reflux, the mixture was allowed to cool to room temperature, and any excess methanol was subsequently distilled off. The resultant product was rendered alkaline by the addition of a 10% NaOH solution in cold water. A brown-colored product identified as C1 was obtained, filtered, and subsequently washed with cold water. To purify the product further, it was recrystallized from a 1:1 mixture of methanol and water.

4.3.2. Synthesis of 3-(3,4-Dihroxyphenyl)acrylohydrazide (C2)

A solution was prepared by dissolving 0.078 mol of compound C1 in a mixture of hydrazine hydrate (0.233 mol) in methanol (50 mL). The resulting solution was then refluxed for a duration of 3 h. A solid, sticky, brown-colored residue was obtained, identified as C2. The end of the reaction was verified by conducting a single-spot TLC. The product was filtered by suction filtration, followed by washing with methanol, and, ultimately, dried.

4.3.3. Synthesis of 4-[2-(4-Amino-5-mercapto-4H-triazol-3-yl)vinyl] benzene-1,2-diol (CTh)

Carbon disulfide (0.031 mol) and compound C2 were added to a methanol solution of NaOH (0.031 mol) while stirring continuously. The reaction mixture was stirred at room temperature for a period of 6 h. The precipitates formed during the reaction were filtered using suction, washed with water, and then dried under vacuum. The product obtained from the initial step was directly utilized in the subsequent step without undergoing additional purification. This product was subjected to a treatment with 0.062 mol of hydrazine hydrate and water (40 mL), and the resulting mixture was refluxed for a period of 2 h. During the refluxing, hydrogen sulfide gas evolved, and a light brown homogeneous solution was obtained. The completion of the reaction was verified by conducting a single-spot TLC technique. The resulting solution was diluted with ice-cold water. The solution was subsequently acidified with dilute hydrochloric acid, leading to the formation of light brown precipitates of compound CTh. The precipitates of CTh were filtered, washed, and subsequently recrystallized with alcohol to yield the purified product.

4.3.4. General Synthetic Procedure for the Synthesis of Schiff Bases

Dissolve compound CTh (0.002 mol) in 10 mL of dimethylformamide (DMF). Different aldehydes (0.005 mol) and 2–3 drops of glacial acetic acid (as a catalyst) were added to the solution of compound CTh. Allow the mixture to stir for 4 h at room temperature. After the completion of the reaction, pour the resulting product into crushed ice to induce precipitation. The precipitates so obtained were filtered by suction, rinsed with cold water, and subsequently subjected to recrystallization in alcohol to obtain purified products.

4.4. Assessment of Biological Activity

4.4.1. In Vitro DHFR Inhibition Assessment

To evaluate the ability of the synthesized derivatives to block DHFR activity, a spectrophotometric assay technique was utilized at 340 nm for 180 seconds to detect the enzyme-catalyzed conversion of dihydrofolic acid to tetrahydrofolic acid. The DHFR assay kit (Sigma-Aldrich, catalog number CS0340) was used in this assay. The kit was tested using recombinant DHFR, NIH 3T3, A431, and CHO cell lines, as well as rat kidney, brain, liver, and muscle tissue extracts. Dihydrofolic acid, NADPH, standard drug solution, assay buffer, and DHFR stock solutions were prepared according to the instructions given in the technical bulletin that came along with the test kit. Except for the assay buffer, all stock solutions were stored on ice.

Each well of a 96-well clear microplate was filled with a calculated amount of assay buffer (190 μL), followed by 6 μL of diluted DHFR, which was thoroughly mixed. Different concentrations of serially diluted synthesized compounds (0.01–10 μM) and diluted NADPH (1.2 μL) were introduced to the wells. The microplate was then covered with parafilm, and the contents were mixed by inversion. Following the addition of 1 μL of substrate dihydrofolic acid, the absorbance was measured at 340 nm over a period of 180 s using an ELISA reader (Gen BioTek Microplate Reader, Winooski, VT, USA) in kinetic mode at room temperature. Plotting the relationship between absorbance and time generates the slope for each test inhibitor sample. The relative inhibition percentages were calculated using Equation (1) [60–62].

$$\%\ \text{relative inhibition} = (\text{slope of [EC]} - \text{slope of [S]}/\text{slope of [EC]}) \times 100 \tag{1}$$

where S is the sample of interest, EC (enzyme control).

4.4.2. Anticancer Activity (MTT Assay)

The potential newly synthesized derivatives were evaluated for their cytotoxic activity on the MCF-7 cell line (sourced from NCCS in Pune, India) using the MTT (3-[4,5-dimethylthiazol-2-yl]-2,5-diphenyl tetrazolium bromide) assay. To perform the assay, 10,000 cells were placed in each well of 96-well plates and left to incubate for 24 h in Dulbecco's Modified Eagle Medium (DMEM). The medium was supplemented with 1% antibiotic solution and 10% FBS (fetal bovine serum), and the incubation occurred at a temperature of 37 °C in an environment containing 5% CO_2. The next day, the cells were exposed to various concentrations of the derivatives, which had been prepared in an incomplete medium. After a 24 h incubation period, MTT solution (603 µM) was introduced into the cell cultures. The cultures were then subjected to additional 2 h incubation. Upon concluding the experiment, the culture supernatant was extracted, and the cellular matrix was dissolved in 100 µL of DMSO. The absorbance of the resulting solution underwent evaluation using an ELISA plate reader (iMark, Biorad, Hercules, CA, USA) at 540 nm. Triplicate measurements were performed to enhance the reliability and reproducibility of the results obtained [63–66].

4.4.3. Antimicrobial Activity

The antibacterial and antifungal activity of potential synthesized compounds was evaluated using the broth microdilution method to determine their minimum inhibitory concentration (MIC). All the synthesized derivatives were tested against bacterial strains *S. aureus* (ATCC FDA209-P), *P. aeruginosa* (ATCC 27853), and *E. coli* (ATCC 25922) and fungal strains *A. niger* (MTCC 1688) and *C. albicans* (ATCC SC5314). Individual MIC (µM) values were determined using serial dilution technique with Mueller–Hinton Broth (MHB) media for bacterial growth and Roswell Park Memorial Institute (RPMI) media for fungal growth. Ampicillin, trimethoprim (TMP), tetracycline, voriconazole, and fluconazole were used as reference drugs to set benchmarks. Bacterial strains were cultured in MHB and incubated at 37 ± 1 °C for 24 h, while the fungal strain (*C. albicans*) was grown in RPMI medium at 37 ± 1 °C for 48 h, and the growth of *A. niger* was observed at 25 ± 1 °C for 7 days in the same RPMI medium. The turbidity of the spore suspension was corrected to 0.5 McFarland by adding distilled water. Stock solutions of synthesized compounds were prepared in DMSO solvent and serially diluted to determine the MIC value. Each spore (0.1 mL) was put into each well holding the finally serially diluted synthesized compound concentration, standards, and controls. Optically clear (MIC-0) endpoint criteria, which defined MIC as resulting in 99.9% suppression of visible bacterial and fungal growth following incubation, were used. colony-forming unit (CFU) was also determined by taking 100 µL cultures from the twelfth well, which were dispensed to the tubes containing 900 µL of saline and serially diluted. Agar plate spotting was carried out for each spore and incubated for 24 h at 37 ± 1 °C (for bacteria spores), 48 h at 37 ± 1 °C (for *C. albicans*), and 7 days at 25 ± 1 °C (for *A. niger*) to determine CFU. To assess reproducibility, duplicate tests were performed on three separate days [67–70].

5. Limitation and Future Scope

This research shows promise for the development of a new class of DHFR inhibitors with potential anti-cancer and anti-microbial applications. The synthesized compounds act potently against DHFR in in vitro assays and are predicted to have good drug-like properties. The predicted pharmacokinetic parameters and toxicity data indicate favorable ADME properties, suggesting that the compounds are likely to be well-absorbed, distributed effectively, metabolized efficiently, and excreted without significant toxicity concerns in the tested models. Additional research is imperative in order to validate the efficacy and safety of these compounds in living organisms. Conducting further testing in animal models (in vivo) and eventually progressing to clinical trials is essential to establish their safety profile and efficacy, thereby confirming their status as safe compounds.

Future research can explore these compounds in greater detail. We can investigate their activity in different animal models and test them against other diseases, such as tuberculosis and malaria. Additionally, by incorporating different heterocyclic rings instead of 1,2,4-triazole, we can design and synthesize novel caffeic acid derivatives. This will provide a clearer understanding of the relationship between a structure of compound and its biological activity. Overall, this study identifies a promising new class of DHFR inhibitors that warrants further exploration to identify optimal drug candidates.

6. Conclusions

Novel caffeic acid derivatives were designed, synthesized, and characterized, and their potential to inhibit DHFR, along with their antimicrobial and anticancer properties, was evaluated. To mitigate last-stage failures, it is crucial to investigate the initial pharmacokinetic parameters. An ADMET prediction was conducted, which indicates that all the synthesized compounds have promising drug-like characteristics. All the synthesized derivatives exhibited substantial DHFR inhibition, aligning well with the docking results. Among them, CTh7 showcased the highest DHFR inhibitory activity with an IC_{50} value of 0.15 μM. Enzyme inhibition assays confirmed that the newly synthesized compounds exert their action through DHFR-mediated mechanisms. Compound CTh7 demonstrated remarkable anticancer activity against the breast tumor cell line MCF7, boasting an IC_{50} value of 8.53 μM. The most active DHFR inhibitor and anticancer compound, CTh7, displayed excellent docking scores and binding energy (−9.9, −70.38 kcal/mol), outperforming methotrexate. It bound to the same site where MTX binds. It effectively interacted with key residues such as Glu 30, Phe34, Tyr121, Ile16, Val115, and Phe31. The voluminous and hydrophobic structure of CTh7 allowed it to snugly fit into the hydrophobic pocket of the modeled protein. Synthesized compounds displayed significant inhibitory potency against bacterial strains and moderate activity against fungi. Specifically, compounds CTh3, CTh4, and CTh6 demonstrated excellent antibacterial effects, while compound CTh4 also exhibited strong antifungal activity against *C. albicans*. For the most active antimicrobial agent, a CTh3 docking analysis revealed that it forms two hydrogen bonds with Thr 121 and Asn 18, a π–cation bond with Phe 92, and a salt bridge with the polar residue Asp 27. This interaction pattern resembled that of the co-crystallized ligand trimethoprim. CTh3 exhibited excellent docking scores and binding energy (−9.8, −71.12 kcal/mol), surpassing the standard trimethoprim. Overall, this study has discovered a potentially impactful structural class of compounds with antimicrobial and anticancer properties. These compounds could act as pivotal molecules for continued research and development in these respective fields.

Author Contributions: Conceptualization, R.S. and A.K.; methodology, R.S., R.P. and D.K.; validation, P.R.; formal analysis, R.S., R.P. and D.K.; investigation, R.S.; data curation, R.S.; writing—original draft preparation, R.S., A.K. and E.K.A.; writing—review and editing, P.R., A.K., E.K.A. and E.S.-S.; visualization, R.S., E.K.A. and E.S.-S. All authors have read and agreed to the published version of the manuscript.

Funding: This research received no external funding.

Institutional Review Board Statement: Not applicable.

Informed Consent Statement: Not applicable.

Data Availability Statement: The datasets produced during this study are accessible upon request from the corresponding authors. Additional inquiries can be directed to the corresponding author.

Acknowledgments: We express our sincere gratitude to the Schrödinger team, particularly Vinod Devaraji, for their invaluable assistance and support in providing necessary help.

Conflicts of Interest: The authors declare no conflict of interest.

References

1. Van Vo, G.; Nguyen, T.P.; Nguyen, H.T.; Nguyen, T.T. In silico and in vitro studies on the anti-cancer activity of artemetin, vitexicarpin and penduletin compounds from Vitex negundo. *Saudi Pharm. J.* **2022**, *30*, 1301–1314.
2. World Health Organization. Cancer. 2022. Available online: https://www.who.int/health-topics/cancer#tab=tab_1,2018 (accessed on 19 November 2022).
3. World Health Organization. Antibiotic Resistance. 2022. Available online: https://www.who.int/news-room/fact-sheets/detail/antimicrobial-resistance (accessed on 19 November 2022).
4. Salam, M.A.; Al-Amin, M.Y.; Salam, M.T.; Pawar, J.S.; Akhter, N.; Rabaan, A.A.; Alqumber, M.A. Antimicrobial resistance: A growing serious threat for global public health. *Healthcare* **2023**, *11*, 1946. [CrossRef]

Majumder, M.A.A.; Rahman, S.; Cohall, D.; Bharatha, A.; Singh, K.; Haque, M.; Gittens-St Hilaire, M. Antimicrobial stewardship: Fighting antimicrobial resistance and protecting global public health. *Infect. Drug Resist.* **2020**, *13*, 4713–4738. [CrossRef] [PubMed]

Aragaw, W.W.; Negatu, D.A.; Bungard, C.J.; Dartois, V.A.; Marrouni, A.E.; Nickbarg, E.B.; Olsen, D.B.; Warrass, R.; Dick, T. Pharmacological validation of dihydrofolate reductase as a drug target in Mycobacterium abscessus. *Antimicrob. Agents Chemother.* **2024**, *68*, e00717–e00723. [CrossRef] [PubMed]

Ducker, G.S.; Rabinowitz, J.D. One-Carbon Metabolism in Health and Disease. *Cell Metab.* **2017**, *25*, 27–42. [CrossRef] [PubMed]

Brown, P.M.; Pratt, A.G.; Isaacs, J.D. Mechanism of action of methotrexate in rheumatoid arhritis, and the search for biomarkers. *Nat. Rev. Rheumatol.* **2016**, *12*, 731–742. [CrossRef] [PubMed]

Sehrawat, R.; Rathee, P.; Khatkar, S.; Küpeli Akkol, E.; Khayatkashani, M.; Nabavi, S.M.; Khatkar, A. Dihydrofolate Reductase (DHFR) Inhibitors: A Comprehensive Review. *Curr. Med. Chem.* **2023**, *31*, 799–824. [CrossRef]

Foye, W.O.; Lemke, T.L.; Williams, D.A. *Principles of Medicinal Chemistry*, 4th ed.; Lea & Febiger: Philadelphia, PA, USA, 1974.

He, J.; Qiao, W.; An, Q.; Yang, T.; Luo, Y. Dihydrofolate reductase inhibitors for use as antimicrobial agents. *Eur. J. Med. Chem.* **2020**, *195*, 112268. [CrossRef]

Gibson, M.W.; Dewar, S.; Ong, H.B.; Sienkiewicz, N.; Fairlamb, A.H. *Trypanosoma brucei* DHFR-TS Revisited: Characterisation of a Bifunctional and Highly Unstable Recombinant Dihydrofolate Reductase-Thymidylate Synthase. *PLoS Negl. Trop. Dis.* **2016**, *10*, e0004714. [CrossRef]

El-Gazzar, Y.I.; Georgey, H.H.; El-Messery, S.M.; Ewida, H.A.; Hassan, G.S.; Raafat, M.M.; Ewida, M.A.; El-Subbagh, H.I. Synthesis, biological evaluation and molecular modeling study of new (1,2,4-triazole or 1,3,4-thiadiazole)-methylthio-derivatives of quinazolin-4(3H)-one as DHFR inhibitors. *Bioorg. Chem.* **2017**, *72*, 282–292. [CrossRef]

Banerjee, D.; Mayer-Kuckuk, P.; Capiaux, G.; Budak-Alpdogan, T.; Gorlick, R.; Bertino, J.R. Novel aspects of resistance to drugs targeted to dihydrofolate reductase and thymidylate synthase. *Biochim. Biophys. Acta (BBA) Mol. Basis Dis.* **2002**, *1587*, 164–173. [CrossRef]

McIvor, R.S. Drug-resistant dihydrofolate reductases: Generation, expression and therapeutic application. *Bone Marrow Transplant.* **1996**, *18* (Suppl. S3), S50–S54. [PubMed]

Wróbel, A.; Arciszewska, K.; Maliszewski, D.; Drozdowska, D. Trimethoprim and other nonclassical antifolates an excellent template for searching modifications of dihydrofolate reductase enzyme inhibitors. *J. Antibiot.* **2020**, *73*, 5–27. [CrossRef] [PubMed]

Reeve, S.M.; Si, D.; Krucinska, J.; Yan, Y.; Viswanathan, K.; Wang, S.; Holt, G.T.; Frenkel, M.S.; Ojewole, A.A.; Estrada, A.; et al. Toward broad spectrum dihydrofolate reductase inhibitors targeting trimethoprim resistant enzymes identified in clinical isolates of methicillin resistant Staphylococcus aureus. *ACS Infect. Dis.* **2019**, *5*, 1896–1906. [CrossRef]

Sehrawat, R.; Rathee, P.; Khatkar, S.; Küpeli Akkol, E.; Khatkar, A. In silico and in vitro analysis of phenolic acids for identification of potential dhfr inhibitors as antimicrobial and anticancer agents. *Curr. Protein Pept. Sci.* **2024**, *25*, 44–58. [CrossRef] [PubMed]

Sehrawat, R.; Rathee, P.; Küpeli Akkol, E.; Khatkar, S.; Lather, A.; Redhu, N.; Khatkar., A. Phenolic acids-versatile natural moiety with numerous biological applications. *Curr. Top. Med. Chem.* **2022**, *22*, 1472–1484. [CrossRef]

Espíndola, K.M.M.; Ferreira, R.G.; Narvaez, L.E.M.; Silva Rosario, A.C.R.; Da Silva, A.H.M.; Silva, A.G.B.; Vieira, A.P.O.; Monteiro, M.C. Chemical and pharmacological aspects of caffeic acid and its activity in hepatocarcinom. *Front. Oncol.* **2019**, *9*, 541. [CrossRef] [PubMed]

Pavlíková, N. Caffeic acid and diseases—Mechanisms of action. *Int. J. Mol. Sci.* **2022**, *24*, 588. [CrossRef]

Kępa, M.; Miklasińska-Majdanik, M.; Wojtyczka, R.D.; Idzik, D.; Korzeniowski, K.; Smoleń-Dzirba, J.; Wąsik, T.J. Antimicrobial potential of caffeic acid against *Staphylococcus aureus* clinical strains. *BioMed Res. Int.* **2018**, *2018*, 7413504. [CrossRef]

Khan, F.; Bamunuarachchi, N.I.; Tabassum, N.; Kim, Y.M. Caffeic acid and its derivatives: Antimicrobial drugs toward microbial pathogens. *J. Agric. Food Chem.* **2021**, *69*, 2979–3004. [CrossRef]

Doshi, H.; Thakkar, S.; Khirsariya, P.; Thakur, M.; Ray, A. 6-Tosyl-4,5,6,7-tetrahydrothieno [2,3-c] pyridine-3-carboxamide analogues: Synthesis, characterization, MO calculation, and antibacterial activity. *Appl. Biochem. Biotechnol.* **2015**, *175*, 1700–1709. [CrossRef] [PubMed]

Sztanke, K.; Tuzimski, T.; Rzymowska, J.; Pasternak, K.; Kandefer-Szerszen, M. Synthesis, determination of the lipophilicity, anticancer and antimicrobial properties of some fused 1,2,4-triazole derivatives. *Eur. J. Med. Chem.* **2008**, *43*, 404–419. [CrossRef] [PubMed]

Popiolek, L.; Kosikowska, U.; Mazur, L.; Dobosz, M.; Malm, A. Synthesis and antimicrobial evaluation of some novel 1,2,4-triazole and 1,3,4-thiadiazole derivatives. *Med. Chem. Res.* **2013**, *23*, 3134–3147. [CrossRef] [PubMed]

Vijesh, A.; Isloor, A.; Shetty, P.; Sundershan, S.; Kun Fun, H. New pyrazole derivatives containing 1,2,4-triazoles and benzoxazoles as potent antimicrobial and analgesic agents. *Eur. J. Med. Chem.* **2013**, *62*, 410–415. [CrossRef] [PubMed]

Salerno, L.; Pittala, V.; Romeo, G. Evaluation of novel aryloxyalkyl derivatives of imidazole and 1,2,4-triazole as heme oxygenase-1 (HO-1) inhibitors and their antitumor properties. *Bioorg. Med. Chem.* **2013**, *21*, 5145–5153. [CrossRef] [PubMed]

Kulabas, N.; Tatar, E.; Ozakpınar, O.B. Synthesis and antiproliferative evaluation of novel 2-(4H–1,2,4- triazole-3-ylthio)acetamide derivatives as inducers of apoptosis in cancer cells. *Eur. J. Med. Chem.* **2016**, *121*, 58–70. [CrossRef] [PubMed]

Al-Abdullah, E.; Asiri, H.; Lahsasni, S.; Habib, E.; Ibrahim, T.; El-Emam, A. Synthesis, antimicrobial, and anti-inflammatory activity, of novel S-substituted and Nsubstituted 5-(1-adamantyl)-1,2,4-triazole-3-thiols. *Drug Des. Dev. Ther.* **2014**, *8*, 505–518.

31. Liu, J.; Liu, Q.; Yang, X. Design, synthesis, and biological evaluation of 1,2,4-triazole bearing 5-substituted biphenyl-2-sulfonamide derivatives as potential antihypertensive candidates. *Bioorg. Med. Chem.* **2013**, *21*, 7742–7775. [CrossRef]
32. Nair, V.P.; Anang, S.; Subramani, C. Endoplasmic reticulum stress induced synthesis of a novel viral factor mediates efficient replication of genotype-1 hepatitis E virus. *PLoS Pathog.* **2016**, *12*, e1005521. [CrossRef]
33. Fischer, J.; Ganellin, C.R.; Rotella, D.P. *Analogue-Based Drug Discovery III*, 3rd ed.; Wiley: Hoboken, NJ, USA, 2012.
34. Shi, L.; Ge, H.; Tan, S. Synthesis and antimicrobial activities of Schiff bases derived from 5-chloro-salicylaldehyde. *Eur. J. Med. Chem.* **2007**, *42*, 558–564. [CrossRef]
35. Sinha, D.; Tiwari, A.; Singh, S. Synthesis, characterization and biological activity of Schiff base analogues of indole-3-carboxaldehyde. *Eur. J. Med. Chem.* **2008**, *43*, 160–165. [CrossRef] [PubMed]
36. Karthikeyan, M.S.; Prasad, D.J.; Poojary, B.; Bhat, K.S.; Holla, B.S.; Kumari, N.S. Synthesis and biological activity of Schiff and Mannich bases bearing 2,4-dichloro-5-fluorophenyl moiety. *Bioorg. Med. Chem.* **2006**, *14*, 7482–7489. [CrossRef] [PubMed]
37. Patole, J.; Shingnapurkar, D.; Padhyea, S.; Ratledge, C. Schiff base conjugates of paminosalicylic acid as antimycobacterial agents. *Bioorg. Med. Chem. Lett.* **2006**, *16*, 1514–1517. [CrossRef] [PubMed]
38. Hearn, M.J.; Cynamon, M.H. Design and synthesis of antituberculars: Preparation and evaluation against *Mycobacterium tuberculosis* of an isoniazid Schiff base. *J. Antimicrob. Chemoter.* **2004**, *53*, 185–191. [CrossRef] [PubMed]
39. Vicini, P.; Geronikaki, A.; Incerti, M.; Busonera, B.; Poni, G.; Cabras, C.A.; La Colla, P. Synthesis and biological evaluation of benzo[d]isothiazole, benzothiazole and thiazole Schiff bases. *Bioorg. Med. Chem.* **2003**, *11*, 4785–4789. [CrossRef] [PubMed]
40. Bhandari, S.V.; Bothara, K.G.; Raut, M.K.; Patil, A.A.; Sarkate, A.P.; Mokale, V.J. Design, synthesis and evaluation of antiinflammatory, analgesic and ulcerogenicity studies of novel S-substituted phenacyl-1,3,4-oxadiazole-2-thiol and Schiff bases of diclofenac acid as nonulcerogenic derivatives. *Bioorg. Med. Chem.* **2008**, *16*, 1822–1831. [CrossRef] [PubMed]
41. Sridhar, S.K.; Pandeya, S.N.; Stables, J.P.; Ramesh, A. Anticonvulsant activity of hydrazones, Schiff and Mannich bases of isatin derivatives. *Eur. J. Pharm. Sci.* **2002**, *16*, 129–132. [CrossRef]
42. Riyadh, S.M.; El-Motairi, S.A.; Ahmed, H.E.; Khalil, K.D.; Habib, E.S.E. Synthesis, biological evaluation, and molecular docking of novel thiazoles and [1,3,4] thiadiazoles incorporating sulfonamide group as DHFR Inhibitors. *Chem. Biodivers.* **2018**, *15*, e1800231. [CrossRef] [PubMed]
43. Katsila, T.; Spyroulias, G.A.; Patrinos, G.P.; Matsoukas, M.T. Computational approaches in target identification and drug discovery. *Comput. Struct. Biotechnol. J.* **2016**, *14*, 177–184. [CrossRef]
44. Mali, G.; Shaikh, B.A.; Garg, S.; Kumar, A.; Bhattacharyya, S.; Erande, R.D.; Chate, A.V. Design, synthesis, and biological evaluation of densely substituted dihydropyrano [2, 3-c] pyrazoles via a taurine-catalyzed green multicomponent approach. *ACS Omega* **2021**, *6*, 30734–30742. [CrossRef]
45. Thakkar, S.S.; Thakor, P.; Doshi, H.; Ray, A. 1,2,4-Triazole and 1,3,4-oxadiazole analogues: Synthesis, MO studies, in silico molecular docking studies, antimalarial as DHFR inhibitor and antimicrobial activities. *Bioorg. Med. Chem.* **2017**, *25*, 4064–4075. [CrossRef] [PubMed]
46. Wang, M.; Yang, M.; Yuan, M.; Xue, L.; Li, H.; Tian, C.; Wang, X.; Liu, J.; Zhang, Z. Synthesis and antiproliferative activity of a series of novel 6-substituted pyrido[3,2-d]pyrimidines as potential nonclassical lipophilic antifolates targeting dihydrofolate reductase. *Eur. J. Med. Chem.* **2017**, *128*, 88–97. [CrossRef] [PubMed]
47. Ikram, M.; Butt, A.R.; Fatima, A.; Shahzadi, I.; Haider, A.; Ul-Hamid, A.; Nabgan, W. Graphitic-carbon nitride and poly acrylic acid doped vanadium oxide for efficient catalytic and antimicrobial activity: In silico molecular docking studies. *J. Photochem. Photobiol. A Chem.* **2023**, *443*, 114835. [CrossRef]
48. Aouidate, A.; Ghaleb, A.; Ghamali, M.; Chtita, S.; Choukrad, M.; Sbai, A.; Bouachrine, M.; Lakhlifi, T. Combined 3D-QSAR and Molecular Docking Study on 7,8-dialkyl-1,3-diaminopyrrolo-[3,2-f] Quinazoline Series Compounds to understand the binding mechanism of DHFR Inhibitors. *J. Mol. Struct.* **2017**, *1139*, 319–327. [CrossRef]
49. Bouhidel, A.; Cherouana, P.; Durand, A.; Doudouh, F.; Morini, B.; Guillot, S.; Dahaoui, S. Synthesis, spectroscopic characterization, crystal structure, Hirshfeld surface analysis and antimicrobial activities of two triazole Schiff bases and their silver complexes. *Inorganica Chim. Acta* **2018**, *482*, 34–47. [CrossRef]
50. Kumar, S.; Dhankhar, S.; Arya, V.P.; Yadav, S.; Yadav, J.P. Antimicrobial activity of *Salvadora oleoides* Decne. against some microorganisms. *J. Med. Plants Res.* **2012**, *6*, 2754–2760.
51. *Maestro*, version 12; Schrodinger, LLC.: New York, NY, USA, 2020.
52. Available online: https://www.rcsb.org/structure/2W9S (accessed on 11 November 2021).
53. Halgren, T.A.; Murphy, R.B.; Friesner, R.A.; Beard, H.S.; Frye, L.L.; Pollard, W.T.; Banks, J.L. Glide: A new approach for rapid accurate docking and scoring. Enrichment factors in database screening. *J. Med. Chem.* **2004**, *47*, 1750–1759. [CrossRef]
54. Friesner, R.A.; Banks, J.L.; Murphy, R.B.; Halgren, T.A.; Klicic, J.J.; Mainz, D.T.; Repasky, M.P.; Knoll, E.H.; Shelley, M.; Perry, J.K. et al. Glide: A new approach for rapid, accurate docking and scoring. Method and assessment of docking accuracy. *J. Med. Chem* **2004**, *47*, 1739–1749. [CrossRef] [PubMed]
55. Friesner, R.A.; Murphy, R.B.; Repasky, M.P.; Frye, L.L.; Greenwood, J.R.; Halgren, T.A.; Sanschagrin, P.C.; Mainz, D.T. Extra precision glide: Docking and scoring incorporating a model of hydrophobic enclosure for protein-ligand complexes. *J. Med. Chem* **2006**, *49*, 6177–6196. [CrossRef]
56. *Glide*, version 6.6; Schrödinger, LLC.: New York, NY, USA, 2015.

Genheden, S.; Ryde, U. The MM/PBSA and MM/GBSA methods to estimate ligand-binding affinities. *Expert Opin. Drug Discov.* **2015**, *10*, 449–461. [CrossRef]

Available online: https://admetmesh.scbdd.com/ (accessed on 5 February 2021).

Sharma, K.; Tanwar, O.; Sharma, S.; Ali, S.; Alam, M.M.; Zaman, M.S. Structural comparison of Mtb-DHFR and h-DHFR for design, synthesis and evaluation of selective non-pteridine analogues as antitubercular agents. *Bioorg. Chem.* **2018**, *80*, 319–333. [CrossRef] [PubMed]

Azzam, R.A.; Elsayed, R.E.; Elgemeie, G.H. Design, synthesis, and antimicrobial evaluation of a new series of N-sulfonamide 2-pyridones as dual inhibitors of DHPS and DHFR enzymes. *ACS Omega* **2020**, *5*, 10401–10414. [CrossRef] [PubMed]

DHFR Assay. Available online: https://www.sigmaaldrich.com/deepweb/assets/sigmaaldrich/product/documents/427/408/cs0340bul.pdf (accessed on 13 March 2023).

Francesconi, V.; Giovannini, L.; Santucci, M.; Cichero, E.; Costi, M.P.; Naesens, L.; Tonelli, M. Synthesis, biological evaluation and molecular modeling of novel azaspiro dihydrotriazines as influenza virus inhibitors targeting the host factor dihydrofolate reductase (DHFR). *Eur. J. Med. Chem.* **2018**, *155*, 229–243. [CrossRef] [PubMed]

Morgan, D.M.L. Tetrazolium (MTT) assay for cellular viability and activity. *Methods Mol. Biol.* **1998**, *79*, 179–184. [PubMed]

Van Meerloo, J.; Kaspers, G.J.L.; Cloos, J. Cell sensitivity assays: The MTT assay. *Methods Mol. Biol.* **2011**, *731*, 237–245.

Fotakis, G.; Timbrell, J.A. In vitro cytotoxicity assays: Comparison of LDH, neutral red, MTT and protein assay in hepatoma cell lines following exposure to cadmium chloride. *Toxicol. Lett.* **2006**, *160*, 171–177. [CrossRef] [PubMed]

Tihauan, B.; Berca, L.M.; Adascălului, M.; Sanmartin, A.; Nica, S.; Cimponeriu, D.; Duta, D. Experimental in vitro cytotoxicity evaluation of plant bioactive compounds and phytoagents: A review. *Rom. Biotechnol. Lett.* **2020**, *25*, 1832–1842.

Indian Pharmacopoeia, Vol-I. Indian Pharmacopoeia Commission. The Controller of Publications, New Delhi. 2018; p. 37. Available online: https://ipc.gov.in/shop/index.php?route=product/category&path=59 (accessed on 21 March 2023).

Narasimhan, B.; Narang, R.; Judge, V.; Ohlan, S.; Ohlan, R. Synthesis, antimicrobial and QSAR studies of substituted anilides. *Arkivoc* **2007**, *15*, 112–126. [CrossRef]

Lather, A.; Sharma, S.; Khatkar, A. Naringin derivatives as glucosamine-6-phosphate synthase inhibitors based preservatives and their biological evaluation. *Sci. Rep.* **2020**, *10*, 20477. [CrossRef]

Pasrija, R.; Kundu, D. Interaction of curcumin with azoles and polyenes against aspergillus infections. *Res. J. Life Sci. Bioinform. Pharm. Chem. Sci.* **2018**, *4*, 271–279.

antibiotics

MDPI

Article

Enhancing Bioactive Compound Classification through the Synergy of Fourier-Transform Infrared Spectroscopy and Advanced Machine Learning Methods

Pedro N. Sampaio [1,2,*] and Cecília C. R. Calado [3,4]

[1] COPELABS—Computação e Cognição Centrada nas Pessoas, Faculty of Engineering, Lusófona University, Campo Grande, 376, 1749-024 Lisbon, Portugal
[2] GREEN-IT—BioResources for Sustainability Unit, Institute of Chemical and Biological Technology António Xavier, ITQB NOVA, Av. da República, 2780-157 Oeiras, Portugal
[3] ISEL—Instituto Superior de Engenharia de Lisboa, Instituto Politécnico de Lisboa, Rua Conselheiro Emídio Navarro 1, 1959-007 Lisbon, Portugal; cecilia.calado@isel.pt
[4] iBB—Institute for Bioengineering and Biosciences, i4HB—The Associate Laboratory Institute for Health and Bioeconomy, IST—Instituto Superior Técnico, Universidade de Lisboa, Av. Rovisco Pais, 1049-001 Lisbon, Portugal
* Correspondence: pedro.sampaio@ulusofona.pt; Tel.: 351-21-750-55-00

Abstract: Bacterial infections and resistance to antibiotic drugs represent the highest challenges to public health. The search for new and promising compounds with anti-bacterial activity is a very urgent matter. To promote the development of platforms enabling the discovery of compounds with anti-bacterial activity, Fourier-Transform Mid-Infrared (FT-MIR) spectroscopy coupled with machine learning algorithms was used to predict the impact of compounds extracted from *Cynara cardunculus* against *Escherichia coli*. According to the plant tissues (seeds, dry and fresh leaves, and flowers) and the solvents used (ethanol, methanol, acetone, ethyl acetate, and water), compounds with different compositions concerning the phenol content and antioxidant and antimicrobial activities were obtained. A principal component analysis of the spectra allowed us to discriminate compounds that inhibited *E. coli* growth according to the conventional assay. The supervised classification models enabled the prediction of the compounds' impact on *E. coli* growth, showing the following values for accuracy: 94% for partial least squares-discriminant analysis; 89% for support vector machine; 72% for k-nearest neighbors; and 100% for a backpropagation network. According to the results, the integration of FT-MIR spectroscopy with machine learning presents a high potential to promote the discovery of new compounds with antibacterial activity, thereby streamlining the drug exploratory process.

Keywords: antimicrobial; *Cynara cardunculus;* machine learning; MIR-Spectroscopy; PCA; PLS-DA; SVM; KNN; BPN

1. Introduction

As a consequence of the rise in antibiotic resistance, it is imperative to develop fast and affordable systems for discovering bioactive compounds with antimicrobial activity that are also non-cytotoxic to the human host [1]. Antibiotic-resistant bacteria present a formidable and concerning issue in contemporary medicine, which is compounded by the dwindling progress in new antibiotic development and the escalating spread of multi-drug-resistant determinants, exacerbating the problem. Plant extracts are important sources of biologically active compounds and are indispensable for the development and synthesis of many drugs, including antibiotics. *Cynara cardunculus* (Cardoon), for example, has been used in traditional medicine [2], namely, extracts of its rhizomes are characterized by antioxidant and antimicrobial activities [3], while its leaves show diuretic, choleretic, and hepatoprotective properties [4], in addition to in vitro anti-proliferative potential against breast cancer [5,6] and cervical cancer uterus [7]. Despite the promising potential of these plant extracts, evaluating their antimicrobial activity typically involves expensive and time-consuming techniques that demand significant sample quantities [8,9]. The current study investigates how Fourier-Transform Infrared (FTIR) spectroscopy, when coupled with machine learning (ML) and neural network modeling, can overcome these limitations and enhance its capabilities. FTIR spectroscopy, particularly in the mid-infrared region (MIR), captures the vibrational signatures of key biomolecules like proteins, carbohydrates, lipids, and nucleic acids [10]. As a result, these spectra serve as metabolic fingerprints for biological samples, offering high sensitivity and specificity [11]. This technique has been applied to classify bacteria [12], differentiate between cell death mechanisms such as apoptosis and necrosis [13], conduct studies and diagnoses related to cancer [14], monitor the effects of drugs on *Helicobacter pylori* [15,16], and characterize cell metabolism in bioreactor cultures [17,18]. This demonstrates the versatility and effectiveness of FTIR spectroscopy in diverse biomedical and biotechnological applications.

In untargeted analyses, FTIR spectra have been proven instrumental in identifying specific phenotypes. They have been utilized to uncover the antimicrobial properties of novel extracts [19], differentiate the effects of various surfactants on *E. coli* cells [20], compare how *E. coli* responds to diverse stress conditions [21], and develop bioassays for toxicity testing in yeasts [22], among other applications. Compared to various omics techniques, FTIR spectroscopy offers valuable insights with less data intensity and minimal impact on metabolic sensitivity [23]. This underscores its significance and potential for further advancements in research across different fields. Given the significant amount of information present in the obtained spectra, their analysis must be carried out using multivariate statistical methods, such as ML, in which the classification and discrimination methods used allow the development of mathematical models capable of discriminating the samples.

Machine learning (ML) is commonly categorized as a subset of artificial intelligence (AI), relying on mathematical frameworks to enable computers to learn autonomously and discern patterns and structures within both structured and unstructured datasets. This process eliminates the need for explicit programming instructions, thereby empowering machines to adapt and improve their performance over time [24]. In ML, two main types of techniques are used: supervised and unsupervised learning. Concerning unsupervised learning, hidden patterns or intrinsic structures in the input data are identified and subsequently grouped correctly [25]. Unsupervised methods usually serve exploratory purposes, aiding in the development of models that facilitate data clustering in a manner not explicitly specified by the user. Supervised learning methods are employed to construct training models that predict future values of data categories or continuous variables. Examples of supervised techniques include partial least squares-discriminant analysis (PLS-DA), k-nearest neighbors (kNNs), and support vector machines (SVMs). Other techniques include artificial neural networks (ANNs) such as backpropagation networks (BPNs) and counter-propagation networks, which are multilayer networks based on the combinations of the input, output, and clustering layers [26].

The main objective of this study was to show the advantage of FT-MIR spectroscopy and machine learning methods to screen for and, consequently, discover promissory bioactive molecules as antimicrobial agents. The extracts, obtained with different solvents, from the seeds, dry and fresh leaves, and flowers of *C. cardunculus* were evaluated. This study evaluated if the spectra of *Escherichia coli* could be used to predict the impact of rapid exposure to various compounds on cells. For this purpose, diverse unsupervised and supervised machine-learning algorithms were evaluated over the obtained *E. coli* spectral data. This approach serves as a tool in targeted exploration for novel compounds with antimicrobial activity.

2. Results and Discussion

2.1. Analysis of C. cardunculus Extracts

The compounds were extracted with diverse solvents from *C. cardunculus* flowers, seeds, and fresh and dried leaves. The following extracts presented antimicrobial activity against *E. coli*: seeds subjected to extraction with water and ethanol, leaves processed with methanol, dry leaves treated with water, and flowers subjected to extraction with ethanol and methanol (Figure 1). Previously, Ibrahim et al. [27] also observed that seed extracts showed antimicrobial activity against *Escherichia coli*, *Staphylococcus aureus*, *Staphylococcus sprophyticus*, and *Klebsiella pneumoniae* species.

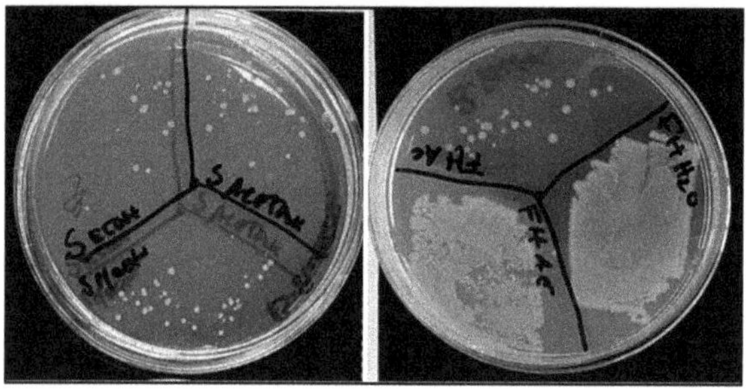

Figure 1. Antimicrobial activity of different assays using *C. cardunculus* extracts in *E. coli* cells in solid culture medium: seeds extracted with ethanol; seeds extracted with methanol; seeds extracted with acetone; fresh leaves extracted with acetone; fresh leaves extracted with ethyl acetate; and fresh leaves extrated with water.

Methanol, ethanol, and water were the most suitable solvents to extract compounds characterized by significant antioxidant activity. The extracts obtained from seeds showed significant antioxidant activity (AA): 75% AA for methanol and 60% AA for water. Fresh leaves also demonstrated significant antioxidant activity (76% AA for methanol and 86% AA for water). The extracts obtained from fresh leaves and seeds using a methanol and water mixture revealed substantial antioxidant activity. The antioxidant level most probably reflects the specific nature of the tissues and their respective functions in terms of plant defense. The highest phenol concentrations were registered for the aqueous extracts from flowers and seeds (8.03 μg gallic acid equivalent (GAE)/g and 3.93 μg GAE/g, respectively), most probably due to the high temperatures used during extraction (~100 °C). The extracts obtained with methanol and ethanol also presented high phenolic compounds, most probably due to the affinity of phenolic compounds with organic solvents. The extracts obtained with ethyl acetate (0.80 μg GAE/g) and acetone (1.22 μg GAE/g) presented low values, suggesting that these solvents presented low affinity for the extraction of these compounds. The present results are in accordance with the results reported by other authors, showing that extracts from *C. cardunculus* L. presented polyphenolic compounds, specifically flavonoids and phenolic acids [28]. Phenolic compounds are the main plant antioxidants and may act through their hydrogen-donor and metal-chelating capacities. Indeed, it was observed that there was a high correlation between the phenol content and antioxidant levels for the extracts obtained with ethanol (r^2 = 0.63) and with methanol (r^2 = 0.96). This was in accordance with Fu et al. [29], who also observed a positive correlation between the content of phenolic compounds and antioxidant activity.

The phenolic contents of plant extracts have been associated with beneficial effects on health, which include anti-allergic, anti-atherogenic, anti-inflammatory, antimicrobial, and antithrombotic effects, as well as acting as cardiovascular protectors and vasodilators [30]. The pronounced lipophilic nature of phenolic compounds enhances their antimicrobial activity, most probably through diverse mechanisms [3,5,31,32]. This antimicrobial activity can affect the permeability of cell membranes. Moreover, interactions with cell membranes can lead to changes in cell wall rigidity, resulting in a loss of integrity and ultimately inhibiting microbial proliferation [33].

2.2. PCA of FT-MIR Spectra of *E. coli*

The FT-MIR spectra of *E. coli* cells incubated with these extracts were pre-processed by different algorithms, such as multiplicative scatter correction (MSC), straight normal variate (SNV), the first derivative, and the second derivative. The impact of these spectral pre-processing methods was evaluated using PCA (Figure 2A). The most effective pre-processing methods for clustering scores in the PCA score plot according to the antimicrobial activity were MSC and the use of the second derivative. The second derivative allowed us to resolve overlapped spectral bands, while MSC minimized the effect of light scattering (Figure 2B).

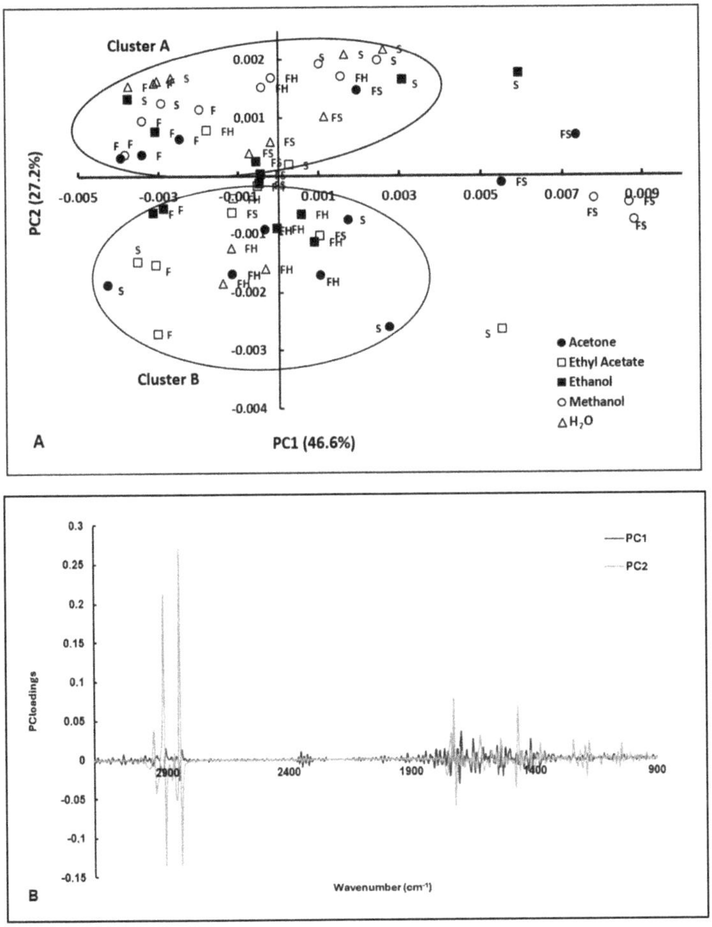

Figure 2. PCA of FT-MIR spectra of *E. coli* exposed to extracts from *C. cardunculus* tissues (S—seeds; F—flowers; FH—fresh leaves; FS—dry leaves) obtained by different solvents (AE—ethyl acetate, EtOH—ethanol; Ac—acetone; MeOH—methanol, and H_2O—water) (**A**), and the corresponding loading vectors (**B**). Spectra were pre-processed by MSC and 2nd derivative. Cluster A is defined by samples characterized by antimicrobial activity. Cluster B is constituted by samples characterized by low antimicrobial activity.

The scores were clustered according to the extracts' antimicrobial activity, as observed according to the conventional assay (based on counting colonies on agar plates) (Figure 2A,B). The clusters of the extracts with antimicrobial activity included extracts from seeds (extracted with water and ethanol), leaves (extracted with methanol), dry leaves (extracted with water), and flowers (extracted with ethanol and methanol). These extracts also presented significant antioxidant activity. According to the results obtained, 14 samples showed a residual antimicrobial activity, while 6 samples showed significant antimicrobial activity. Meanwhile, the PCA loadings highlighted the spectral regions that contributed to the resolution of scores between the two clusters characterized by high and low antimicrobial activity, especially PC2 loading, which had a strong contribution from the following regions: 1000 to 1800 cm^{-1}, and 2700 to 3000 cm^{-1} (Figure 2B). The spectral regions ranging from 1600 to 1700 cm^{-1} primarily arose from amide I and amide II absorptions associated with proteins, typically constituting the most prominent bands in a bacterial infrared spectrum. This prominence is attributed to proteins accounting for approximately half of the bacterial cell composition on average. The peak registered at 1740 cm^{-1} is related to the ester C=O stretching of phospholipids. At wavenumbers around 1080 cm^{-1} and 1240 cm^{-1}, there is a significant presence of the P=O bond characteristic of phosphate groups found in molecules such as nucleotides (like ATP), nucleic acids, phosphorylated proteins, and lipids. Between 1000 and 1100 cm^{-1}, there is a broad contribution from bonds such as C–O, C–C, C–OH, and C–O–C, which are commonly found in carbohydrates. The spectral region between 2700 and 3000 cm^{-1} exhibits a notable intensity in the stretching vibrations of C-H bonds, particularly from lipids. This resonance is particularly pronounced in bacterial membranes (as outlined in Table 1).

Table 1. Vibrational frequencies of some functional groups in biomolecules (adapted from Sales et al. [17]).

Wavenumber (cm^{-1})	Vibrational Mode in the Functional Group	Commonly Assigned Biochemical Component
3300	ν(N-H)	Amide A: peptides, proteins
3100	ν(N-H)	Amide B: peptides, proteins
2960	$\nu_{as}(CH_3)$	Lipids
2929 2870 2850	$\nu_{as}(CH_2)$ $\nu_s(CH_3)$ $\nu_s(CH_2)$	Lipids
1740	ν(C=O)	Phospholipid esters
1655	[80% ν(C=O) and 20% ν(CN)] in O=C-N	Amide I in peptides and proteins
1550	ν(C-C) in O=C-N	Amide II in peptides and proteins
1450	$\nu_{as}(CH_3)$	Lipids, proteins
1395	$\nu_s(CH_3)$	Lipids, proteins
1380	$\nu_s(CH_3)$	Phospholipids, fatty acids, triglyceride
1240	$\nu_s(PO_2^-)$	Phosphodiesters in nucleic acids, phospholipids, and phosphorylated proteins
1160 & 1111	ν_s(C-O)	Ribose in RNA
1078	$\nu_s(PO_2^-)$	Phosphodiesters
1055	ν_s(C-O-P)	Phosphate esters
1032	def(C-OH)	Glycogen

Vibration type: ν_s—symmetric stretching; ν_{as}—asymmetric stretching; def—deformations.

2.3. Analysis of Spectral Bands

The spectral second derivative presents bands that, on average, are different between *E. coli* subjected to compounds with antimicrobial or low antimicrobial activity (Figure 3A,B). To further identify spectral bands that exhibit significant differences between bacteria exposed to inhibitory and non-inhibitory compounds, several ratios between the bands, as highlighted by the PCA loading vector, were examined. Twelve ratios between the spectral bands were analyzed, where "A" followed by a number represents the absorbance at that wavenumber (Table 2). From the twelve ratios evaluated, the following nine ratios were statistically different at the 5% significance level between the two clusters: A2856/A1705, A1740/A1656, A1740/1545, A2847/1545, A1617/A1545, A1476/1545, 1215/1179, A1215/1545, and A1244/1230. These ratios reflect the compounds' impact on bacterial metabolism related to protein synthesis (as indicated by the ratio between amides I and II, i.e., A1617/A1545, and various ratios involving the amide I peak). Additionally, changes in the composition of the cell membrane are highlighted by the CH_3/CH_2 ratio (i.e., A2912/A2856), as well as the ratios incorporating the bands at 1740 cm^{-1} and 1179 cm^{-1} corresponding to carbonyl vibrations of esters of phospholipids, and the bands at 2856 cm^{-1} and 1476 cm^{-1} originating from methyl and methylene groups. Figure 4 presents the box plots for the ratios of spectral bands that are statistically different. These findings underscore the remarkable sensitivity of FT-MIR spectroscopy in capturing the influence of compounds on *E. coli* metabolism.

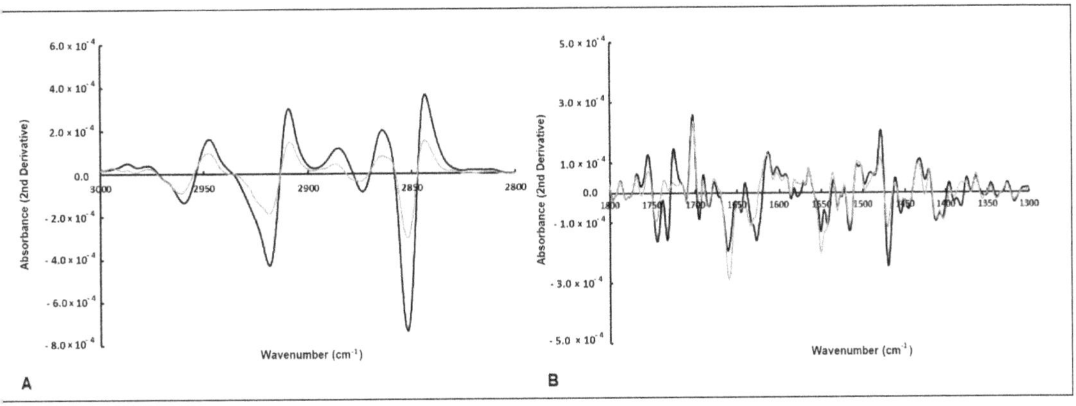

Figure 3. Average of second-derivative spectra of *E. coli* exposed to antimicrobial (gray line) and non-antimicrobial (black line) compounds. The arrows highlight spectral bands that are different between the two bacterial populations (**A**,**B**).

Table 2. Ratios between spectral bands of *E. coli* exposed to biocompounds with and without antimicrobial activities.

Ratio between Bands (cm^{-1})	Average		Standard Deviation		*p*-Value
	Antimicrobial	Non-Antimicrobial	Antimicrobial	Non-Antimicrobial	
A2912/2856	1.359	1.356	0.114	0.162	0.472
A2912/1740	3.921	4.121	1.086	1.114	0.263
A2856/1705	2.601	1.334	0.753	0.248	**<0.0001**
A1740/1656	0.261	0.114	0.089	0.038	**<0.0001**
A1740/1545	0.434	0.218	0.151	0.082	**<0.0001**
A2847/1545	0.977	0.537	0.253	0.124	**<0.0001**
A2847/1740	2.395	2.720	0.657	0.966	0.102
A1617/1545	1.129	1.035	0.124	0.131	**0.008**
A1476/1545	0.304	0.146	0.167	0.081	**<0.0001**
A1215/1179	1.269	2.950	0.605	2.314	**0.004**
A1215/1545	0.942	0.708	0.174	0.082	**<0.0001**
A1244/1230	1.312	1.158	0.180	0.027	**<0.0001**

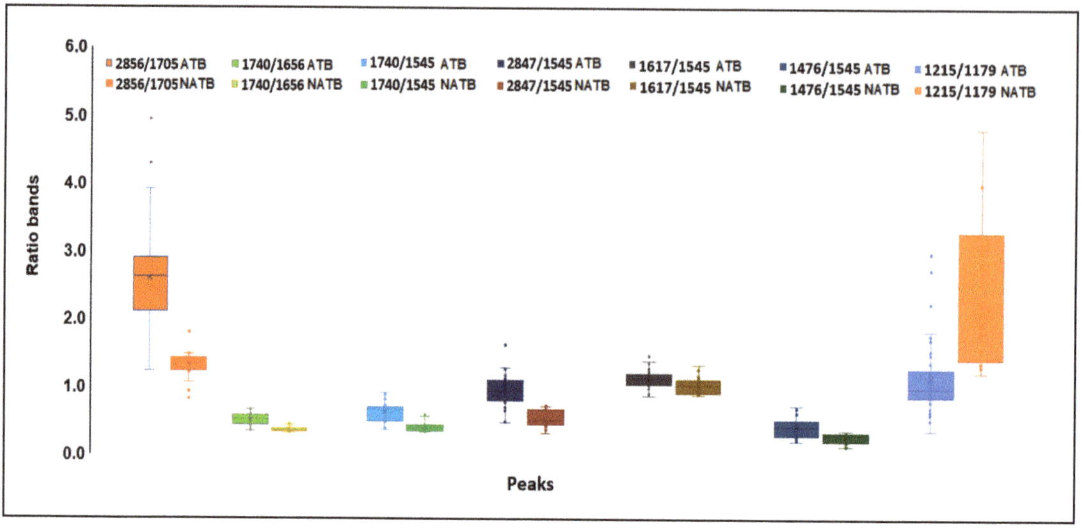

Figure 4. Box plots of ratios of spectral bands of *E. coli* cells exposed to compounds with (A) and without (NA) inhibitory activities.

2.4. Classification Models

To assess the feasibility of predicting the impact of compounds on *E. coli* inhibition, based on the spectral data from bacterial cells, several supervised classification machine learning algorithms were evaluated, including PLS-DA, k-NN, SVM, and BPN. For all models, 70% and 30% of the spectra were used as the training data (n = 42) and test data (n = 18), respectively (Table 2). The calibration and cross-validation performances of the tested models were compared in terms of classification parameters, such as non-error rate (NER) and accuracy, with an analysis of confusion matrices, thus providing a comprehensive overview and a rational means of selecting the best model.

2.4.1. PLS-DA Model

The most appropriate PLS-DA model was developed through the implementation of the spectra after MSC pre-processing, explaining a total variance of 99% for 11 LVs (Table 3). The error for the validation procedure was 11%. The model presented sensibility, specificity, precision, and accuracy of 86%, 92%, 96%, and 88%, respectively.

Table 3. Performance of the prediction models of the impact of compounds on inhibiting *E. coli*, based on the FT-MIR spectra of *E. coli*, after diverse pre-processing methods. Data are shown for the training and cross-validation dataset.

Model	Calibration			Cross-Validation			Test		
	NER	ER	Accuracy	NER	ER	Accuracy	NER	ER	Accuracy
PLS-DA	100	0	100	88	22	86	92	8	89
PLS-DA msc	100	0	100	89	11	88	96	4	94
PLS-DA snv	97	3	95	89	11	88	92	8	89
PLS-DA 1st	100	0	100	78	22	81	50	50	72
PLS-DA 2nd	100	0	100	50	50	100	0	100	0
kNN	70	30	76	69	31	69	52	48	67
kNN msc	82	18	83	93	7	90	65	35	67
kNN snv	79	21	83	85	15	86	65	35	67
kNN 1st	87	13	88	91	9	90	50	50	72
kNN 2nd	74	26	76	69	31	71	50	50	28

Table 3. *Cont.*

Model	Calibration			Cross-Validation			Test		
	NER	ER	Accuracy	NER	ER	Accuracy	NER	ER	Accuracy
SVM	77	23	83	58	42	55	72	28	78
SVM msc	91	9	93	80	20	81	92	8	89
SVM snv	87	13	90	80	20	81	92	8	89
SVM 1st	92	8	95	79	21	83	50	50	28
SVM 2nd	100	0	100	68	32	79	50	50	28
BPN:1:10	100	0	100	60	40	62	68	32	72
BPN:1:10 msc	96	4	95	90	10	90	81	19	71
BPN:1:10 snv	98	2	98	77	23	79	68	32	72
BPN:1:10 1st	100	0	100	73	27	80	62	38	72
BPN:1:10 2nd	85	15	87	51	49	52	47	53	50
BPN:1:20	100	0	100	68	32	71	100	0	100
BPN:1:20 msc	100	0	100	78	22	81	86	14	89
BPN:1:20 snv	100	0	100	86	14	86	81	19	87
BPN:1:20 1st	100	0	100	60	40	74	67	33	88
BPN:1:20 2nd	96	4	94	63	37	69	43	57	53
BPN:2:10	98	2	97	74	26	79	82	18	83
BPN:2:10 msc	98	2	98	86	14	87	78	22	76
BPN:2:10 snv	100	0	100	77	23	80	72	28	78
BPN:2:10 1st	84	16	85	72	25	82	75	25	80
BPN:2:10 2nd	100	0	100	62	38	60	66	34	78
BPN:2:20	100	0	100	60	40	64	50	50	72
BPN:2:20 msc	100	0	87	87	13	88	82	18	83
BPN:2:20 snv	100	0	100	72	28	78	68	32	72
BPN:2:20 1st	100	0	100	53	47	58	58	42	75
BPN:2:20 2nd	100	0	100	58	42	68	53	47	50

NER—non-error rate; ER—error rate.

The PLS-DA model's performance could be affected by outliers, as predicted by several parameters such as leverages, Q residuals, and Hotelling's T2. The samples exhibited low residual Q and Hotelling's T2 values, suggesting minimal variance between the actual and predicted outputs of the model [34]. Figure 5A illustrates the response of cluster A (exhibiting antimicrobial activity) for the training samples. The classification threshold to distinguish cluster A (antimicrobial activity) from cluster B (low antimicrobial activity) was set at 0.210 (depicted by the dotted red line).

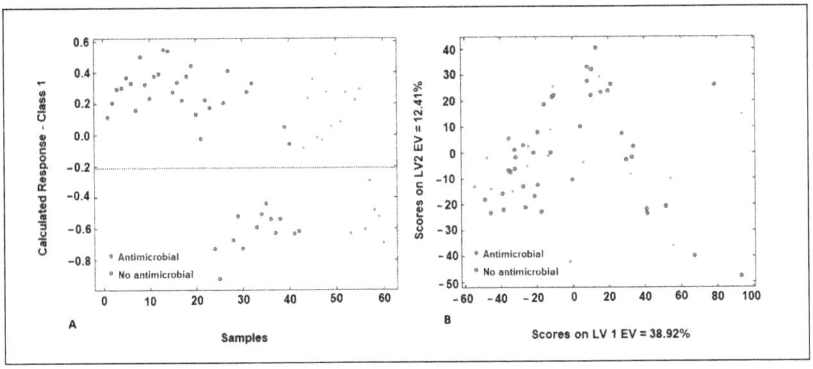

Figure 5. Representation of samples in the group with antimicrobial activity (blue) and the group without antimicrobial activity (red) of plant extracts (**A**) for the PLS-DA model developed using *E. coli* spectra with MSC pre-processing. Score plot obtained during the development of the PLS-DA model (**B**).

Misclassified samples were evident in our analysis, with samples from cluster A being mistakenly categorized in cluster B (false negatives), and samples from cluster B erroneously assigned to cluster A (false positives) (Figure 5A). The score plot indicates a discernible pattern in the distribution of samples, with a noticeable degree of overlap between them. The first latent variable explains 39% of the variance, suggesting it captures a significant portion of the underlying structure in the data. However, the second latent variable explains only 12% of the variance, indicating that there is still substantial variability not accounted for by the model (Figure 5B). In Figure 5B, it is evident that samples belonging to cluster A, which are characterized by antimicrobial action, exhibit higher scores for the first latent variable (LV) compared to samples from cluster B, which have low antimicrobial activity. This suggests that the first LV effectively discriminates between the two clusters based on their antimicrobial properties. Concerning the second LV, samples from cluster B are represented by negative scores, while samples from cluster A exhibit positive scores. This distinction demonstrates a significant resolution between the two types of plant extracts, further highlighting the efficacy of the model in distinguishing between them based on their characteristics. In the validation step, it was observed that the samples utilized in the test phase for both groups were placed close to the samples utilized in the training set. This finding indicates that the model exhibits high accuracy, thereby adding value to the classification process facilitated by the PLS-DA technique. The studies conducted by Pellegrini et al. (2017) showed the advantages of PLS-DA models as classification models for different essential oils extracted from *B. latifolia*, *Buddleja globosa*, *Solidago chilensis*, and *Aloysia polystachia*, which showed antimicrobial activity against *Paenibacillus larvae* [35].

2.4.2. K-Nearest Neighbor (KNN)

The KNN-optimized models, developed based on the spectra with MSC pre-processing, showed a high accuracy value (90%) and a low classification error (12%), with a cross-validation error rate of 20% (Table 3). The training model was characterized by an error of 13%, and sensitivity, specificity, and accuracy of 90%, 85%, and 88%, respectively, in predicting antimicrobial activity. For the cross-validation dataset, we obtained an error of 10% and sensitivity and specificity of 92% and 90%, respectively, in predicting antimicrobial activity (Table 3). For the test dataset, the accuracy, specificity, and sensitivity in predicting antimicrobial activity were 72%, 100%, and 0%, respectively. The performance of KNN revealed that for the classification of samples characterized by antimicrobial activity, the classification rate was 88% and 90% for the training and validation process.

2.4.3. Support Vector Machine (SVM)

The SVM model, based on the *E. coli* spectra with MSC pre-processing, was developed after the fine-tuning of several parameters and was characterized by a radial basis function SVM-kernel, with C = 100. The model showed significant fitting accuracy (90%), with a 12% classification error and a 20% cross-validation error rate (Table 3). The training model was characterized by excellent accuracy (93%) and a low calibration error (9%). The sensitivity and specificity for cluster A (antimicrobial activity) were 97% and 85%, respectively. In terms of the calibration model, the specificity and sensitivity for cluster A were 80% and 85%, respectively. In terms of cross-validation, the accuracy and validation error were, respectively, 81% and 20%, being characterized by excellent specificity and sensitivity (92% and 93%) in predicting cluster A (Table 3). In terms of prediction, the model presented an accuracy value of 89%, with values of 80% and 85% for specificity and sensitivity, respectively, in predict clustering A. The SVM model combined with the MIR spectral data, therefore, was a robust model for predicting antimicrobial activity.

2.4.4. Backpropagation Network (BPN)

The success of an artificial neural network depends on the size and quality of the training data, as well as its structure and the learning algorithms used. The samples were randomly assigned to a training set (65%) and a test set (35%). To optimize the network, an appropriate learning rate and different numbers of input nodes (10 and 20) and hidden layer neurons (1 and 2) were evaluated. According to the test set values obtained for the different models, the BPN algorithms (1:10; 1:20; 2:10, and 2:20) presented the best results when compared to the other algorithms (Table 3). Based on the MSE value, the model developed with one layer and 20 nodes was more suitable for the classification process. For each condition, 20 neurons per layer were considered, with a learning rate = 0.01, 1000 iterations, and a momentum term of 0.5, which allowed for faster convergence with the use of smaller learning coefficients. A two-layer BP model was developed at last, with the sigmoid transfer function. The scores of the first 15 PCs were applied as the input variables in the BP-ANN models. The epoch number was set at 1000, and after the multi-training step, the optimal number of nodes in the hidden layer and the learning rate were defined as 20 and 0.01, respectively. The optimized BPN model, based on the spectra with MSC pre-processing, showed an excellent fitting accuracy of 90%, a low classification error of 12%, and a cross-validation error rate of 20% (Table 3). The training model was characterized by an accuracy of 100% and a calibration error of 22%. Sensitivity was 100% and specificity was 100% in predicting the antimicrobial effect (Table 3). For the test dataset, the error rate was 14% and the accuracy value was 89%, showing sensitivity of 92% and specificity of 80% for cluster A.

The BPN 1:20 model was the one with the best classification results in terms of the test conditions, followed by the BPN model with two layers and 10 nodes and the BPN model with two layers and 20 nodes. Based on these results, the BPN algorithms provide the best guarantee for the development of classification models to predict extracts with antimicrobial activity (Figure 6). These neural networks have an advantage over conventional mathematical methods for modeling complex biological systems because (1) they are based only on an actual measured set of input and output variables, without requiring prior information about the interrelationship between the variables, and (2) they possess strong generalization and prediction ability. The principal advantages of artificial neural network (ANN) techniques include their capacity for data learning and generalization, fault tolerance, and an innate ability to process contextual information, alongside their rapid computational capabilities [36]. According to Badura et al. (2021), ANNs serve as excellent tools to support researchers in laboratory work, enabling the estimation of whether a tested compound possesses the desired antimicrobial activity [37]. It is also noteworthy that the primary advantage of ANNs is their ability to streamline the work of researchers who aim to uncover the intricate relationships among the physicochemical properties of compounds. This approach enables researchers to target the chemical synthesis of compounds with the desired activity by leveraging theoretical models obtained through computational chemistry [37]. According to the studies conducted by Cabrera et al. (2010), artificial neural networks serve as reliable, fast, and cost-effective tools, paving the way for predicting the antioxidant activity of essential oils [38].

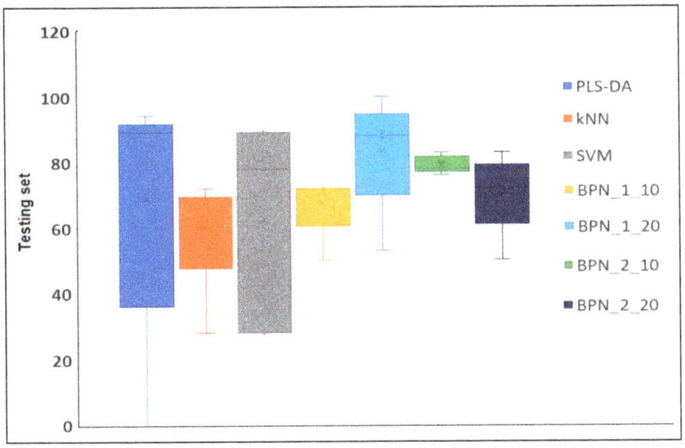

Figure 6. Comparison of test parameters related to different classification methods being studied (PLS-DA, kNN, SVM, BPN:1:10; BPN:1:20; BPN:2:10; and BPN:2:20 models).

3. Materials and Methods

3.1. Extraction Process

The plant materials, including the flowers, seeds, and fresh and dried leaves of *C. cardunculus*, were subjected to maceration using a mortar and pestle. Subsequently, the macerated materials were transferred to a flask containing the respective solvent (water, ethanol, acetone, methanol, and ethyl acetate) at a concentration of 5% (*w*/*v*). The mixture was stirred at room temperature for 16 h. The extracts were then filtered under vacuum, and the solvents were evaporated using a rotavapor at 60 °C. The resulting dry extract was weighed, and a suitable volume of H_2O was added to achieve a final concentration of 10 mg/mL. In the case of aqueous extraction, the material was boiled at approximately 100 °C for 10 min (Figure 7).

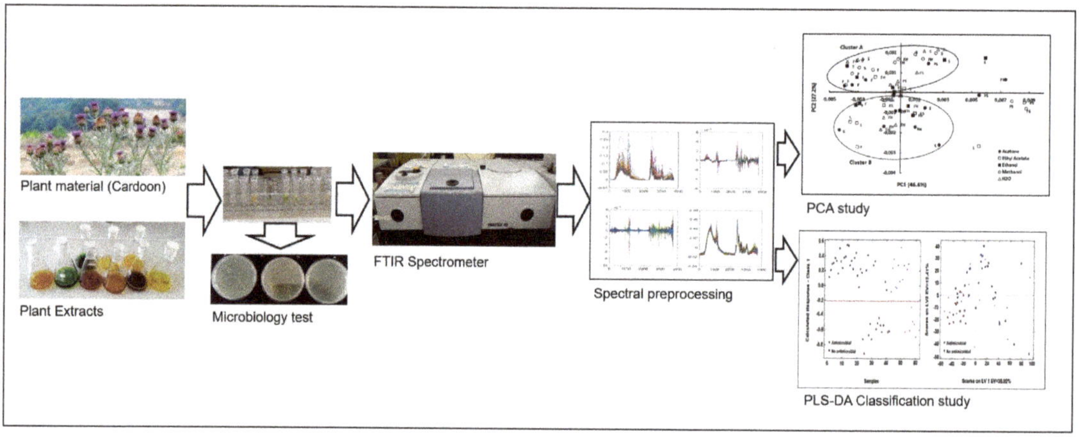

Figure 7. Experimental procedure carried out from selecting the plant material to extract classification based on *E. coli* FTIR spectra and machine learning methodology.

3.2. Antioxidant Activity

The antioxidant activity was determined according to the procedure developed by Mensor et al. [39], with slight adaptations as follows: the samples were diluted in methanol at concentrations of 10, 50, and 200 µg/mL. A total of 1 mL of alcoholic solution of 2,2-diphenyl-1-picryl-hydrazyl (DPPH) (50 µg/mL) was added to 2.5 mL of the sample. The mixture was maintained in the dark for 30 min. Absorbance was evaluated using a spectrophotometer at 518 nm (Shimadzu, Japan). The percentage of free radical scavenging (%FRS) was evaluated according to Equation (1): *Ac*–control absorbance; *As*–sample absorbance. The blank was prepared by replacing DPPH with ethanol in the reaction mixture. The negative control was prepared by adding 1 mL of DPPH in 2.5 mL of ethanol.

$$FRS\ (\%) = \frac{(Ac - As)}{Ac} \times 100 \quad (1)$$

3.3. Total Phenols

The spectrophotometric determination of phenolic compounds was conducted according to the method developed by Slinkard and Singleton [40]. The calibration curve was obtained using different concentrations of gallic acid (0–1000 mg/L). Each sample (1 mL) was transferred to a 25 mL flask, while the blank was prepared with distilled H_2O. Subsequently, 1 mL of Folin–Ciocalteu reagent was added, and the mixture was stirred and then left to rest for 5 min. To this mixture, 7 mL of (0.2 g/L) Na_2CO_3 solution was added, and the resulting mixture was kept in the dark for 2 h. The absorbance of the samples was determined at 765 nm, and the values were consequently converted to gallic acid equivalent (GAE) per gram of sample.

3.4. Antimicrobial Activity

The extracts obtained from different tissues of *C. cardunculus*, using diverse solvents, were tested on agar plates against *E. coli* cells. The bacteria were grown at 37° C with a stirrer (200 rpm) for 16 h. Then, 100 µL of the extracts was incubated with a similar volume of culture medium containing a specific concentration of cells. After an incubation period (2 h at 37 °C), 100 µL of this mixture was plated on a solid culture medium containing 2% (*w*/*v*) of agar. The plates were incubated at 37 °C for 16 h before counting the number of colonies.

3.5. FTIR Spectral Analysis

3.5.1. Acquisition of Spectra

The samples containing *E. coli* cells were subsequently diluted to obtain a similar optical density (OD = 0.2) for each assay. All assays were performed in triplicate. The plant extracts (100 µL) were incubated with the cells in the culture medium for 2 h at 37 °C. After that, the microtubes were centrifuged at 10,000 rpm for 3 min. The pellet was resuspended in 100 µL of 0.9% (*w*/*v*) NaCl. Then, 30 µL of these final suspensions was pipetted, in triplicate, into Zn-Se plates and, subsequently, dried under a vacuum system for 3.5 h at room temperature. The spectra were obtained with an HTS-XT system (Vertex, Bruker Optics, Mannheim, Germany) equipped with an FTIR spectrometer (Bruker Optics, Mannheim, Germany) between 4000 and 400 cm^{-1} with a resolution of 4 cm^{-1} (Figure 7).

3.5.2. Spectral Pre-Processing

The spectra were pre-processed by baseline correction, multiplicative scatter correction (MSC), standard normal variate (SNV), and derivatives (1st and 2nd derivatives).

3.6. Chemometric Methods

Principal Component Analysis

The compression and classification of data can be performed using the principal component analysis (PCA) technique. The goal is to reduce the dimensionality of a dataset (sample) by identifying a new set of variables that is smaller than the original set but still retains most of the information. This method groups the samples according to the existing information so that individuals in a group are as similar to each other as possible and as different from the remaining groups as possible (Figure 7).

3.7. Classification Methods

3.7.1. Partial Least Squares-Discriminant Analysis (PLS-DA)

Partial least squares-discriminant analysis (PLS-DA) is a linear classification tool used to calculate predictive models. It employs the partial least squares regression algorithm to identify latent variables that exhibit maximum covariance between the predictor variables and the class labels [34,41]. The PLS-DA models were calibrated using a total of 42 samples that were randomly selected: 29 samples with antimicrobial effect and 13 samples with no antimicrobial effect (Figure 7). Cross-validation was conducted using randomly selected training samples, with 5 data splits comprising 20% of the calibration samples each. This process was repeated for 20 iterations to ensure the robustness and reliability of the results. The test samples were constituted by 18 samples that were randomly selected: 13 samples with antimicrobial effect and 5 samples with no antimicrobial effect. Prediction was performed with the samples of the testing set. The evaluation of the prediction ability of the models was conducted based on their sensitivity (*Sn*), specificity (*Sp*), and accuracy, defined according to the true positives (*tp*), true negatives (*tn*), false positives (*fp*), and false negatives (*fn*), as follows:

Sensitivity (Sn):

$$Sn = \frac{tp}{tp + fp} \times 100\% \quad (2)$$

Specificity (Sp):

$$Sp = \frac{tn}{tn + fp} \times 100\% \quad (3)$$

Accuracy (Acc):

$$Acc = \frac{tp + tn}{tp + fn + tn + fp} \times 100\% \quad (4)$$

3.7.2. K-Nearest Neighbors

The K-nearest neighbors (KNN) algorithm is a supervised machine learning algorithm in which classification rules are derived from the neighborhood of training-set objects. Here, 'k' represents the number of neighbors surrounding the new data point, and classification is based on the majority class among its nearest neighbors. Distance functions play a crucial role in computing the distance between the new data point and its 'k' neighbors [42]. The neighborhoods are determined by calculating either Euclidean distances or correlation coefficients between the unknown object and the objects in the training set. For a training set comprising n samples, n distances or correlations are computed. The unknown object is assigned to the class to which the majority of the k objects with the smallest Euclidean distances or the highest correlations among the training samples belong.

3.7.3. Support Vector Machine

Support vector machine (SVM) relies on identifying a hyperplane that effectively divides a dataset into two distinct groups. However, when dealing with non-linear data, SVM may struggle with separating the data using linear hyperplanes. In such cases, alternative methods, known as kernel functions, are employed. Kernel functions are mathematical functions that take data as input and transform them into a higher-dimensional space, allowing SVM to effectively classify non-linear data by finding the optimal separating hyperplane in the transformed feature space [43]. Kernel functions map data to a higher dimension, enabling SVM classifiers to achieve optimal performance. The parameters and kernels are fine-tuned to enhance SVM classifier performance. The regularization parameter 'C' balances between minimizing training error and model complexity. Additionally, the kernel parameter 'γ' determines the width of the kernel function and the degree of generalization. These parameters are optimized using a grid-search procedure in SVM. In this study, the optimal conditions were determined by testing a regularization parameter 'C' value of 100 and utilizing a linear kernel function.

3.7.4. Backpropagation Network (BPN)

A backpropagation network (BPN) is highly regarded in data processing due to its ability to model non-linear processes, leverage data-driven features for powerful parallel computing, and demonstrate good fault tolerance and adaptability [44]. A backpropagation network (BPN) utilizes a gradient descent-based delta learning rule, commonly known as backpropagation, to train an artificial neural network [45]. This systematic method efficiently adjusts the weights in the network with functional units to analyze a set of input–output patterns. It aims to minimize the total squared error of the output, thus enabling the trained supervised learning network to effectively balance its ability to accurately respond to input patterns. A BPN algorithm minimizes prediction errors and yields satisfactory results by adjusting each weight of the network. This method is employed when establishing an artificial neural network (ANN) model [46]. During the optimization step, the BPN consisted of either 10 or 20 neurons distributed across one or two hidden layers. In the modeling process, 70% of the total samples were allocated and utilized as the training dataset for constructing the model, comprising 42 samples. To mitigate the risk of overfitting, 30% of the samples were set aside and employed as the validation dataset for implementing the early stopping methodology, consisting of 18 samples. Based on the preliminary experiment, the divider and function were selected to randomly partition the sample data in this study. The Mean Square Error (*MSE*) serves as a commonly used metric in the field of machine learning and neural networks, including BPNs. *MSE* is a measure of the average squared difference between the actual and predicted values. In the context of a backpropagation network (BPN) model during the training process, *MSE* quantifies the extent to which the model's predictions deviate from the actual target values. The formula for *MSE* is typically expressed as shown in Equation (5):

$$MSE = \frac{1}{N}\sum_{i=1}^{N}(y_i - \hat{y}_i)^2 \tag{5}$$

where N is the number of data points in the dataset; y_i represents the actual target value for the ith data point; and $\hat{y}_i$.*MSE* is calculated by taking the average of the squared difference between the actual and predicted values for all data points. The use of the squared difference helps to penalize larger errors more significantly than smaller errors. A BPN neural network is a kind of multilayer feed-forward neural network. The specific topological structure of the BP neural network is shown in Figure 8. The BP-ANN model was developed using the ANN tool of MATLAB 2023a software.

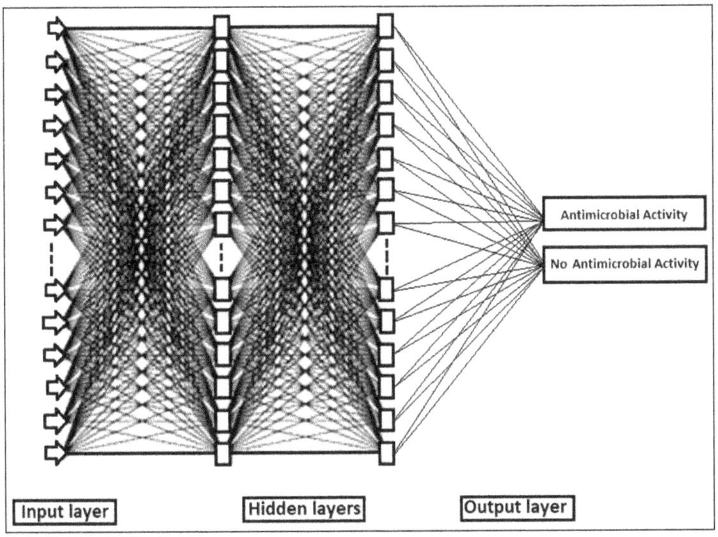

Figure 8. Topological structure of neural network used for the identification of antimicrobial activity.

Neurons are organized into layers within a neural network. The layers situated between the input layer, where the input signal is applied, and the output layer, which generates the output signal, are referred to as the hidden layers. The input layer comprises a total of 1154 absorbances of the entire spectrum. Each hidden layer consists of 20 nodes. The output layer represents the classification results of the extracts, which were categorized as either having antimicrobial activity or no antimicrobial activity. During the training process, the network learns each time a training pattern (comprising input and output data) is presented to it. This learning occurs through the modification of connection weights, and the process is repeated several times (iterations or epochs) until a satisfactory level of error is attained. The MLP parameters were trained using backpropagation to minimize the MSE, thereby optimizing the weights and biases for each neuron. Subsequently, the output layer yielded the result of the hidden layers, which was indicative of the antimicrobial activity. The training process was conducted with a learning rate of 0.001 for 500 epochs, each comprising 1000 batches. Each batch contained 1154 absorbances. At each epoch, the

MLP was evaluated using a validation set of the same size as the training set. The MLP parameters corresponding to each epoch were recorded.

3.8. Other Statistical Analysis

Student's *t*-test was performed using a data analysis tool package (Microsoft Office 365 Excel®) to evaluate differences in the ratios between the bands, at a 5% significance level. Spectral pre-processing and processing were conducted with a classification toolbox (PLS-DA, SVM, k-NN, and BPN) developed by Milano Chemometrics and QSAR Research Group (http://michem.disat.unimib.it/chm (accessed on 1 September 2023)), using MATLAB® 2023a (The MathWorks, Natick, MA, USA).

4. Conclusions

The results obtained in this study suggest that the integration of FT-MIR spectroscopy with machine learning methods represents a promising strategy to evaluate the impact of biocompounds obtained from plant extracts as antimicrobial agents against bacterial cells. This holistic approach holds the potential for efficiently classifying and discovering new bioactive molecules, thereby streamlining the drug exploratory process. It is possible to develop a model based on a backpropagation network for plant extract classification based on the spectra of *E. coli* after rapid exposure to antimicrobial compounds. The combination of FTIR spectroscopy and machine learning techniques creates the opportunity for preliminary research of promising antimicrobial compounds, saving time and resources compared to tests that require greater logistics in terms of consumables and laboratory techniques. Data analysis employing artificial neural networks offers a powerful method to optimize and minimize labor costs by streamlining the synthesis process, focusing only on compounds with anticipated desired properties. This approach serves as an invaluable tool in guiding the design of synthesis procedures and subsequent biological experiments, particularly in the targeted exploration for novel compounds with significant potential as antimicrobial agents.

Author Contributions: Conceptualization, P.N.S. and C.C.R.C.; Methodology, P.N.S. and C.C.R.C.; Software, P.N.S.; Validation, P.N.S.; Formal Analysis, P.N.S. and C.C.R.C.; Investigation, P.N.S. and C.C.R.C.; Writing—Original Draft Preparation, P.N.S.; Writing—Review and Editing, P.N.S. and C.C.R.C.; Visualization, P.N.S. and C.C.R.C.; Supervision, C.C.R.C.; Project Administration, P.N.S. and C.C.R.C.; Funding Acquisition, P.N.S. and C.C.R.C. All authors have read and agreed to the published version of the manuscript.

Funding: Funding was provided by ILIND–Lusófona University through the PLABIA–Research Platform for Bioactive Compounds through Artificial Intelligence.

Institutional Review Board Statement: Not applicable.

Informed Consent Statement: Not applicable.

Data Availability Statement: Data are contained within the article.

Acknowledgments: We thank Paulo Barracosa from the Instituto Politécnico de Viseu and César Garcia from Botanical Garden–Faculty of Sciences of the University of Lisbon.

Conflicts of Interest: The authors declare no conflicts of interest.

References

1. Chinemerem, D.N.; Ugwu, M.C.; Oliseloke, A.C. Antibiotic resistance: The challenges and some emerging strategies for tackling a global menace. *J. Clin. Lab. Anal.* **2022**, *36*, e24655. [CrossRef]
2. Petropoulos, S.; Fernandes, Â.; Pereira, C.; Tzortzakis, N.; Vaz, J.; Soković, M.; Ferreira, I.C.R. Bioactivities, chemical composition and nutritional value of *Cynara cardunculus* L. seeds. *Food Chem.* **2019**, *289*, 404–412. [CrossRef] [PubMed]
3. Falleh, H.; Ksouri, R.; Chaieb, K.; Karray-Bouraoui, N.; Trabelsi, N.; Boulaaba, M.; Abdelly, C. Phenolic composition of *Cynara cardunculus* L. organs, and their biological activities. *Comptes Rendus Biol.* **2008**, *331*, 372–379. [CrossRef] [PubMed]
4. Paris, R.R.; Moyse, H. *Médicale Massons & Cie*; Masson: Paris, France, 1971.

Velez, Z.; Campinho, M.A.; Guerra, Â.R.; García, L.; Ramos, P.; Guerreiro, O.; Felício, L.; Schmitt, F.; Duarte, M. Biological characterization of *Cynara cardunculus* L. methanolic extracts: Antioxidant, anti-proliferative, anti-migratory and anti-angiogenic activities. *Agriculture* **2012**, *2*, 472–492. [CrossRef]

Mileo, A.M.; Di Venere, D.; Linsalata, V.; Fraioli, R.; Miccadei, S. Artichoke polyphenols induce apoptosis and decrease the invasive potential of the human breast cancer cell line MDA-MB231. *J. Cell Physiol.* **2012**, *227*, 3301–3309. [CrossRef] [PubMed]

Pais, M.S.; Sampaio, P.; Soares, R. Pharmaceutical Composition Containing the Enzyme Cyprosin B an Aspartic Peptidase from *Cynara cardunculus* and Its Inclusion in anti-Tumorals Formulations. WO 2009/040778 A3, 16 July 2009.

O'Neill, J. Antimicrobial Resistance: Tackling a Crisis for the Health and Wealth of Nations. *Green Sustain. Chem.* **2014**, *11*, 1–8.

Mohr, K.I. History of Antibiotics Research. *Curr. Top. Microbiol. Immunol.* **2016**, *398*, 237–272.

Rosa, F.; Sales, K.C.; Cunha, B.R.; Calado, C.R.C. A comprehensive high-throughput FTIR spectroscopy-based method for evaluating the transfection event: Estimating the transfection efficiency and extracting associated metabolic responses. *Anal. Bioanal. Chem.* **2015**, *407*, 8097–8108. [CrossRef]

O'Connell, K.M.; Hodgkinson, J.T.; Sore, H.F.; Welch, M.; Salmond, G.P.; Spring, D.R. Combating multidrug-resistant bacteria: Current strategies for the discovery of novel antibacterials. Angew. *Chem. Int. Ed. Engl.* **2013**, *20120*, 10706–10733. [CrossRef]

Baldauf, N.A.; Rodriguez-Romo, L.A.; Männig, A.; Yousef, A.E.; Rodriguez-Saona, L.E. Effect of selective growth media on the differentiation of *Salmonella enterica* serovars by Fourier-transform mid-infrared spectroscopy. *J. Microbiol. Methods.* **2007**, *68*, 106–114. [CrossRef] [PubMed]

Jamin, N.; Miller, L.; Moncuit, J.; Fridman, W.H.; Dumas, P.; Teillaud, J.L. Chemical heterogeneity in cell death: Combined synchrotron IR and fluorescence microscopy studies of single apoptotic and necrotic cells. *Biopolymers* **2003**, *72*, 366–373. [CrossRef] [PubMed]

Hughes, C.; Liew, M.; Sachdeva, A.; Bassan, P.; Dumas, P.; Hart, C.A.; Gardner, P. SR-FTIR spectroscopy of renal epithelial carcinoma side population cells displaying stem cell-like characteristics. *Analyst* **2010**, *135*, 3133–3144. [CrossRef] [PubMed]

Scholz, T.; Lopes, V.V.; Calado, C.R.C. High-throughput analysis of the plasmid bioproduction process in *Escherichia coli* by FTIR spectroscopy. *Biotechnol. Bioeng.* **2012**, *109*, 2279–2285. [CrossRef] [PubMed]

Sampaio, P.N.; Calado, C.R.C. Potential of FTIR-spectroscopy for drug screening against *Helicobacter pylori*. *Antibiotics* **2020**, *9*, 897. [CrossRef] [PubMed]

Sales, K.K.; Rosa, F.; Cunha, B.R.; Sampaio, P.N.; Lopes, M.B.; Calado, C.R.C. Metabolic profiling of recombinant *Escherichia coli* cultivations based on high-throughput FT-MIR spectroscopic analysis. *Biotechnol. Progr.* **2017**, *33*, 285–298. [CrossRef]

Sampaio, P.N.; Sales, K.K.; Rosa, F.; Lopes, M.B.; Calado, C.R.C. High-throughput FTIR-based bioprocess analysis of recombinant cyprosin production. *J. Ind. Microbiol. Biotechnol.* **2018**, *44*, 49–61. [CrossRef] [PubMed]

Alvarez-Ordonez, A.; Carvajal, A.; Arguello, H.; Martínez-Lobo, F.J.; Naharro, G.; Rubio, P. Antibacterial activity and mode of action of a commercial citrus fruit extract. *J. Appl. Microbiol.* **2013**, *115*, 50–60. [CrossRef] [PubMed]

Corte, L.; Tiecco, M.; Roscini, L.; De Vincenzi, S.; Colabella, C.; Germani, R.; Tascini, C.; Cardinali, G. FTIR metabolomic fingerprint reveals different modes of action exerted by structural variants of N-alkyltropinium bromide surfactants on *Escherichia coli* and *Listeria innocua* cells. *PLoS ONE* **2015**, *10*, e0115275. [CrossRef] [PubMed]

Moen, B.; Janbu, A.O.; Langsrud, S.; Langsrud, Ø.; Hobman, J.L.; Constantinidou, C.; Kohler, A.; Rudi, K. Global responses of *Escherichia coli* to adverse conditions determined by microarrays and FT-IR spectroscopy. *Can. J. Microbiol.* **2009**, *55*, 714–728. [CrossRef] [PubMed]

Corte, L.; Rellini, P.; Roscini, L.; Fatichenti, F.; Cardinali, G. Development of a novel, FTIR (Fourier transform infrared spectroscopy) based, yeast bioassay for toxicity testing and stress response study. *Anal. Chim. Acta* **2010**, *659*, 258–265. [CrossRef] [PubMed]

Ribeiro da Cunha, B.; Fonseca, L.P.; Calado, C.R.C. A phenotypic screening bioassay for Escherichia coli stress and antibiotic responses based on Fourier-transform infrared (FTIR) spectroscopy and multivariate analysis. *J. Appl. Microbiol.* **2019**, *127*, 1776–1789. [CrossRef]

Sarker, I.H. Machine Learning: Algorithms, Real-World Applications and Research Directions. *Sn Comput. Sci.* **2021**, *2*, 160. [CrossRef] [PubMed]

Vamathevan, J.; Clark, D.; Czodrowski, P. Applications of machine learning in drug discovery and development. *Nat. Rev. Drug Discov.* **2019**, *18*, 463–477. [CrossRef] [PubMed]

Vora, N.; Tambe, S.S.; Kulkarn, B.D. Counter-propagation neural networks for fault detection and diagnosis. *Comput. Chem. Eng.* **1997**, *21*, 2. [CrossRef]

Ibrahim, D.; Lee, C.C.; Yenn, T.W.; Zakaria, L.; Sheh-Hong, L. Effect of the extract of endophytic fungus, *Nigrospora sphaerica* CL-OP 30, against the growth of methicillin-resistant *Staphylococcus aureus* (MRSA) and *Klebsiella pneumonia* cells. *Trop. J. Pharm. Res.* **2015**, *14*, 2091–2097. [CrossRef]

Ramos, P.A.B.; Santos, S.A.O.; Guerra, A.R.; Guerreiro, O.; Freire, C.S.R.; Rocha, S.M.; Duarte, M.F.; Silvestre, A.J.D. Phenolic composition and antioxidant activity of different morphological parts of *Cynara cardunculus* L. var. *altilis (DC). Ind. Crops Prod.* **2014**, *61*, 460–471. [CrossRef]

Fu, L.; Xu, B.T.; Xu, X.R.; Gan, R.Y.; Zhang, Y.; Xia, E.Q.; Li, H.B. Antioxidant capacities and total phenolic contents of 62 fruits. *Food Chem.* **2011**, *129*, 345–350. [CrossRef]

Da Silva, T.B.C.; Souza, V.K.; Lyra, L. Determination of the phenolic content and antioxidant potential of crude extracts and isolated compounds from leaves of *Cordia multispicata* and *Tournefortia bicolor*. *Pharm. Biol.* **2010**, *48*, 63–69. [CrossRef]

31. Valentão, P.; Fernandes, E.; Carvalho, F.; Andrade, P.B.; Seabra, R.M.; Bastos, M.L. Antioxidative properties of cardoon (*Cynar cardunculus* L.) infusion against superoxide radical, hydroxyl radical, and hypochlorous acid. *J. Agric. Food Chem.* **2002**, *50*, 49 [CrossRef] [PubMed]
32. Pauli, A.; Knobloch, K. Inhibitory effects of essential oil components on growth of food-contaminating fungi. *Z. Lebensm. Unter Forsch.* **1987**, *185*, 10–13. [CrossRef]
33. Ikigai, H.; Nakae, T.; Hara, Y.; Shimamura, T. Bactericidal catechins damage the lipid bilayer. *Biochim. Biophys. Acta* **1993**, *114* 132–136. [CrossRef]
34. Ballabio, D.; Consonni, V. Classification tools in chemistry. Part 1: Linear models PLS-DA. *Anal. Methods* **2013**, *5*, 3790–379 [CrossRef]
35. Pellegrini, M.C.; Alonso-Salces, R.M.; Umpierrez, M.L.; Rossini, C.; Fuselli, S.R. Chemical Composition, Antimicrobial Activit and Mode of Action of Essential Oils against *Paenibacillus larvae*, Etiological Agent of American Foulbrood on *Apis mellifera*. *Chen Biodivers.* **2017**, *14*, e1600382. [CrossRef] [PubMed]
36. Jain, A.K.; Mao, J.; Mohiuddin, K.M. Artificial Neural Networks: A Tutorial. *IEEE Comput.* **1996**, *29*, 31–44. [CrossRef]
37. Badura, A.; Krysiński, J.; Nowaczyk, A.; Buciński, A. Prediction of the antimicrobial activity of quaternary ammonium salt against *Staphylococcus aureus* using artificial neural networks. *Arab. J. Chem.* **2021**, *14*, 103233. [CrossRef]
38. Cabrera, A.C.; Prieto, J.M. Application of artificial neural networks to the prediction of the antioxidant activity of essential oils i two experimental in vitro models. *Food Chem.* **2010**, *118*, 141–146. [CrossRef]
39. Mensor, L.L.; Menezes, F.S.; Leitão, G.G.; Reis, A.; dos Santos, T.C.; Coube, C.S.; Leitão, S.G. Screening of Brazilian plant extract for antioxidant activity by the use of DPPH free radical method. *Phytother. Res.* **2001**, *15*, 127–130. [CrossRef] [PubMed]
40. Slinkard, K.; Singleton, V.L. Total Phenol Analysis: Automation and comparison with manual methods. *Am. J. Enol. Vitic.* **197** *28*, 49–55. [CrossRef]
41. Barker, M.; Rayens, W. Partial least squares for discrimination. *J. Chemom.* **2003**, *17*, 166–173. [CrossRef]
42. Deconinck, E.; Sokeng Djiogo, C.A.; Courselle, P. Chemometrics and chromatographic fingerprints to classify plant foo supplements according to the content of regulated plants. *J. Pharm. Biomed. Anal.* **2017**, *143*, 48–55. [CrossRef]
43. Westerhuis, J.A.; Hoefsloot, H.C.J.; Smit, S.; Vis, D.J.; Smilde, A.K.; van Velzen, E.J.; van Duijnhoven, J.P.M.; van Dorsten, F.A Assessment of PLS-DA cross validation. *Metabolomics* **2008**, *1*, 81–89. [CrossRef]
44. Berry, M.J.; Linoff, G. *Data Mining Techniques: For Marking, Sales and Customer Support*; Wiley: Hoboken, NJ, USA, 2001.
45. Russell, S.; Norvig, P. *Artificial Intelligence: A Modern Approach*; Prestige Hall: Saddle River, NJ, USA, 2003.
46. Tu, J.V. Advantages and disadvantages of using artificial neural networks versus logistic regression for predicting medica outcomes. *J. Clin. Epidemiol.* **1996**, *49*, 1225–1231. [CrossRef] [PubMed]

antibiotics

MDPI

Article

Untargeted MS-Based Metabolomic Analysis of Termite Gut-Associated Streptomycetes with Antifungal Activity against *Pyrrhoderma noxium*

Cherrihan Adra [1], Trong D. Tran [1,2], Keith Foster [3], Russell Tomlin [3] and D. İpek Kurtböke [1,*]

[1] School of Science, Technology and Engineering, University of the Sunshine Coast, Maroochydore BC, QLD 4558, Australia; cherrihan.adra@research.usc.edu.au (C.A.); ttran1@usc.edu.au (T.D.T.)
[2] Centre for Bioinnovation, University of the Sunshine Coast, Maroochydore BC, QLD 4558, Australia
[3] Brisbane City Council, Program, Planning and Integration, Brisbane Square, Level 10, 266 George Street, Brisbane, QLD 4000, Australia; kfosteradsl@bigpond.com (K.F.); russell.tomlin@brisbane.qld.gov.au (R.T.)
* Correspondence: ikurtbok@usc.edu.au

Abstract: *Pyrrhoderma noxium* is a plant fungal pathogen that induces the disease of brown root rot in a large variety of tree species. It is currently infecting many of the amenity trees within Brisbane City of Queensland, Australia. Steering away from harmful chemical fungicides, biological control agents offer environmentally friendly alternatives. Streptomycetes are known for their production of novel bioactive secondary metabolites with biocontrol potential, particularly, streptomycete symbionts isolated from unique ecological niches. In this study, 37 termite gut-associated actinomycete isolates were identified using molecular methods and screened against *P. noxium*. A majority of the isolates belonged to the genus *Streptomyces*, and 15 isolates exhibited strong antifungal activity with up to 98.5% mycelial inhibition of the fungal pathogen. MS/MS molecular networking analysis of the isolates' fermentation extracts revealed several chemical classes with polyketides being among the most abundant. Most of the metabolites, however, did not have matches to the GNPS database, indicating potential novel antifungal compounds in the active extracts obtained from the isolates. Pathway enrichment and overrepresentation analyses revealed pathways relating to polyketide antibiotic production, among other antibiotic pathways, further confirming the biosynthetic potential of the termite gut-associated streptomycetes with biocontrol potential against *P. noxium*.

Keywords: *Streptomyces*; streptomycetes; *Pyrrhoderma noxium*; termites; termite gut symbiosis; antifungal compounds; metabolomics; molecular networking; biological control; mass spectrometry

1. Introduction

Invasive fungal pathogens pose an increasing threat to agriculture and food industries worldwide [1,2]. *Pyrrhoderma noxium* (*P. noxium*), formally known as *Phellinus noxius* [3], is a phytopathogenic fungus belonging to the family *Hymenochaetaceae* and the division Basidiomycota. It induces the disease of brown root rot, although it is classed as a white-rot fungus due to its ability to degrade lignin along with cellulose and hemicellulose [3]. This pathogen has caused much destruction throughout tropical and subtropical regions and is widespread throughout Taiwan [4–6], Japan [7,8], China [9,10], and Australia [11–14]. It is able to infect over 250 tree species, including herbaceous, broad-leaved, fruit, ornamental, and coniferous woody trees [15,16] within forests, landscapes, and orchards [17]. Infection begins within the roots and eventually extends to the basal stem [17]. Early stages of the disease are largely asymptomatic; therefore, diagnosis and treatment are usually hindered, and the tree health is too far declined for intervention [18]. Symptoms can include chlorosis, dieback, defoliation, and decay of stem and roots [19,20].

In response to the pressing need for sustainable and eco-friendly solutions to combat invasive fungal pathogens and reduce reliance on chemical fungicides, biological control methods have gained significant interest [21]. These methods involve the use of natural antagonists, such as functional microorganisms, to minimise the harmful effects of chemical agents used in conventional agriculture [22,23]. Among the diverse sources of natural antagonists, actinomycetes have emerged as prominent candidates as they are one of the largest producers of bioactive secondary metabolites and new antibiotics, with a recent focal shift on rare actinomycetes as an increasing target source for new secondary metabolites [24,25]. In fact, rare members of the genus *Streptomyces* persist to be the largest producer of new innovative metabolites [26]. Between 2015 and 2020, a significantly large number of novel *Streptomyces* spp. were isolated from various niches, such as extreme environments, sediments, marine environments, and symbionts [22,24]. Streptomycetes as microbial symbionts are distributed in a wide range of environments, including insect guts, cuticles, and nests [27]. Within this mutualistic relationship, streptomycetes assist insects in the breaking down of food, provide a synthesis of essential metabolites, alter gene descriptions, and combat pathogens and competitors [28]. One such source of these specialised streptomycetes symbionts is the guts of termites. Termites belong to the order Isoptera and have three main components of their alimentary tract: the foregut (stomodeum), the midgut (mesenteron), and the hindgut (proctodeum) [29]. They predominately feed on litter tissue and wood, or in other words, difficult-to-digest lignocelluloses; therefore, they depend on beneficial symbioses with a diversity of microflora for digestion and acquisition of supplementary nutrition [28]. Previous studies have revealed that actinomycetes are among the most dominant bacteria identified within the termite's symbiotic lifestyle [29]. It is therefore possible that these unexplored insect guts may house rare species of actinomycetes with the ability to produce novel NPs [28,30]. To unravel the intricacies of these NPs in understudied streptomycetes, a metabolomic approach has proved to be a powerful tool [31]. Moreover, molecular networking [32] allows for the dynamic organisation of experimental spectra into families/classes based on spectral similarity with established MS/MS databases. It also provides visualisation and annotation of non-targeted mass spectrometry data [33,34].

In this study, 37 termite gut-associated actinomycetes selected from the University of the Sunshine Coast's Microbial Library [35–37] were screened for antifungal activity against phytopathogen *P. noxium*. Their metabolic profiles were investigated using quadrupole time-of-flight (QTOF) mass spectrometry (MS) as well as molecular networking techniques for the purposes of screening and predicting bioactive molecules from the termite gut-associated actinomycetes active extracts and prioritising the best candidate(s) as potential biological control agents against phytopathogen *P. noxium*.

2. Results and Discussion

2.1. Phylogeny of Termite Gut-Associated Actinomycetes Based on the 16S rRNA Oligonucleotide Analysis

From the 37-termite gut-associated actinomycetes obtained from the UniSC's Microbial Library, 32 of them (86%) had almost complete 16S rRNA gene sequences aligned with corresponding sequences of representative type strains of the genus *Streptomyces* (Tables S1 and S2). Five of the isolates (13%) aligned with the genus *Micromonospora* and one strain aligned with the genus *Saccharopolyspora* (Tables S1 and S2). Phylogenetic analysis demonstrated putative sequence identity percentage between 97 and 100% (Figure S1). The dominance of *Streptomyces* isolates within the termite gut aligned with previous studies and demonstrated the crucial role these bacteria play in termite symbiosis [38,39]. One study analysed the phylogeny of 118 actinomycete isolates collected from three types of termite nests (mound, carton, and subterranean nests) and determined *Streptomyces* spp. to be the most dominant genus accounting for approximately 60% of the isolates [40]. Another study completed a 16S analysis of 16 actinomycetes isolates from termite guts and found 9 of them aligned with the genus *Streptomyces* [41].

2.2. Screening for Antifungal Activity from Termite Gut-Associated Actinomycete Isolates against *P. noxium*

All 37 actinomycete isolates underwent a preliminary evaluation for antifungal activity against *P. noxium* using the co-culture assay technique (Table S2). From this, 15 actinomycete isolates demonstrated strong antifungal activity against the fungal pathogen (Figure 1A). The mean inhibition percentages are demonstrated in Figure 1B. The results showed significant differences between the control and co-culture plates (F (14, 60) = 152.880) $p \leq 0.001$) with isolate USC-595B presenting with the strongest antifungal activity with a mean value of 98.5% mycelial inhibition. This was followed by USC-6918 with 94.5% and USC-6919 with 92.0% inhibition, respectively. These active isolates might therefore be a valuable reservoir of bioactive compounds with antifungal properties. As per the phylogenetic analysis, the 15 isolates demonstrating the potent antifungal activity showed clustering within the genus *Streptomyces*.

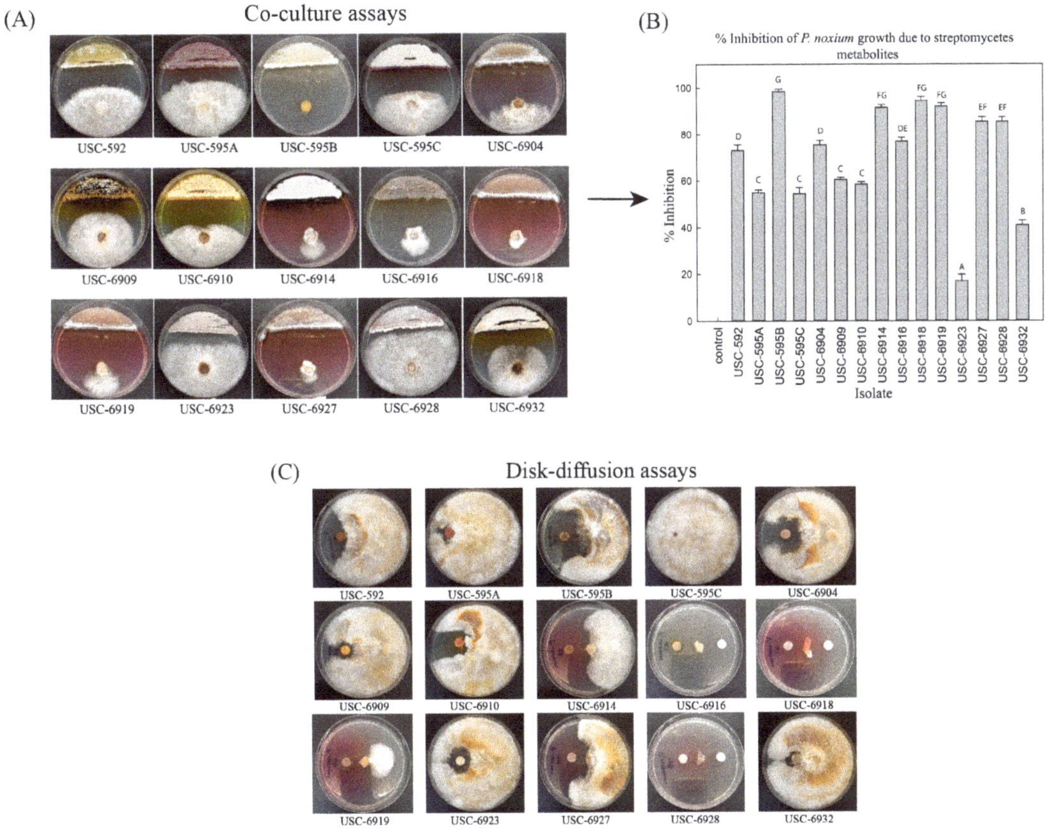

Figure 1. (**A**) Co-culture assays of the selected 15 termite gut-associated *Streptomyces* isolates and the *P. noxium* pathogen. (**B**) Percent inhibition of *P. noxium* mycelia when grown in co-culture with *Streptomyces* isolates. (**C**) Disk-diffusion assays of the *Streptomyces* isolate extracts against *P. noxium* (the left disk contains the active extracts, and the right disk is the control and contains EtOAc).

2.3. Antifungal Activity from Streptomycetes Fermentation Extracts against *P. noxium*

The 15 selected antifungal *Streptomyces* isolates underwent fermentation and extraction of their natural products. The resulting fermentation extracts were further tested for persistent antifungal activity using the Kirby-Bauer disk-diffusion method (Figure 1C).

USC-595B demonstrated persistent strong antifungal activity, and USC-6916 and USC-6928 also displayed a substantial increase in antifungal activity compared with the initial co-culture assays (Figure 2A). The antifungal activity of some of the *Streptomyces* isolates, including USC-6904, USC-6927, USC-6910, and USC-6909, was reduced. This decrease might be due to the absence of antifungal volatile compounds or compound degradation during solvent extraction and evaporation. Based on these disk-diffusion activity assays, isolates presenting with the strongest activities and, therefore, of interest include USC-595B, USC-6904, UCS-6914, USC-6916, USC-6918, USC-6919, USC-6927, and USC-6928. These isolates and their closest related species can be linked to previous studies demonstrating antifungal activities. For instance, USC-595B was found to be most similar to *Streptomyces platensis*, which was identified as a strong antifungal strain [42–44]. Resormycin is an herbicidal and antifungal antibiotic isolated from *S. platensis*, which demonstrated activity against numerous fungal phytopathogens [45]. Wan et al. (2008) also found *S. platensis* to be an effective biocontrol agent for the suppression of many plant-pathogenic fungi, including *Botrytis cinerea*, *Rhizoctonia solani*, and *Sclerotinia sclerotiorum*. Additionally, biosynthetic gene clusters of Phoslactomycins (PLMs), which are polyketide natural products and important antifungals, were identified from *S. platensis* [46]. USC-6904 and USC-6914 were most similar to *S. catenulae*, which demonstrated antifungal activity against the fungal pathogen *P. citriasiana* [47]. USC-6916 was most similar to *S. gelaticus*, which was assessed for antifungal activity against basal stem rot caused by *Ganoderma boninense* using co-culture assays and found to produce up to 100% inhibition. The subsequent crude extracts generated from this strain demonstrated activity of around 87.5% inhibition from a 10% concentration [48]. USC-6918 was most similar to *S. luozhongensis*, which was found to have antifungal activity against *Saccharomyces cerevisiae*, *Aspergillus flavus*, *A. niger*, *Penicillium citrinum*, and *Candida albicans* [49]. USC-6919 was most similar to the thermophilic *S. thermocarboxydus*, which was shown to be a promising biocontrol agent against strawberry anthracnose caused by *Glomerella cingulate* [50]. USC-6927 was most similar to *S. alni* and was found to have antifungal activity against the root rot of grapevine [50]. USC-6928 was most similar to *S. diastaticus*, which was found to produce the macrolide antibiotics oligomycin A and C, which exhibited strong activity against *A. niger*, *Alternaria alternata*, *B. cinerea*, and *Phytophthora capsica* [51]. This further exemplifies the biopotential of these isolates to produce antifungal secondary metabolites for the purposes of biocontrol.

2.4. Metabolite Diversity Analysis of Streptomycetes Fermentation Extracts Using Untargeted MS-Based Metabolomic Approach

Principal component analysis (PCA) of the MS data was generated by comparing the chemical diversity between the 15 streptomycete fermentation extracts (Figure 2A). The PCA demonstrated a clear clustering of samples, indicating spectral similarities between isolates USC-6904, USC-6914, USC-6918, USC-6919, USC-6927, and USC-6928. The remaining isolates had a clear distinction between this cluster but still displayed differences between each other. It can also be noted that USC-6923 displayed the largest metabolic diversity. Additionally, USC-595A and USC-6916 also displayed a degree of similarity in their metabolic profiles. The generated PCA in Figure 2B displays a comparison between the 8 identified active extracts and the remaining 7 non-active extracts. The non-active extracts show a larger chemical diversity; however, there is a great deal of overlap with the active extracts.

2.5. Molecular Networking and Identification of Bioactive Compounds within the Streptomycetes Fermentation Extracts

A molecular network (MN) was generated for the positive and negative electrospray ionisation mode datasets to detect as many compounds as possible. The MN of the positive dataset consisted of 6994 nodes, of which 114 were annotated by the GNPS spectral library database (Figure 3A). The negative dataset comprised 5629 nodes, and 38 nodes were annotated (Figure 3B). Each node represented a metabolite based on its molecular ion signal, and the edges connecting them represented the similarity of MS/MS fragmentation patterns among the nodes. To further enhance chemical structural information within the MN, the in silico structure annotation from the GNPS library search was incorporated into the network using the MolNetEnhancer workflow, which allowed for chemical class annotations using ClassyFire chemical ontology. It is apparent that the termite gut-associated streptomycetes produce structurally diverse compounds both between ionisation modes and between the individual samples (Figure 3). The identified chemical classes include organoheterocyclic compounds, lipids and lipid-like molecules, organic acids and derivatives, phenylpropanoids and polyketides, organic oxygen compounds, benzenoids, organic nitrogen compounds, nucleotides/sides, and lignans and neolignans (Figures S2 and S3). The most abundant chemical classes observed in node clusters include organic oxygen compounds, phenylpropanoids and polyketides, organic acids and derivatives, and lipids and lipid-like molecules. However, a majority of the metabolites visualised in the MN did not have a match to the chemical class annotations due to a lack of annotations from the GNPS library, indicating potential novel metabolites.

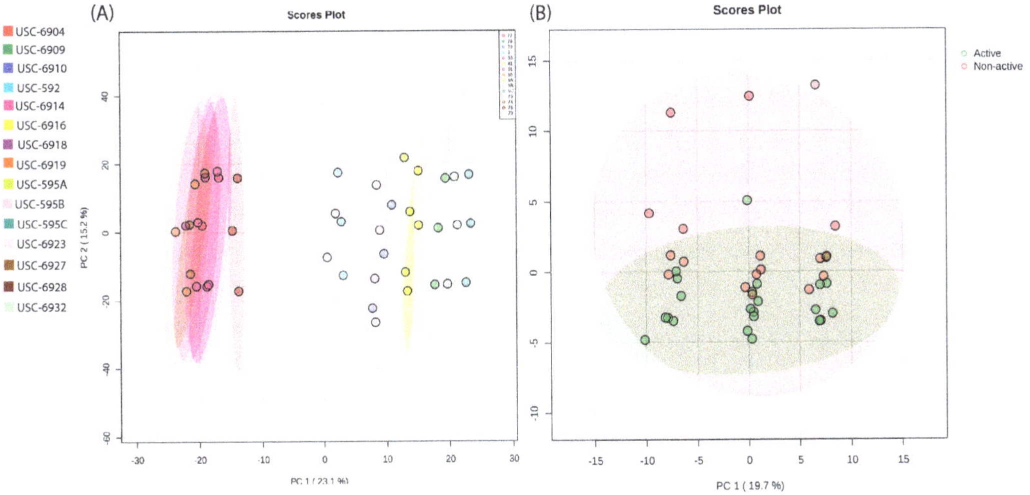

Figure 2. (**A**) Two-dimensional PCA score plot demonstrating the differential production of metabolites between the fifteen *Streptomyces* isolates across the positive and negative ionisation modes (PC1 × PC2) and (**B**) a score plot comparing the active and non-active fermentation extracts.

Streptomycetes are known to synthesise various antimicrobial compounds belonging to chemical classes such as aminoglycosides, peptides, ansamycins, β-lactams, tetracyclines, macrolides, lincosamides, epoxides, polyketides terpenoids, flavones, and amino--coumarins [52,53]. Many of these described chemical classes were also identified within the molecular network of the fermentation extracts of termite gut-associated *Streptomyces* isolates. Polyketides constitute a substantial and diverse group of secondary metabolites that display significant structural variability and versatile biological functions [54]. Beyond their pharmaceutical significance, actinomycete polyketides have also gained much interest in crop protection. Examples of these actinomycete polyketide pesticides include avermectins, milbemycins, and spinosyns, all of which offer a form of protection for plants against various pathogens and have been successfully commercialised and used in the agricultural field [55]. As seen in Figure 3, polyketide metabolites are among the most dominant observed in the metabolome of the termite gut-associated streptomycetes. Accordingly, they were selected for further investigation.

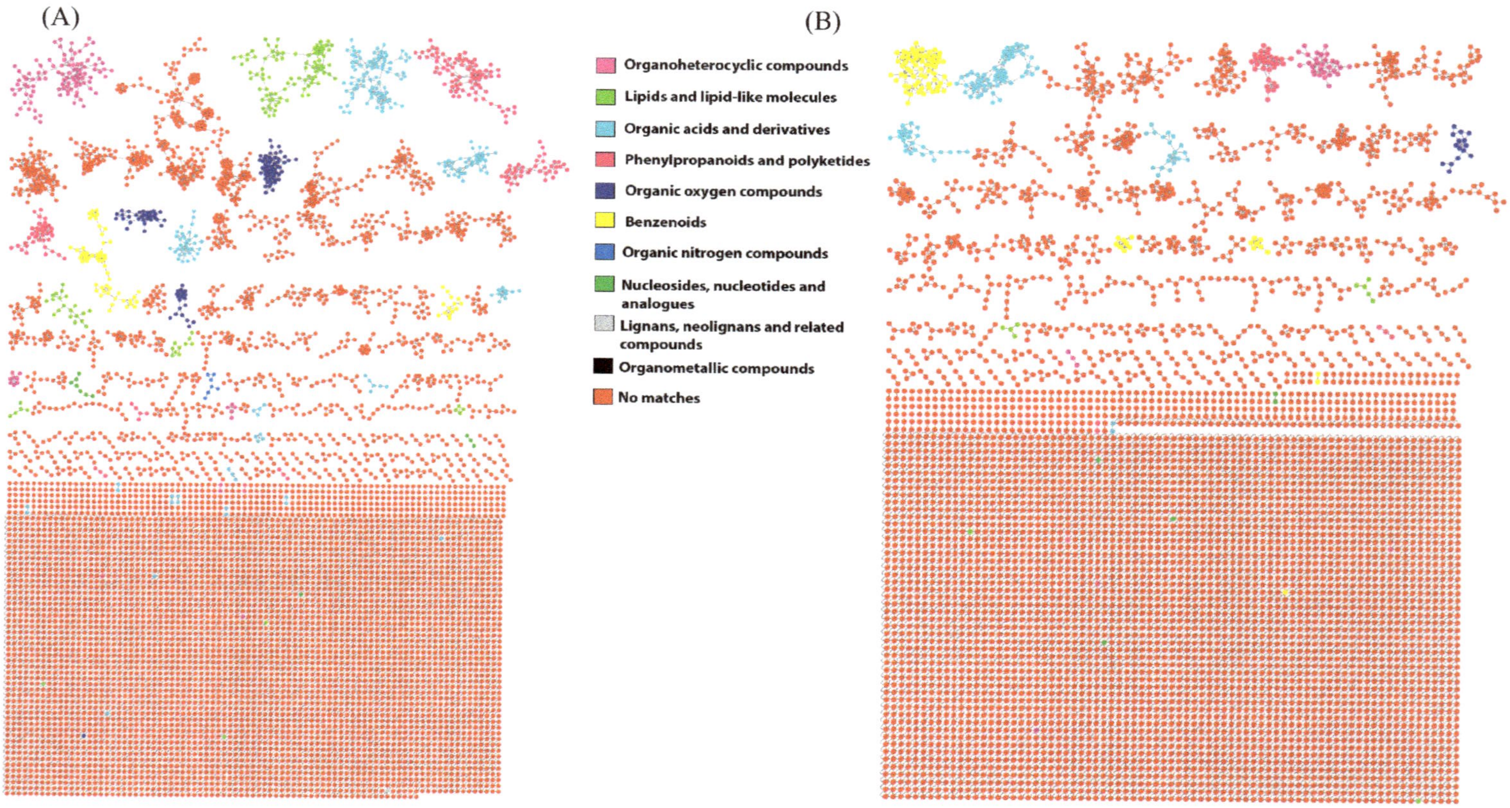

Figure 3. Termite gut-associated streptomycetes molecular network coloured by 11 chemical class terms as indicated by the legend. Chemical class annotation was performed using MolNetEnhancer within the GNPS framework (**A**) class annotated MN in a positive ionisation mode and (**B**) class annotated MN in a negative ionisation mode.

2.5.1. Isoflavonoids

Flavonoids and isoflavonoids are versatile natural products with extensive pharmacological activities. They have extremely low toxicity, making them compounds of interest and a "hotspot" for drug discovery and development [56]. However, research on flavonoids has favoured plant sources, resulting in relatively less focus on microorganisms as potential sources for novel flavonoids [56–58]. Within this class, a cluster of isoflavones, including daidzein (m/z 255.065 $[M+H]^+$) and genistein (m/z 271.061 $[M+H]^+$), were tentatively identified within the GNPS library (Figure 4). One study found that 10% of their 200 tested soil *Streptomyces* isolates produced daidzein and genistein and that genistein increased the production of polyketide antibiotics by 11–42% within its fermentation extracts [59]. The potential discovery of new isoflavonoids may be found in termite-associated streptomycetes as a study found three new isoflavonoid glycosides, which were isolated from termite-associated *Streptomyces* spp. recovered from the cuticle of South African termites, and as such were named termisoflavones A-C. These isolates were also found to produce daidzein and genistein [60]. Derivatives of daidzein and genistein, 7-O-β-D-glucosyl-genistein and 7-O-β-D-glucosyl-daidzein, have also demonstrated considerable antifungal activity against *P. citricarpa*, meaning there may be potential for antifungal activity against *P. noxium* [61]. These two compounds were produced in isolates USC-595C, USC-595A, USC-6928, and USC-6916.

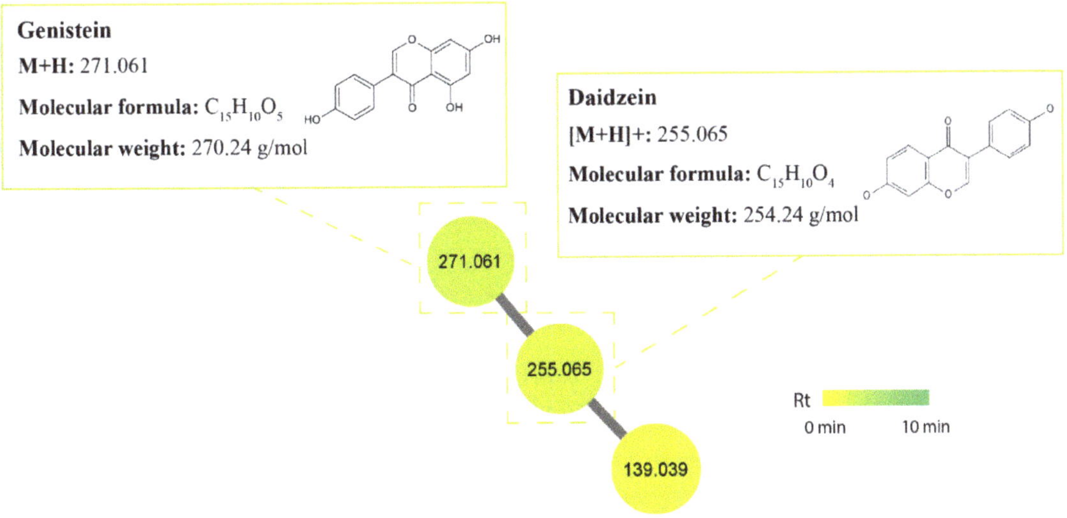

Figure 4. Isoflavonoid metabolite cluster identified in the termite gut-associated streptomycetes molecular network. Nodes represent an individual metabolite; each node displays its parent mass and is coloured based on its retention time.

2.5.2. Macrolides

Macrolides were originally isolated from *Streptomyces* species and are antibiotics consisting of a macrocyclic lactone ring, often 14-, 15- or 16-membered ring compounds with which various amino sugars are attached [62]. The most common macrolides are oleandomycin isolated from *Streptomyces antibioticus*, spiramycin isolated from *S. ambofaciens*, and erythromycin isolated from *S. erythreus*. Recent findings of new macrolide antibiotics from *Streptomyces* species have demonstrated potent antifungal activity. For instance, nine new polyene macrolides were isolates from deep-sea-derived *Streptomyces antibioticus* that demonstrated antifungal activity against *Candida albicans* [63]. Additional four new macrolides named venturicidins G-J were isolated from *Streptomyces* sp. *SN5452*, exhibiting strong inhibition towards *Pyricularia oryzae*, which causes the destructive rice blast disease [53]. A cluster of macrotetrolide antibiotics were also tentatively identified within this class, including nonactin (m/z 737.450 $[M+H]^+$), monactin (m/z 768.495 $[M+NH4]^+$), dinactin (m/z 765.481 $[M+H]^+$), trinactin (m/z 779.494 $[M+H]^+$), bonactin (m/z 418.203 $[M+NH4]^+$), and tetranactin (m/z 793.512 $[M+H]^+$), all found in isolates USC-595A and USC-595C (Figure 5). These macrolides are also known to be produced by a variety of *Streptomyces* spp., particularly by *S. globisporus* [64]. They exhibit antibacterial and antifungal activity as well as acaricidal, insecticidal, coccidiostatic, and anthelminthic activities [65]. One study investigated the bioactive compounds of *S. araugoniae* that proved effective against many phytopathogenic fungi and found a complex of macrotetrolides that included these compounds, all of which presented with antifungal activity against *Botrytis cinerea* [66]. Another study found that dinactin had broad inhibitory activity towards multiple phytopathogenic fungi, including *Colletotrichum gloeosporioides* [67]. A macrolide antibiotic within a separate cluster was identified as oligomycin A (m/z 791.531 [M+H]) and is reported to be produced by some strains of *Streptomyces*. They have a broad spectrum of activities against organisms such as fungi, bacteria, and nematodes. Oligomycin was first described as a new antifungal antibiotic in *Streptomyces diasatochromogenes* with nine isomers identified soon after [68]. A study conducted by Chakraborty (2020) investigated oligomycins produced by streptomycetes and evaluated the in vitro suppressive effects on hyphal growth, condidogenesis, conidial germination, and appressorial formation of the wheat blast fungus *Magnaporthe oryzae* and found strong inhibitory activity across all areas [69]. This oligomycin cluster was found in isolates USC-592, USC-595A, USC-6923, and USC-6928 (Figure 5). A separate phenylpropanoid and polyketide cluster identified as polycyclic tetramate macrolactum (PTM) clifednamide A (m/z 493.271 $[M+H]^+$) is produced by streptomycetes and other bacteria. PTMs are a fast-growing antibiotic natural product family with diverse therapeutic properties, such as antibacterial, antifungal, antiprotozoal, and anticancer [70]. It has also been suggested that the fungicidal action of macrolide antibiotics is their affinity for binding to ergosterol within fungal membranes [71].

2.5.3. Coumarins

Coumarin and its derivatives represent one of the most active classes of compounds possessing biological activities [72], such as antibacterial, antifungal, antiviral, and anticancer [73]. Since the discovery of novobiocin (an aminocoumarin antibiotic) produced by *Streptomyces niveus*, coumarins have attracted more attention as potential antimicrobial agents. The GNPS library displayed a match to 6-methylcoumarin (m/z 161.060 [M+H]) produced by isolates USC-6932 and USC-6904. In a study conducted by Chen (2022), the antifungal activity of coumarins against phytopathogenic fungi was evaluated and found that 6-methylcoumarin presented with the strongest antifungal activity against Valsa mali [74]. Another coumarin was manually annotated as methoxsalen (m/z 217.045 $[M+H]^+$) produced by isolates USC-592 and USC-6932 (Figure 6). Methoxsalen (xanthotoxin) is a photosensitising furocoumarin compound and has been used as the first treatment of mycosis fungoides [75].

2.5.4. Cyclic Peptides (Nocardamine)/Siderophores

Desferrioxamines are the principle siderophores produced by *Streptomyces* species [76]. Studies have shown that streptomycetes are able to use their siderophore production as a means to 'starve' fungi of iron [77] as an antagonistic mechanism. A cluster of siderophore peptides was also identified within the MN, including desferrioxamine E (synonym: nocardamine) (m/z 601.356 $[M+H]^+$) which is a siderophore widely produced by *Streptomyces* species [76] and was found in isolates USC-595A and USC-592. Derivatives of this compound, also found in streptomycetes and identified as dehydroxynocardamine (m/z 585.631 $[M+H]^+$), were found in isolate USC-592 (Figure 7).

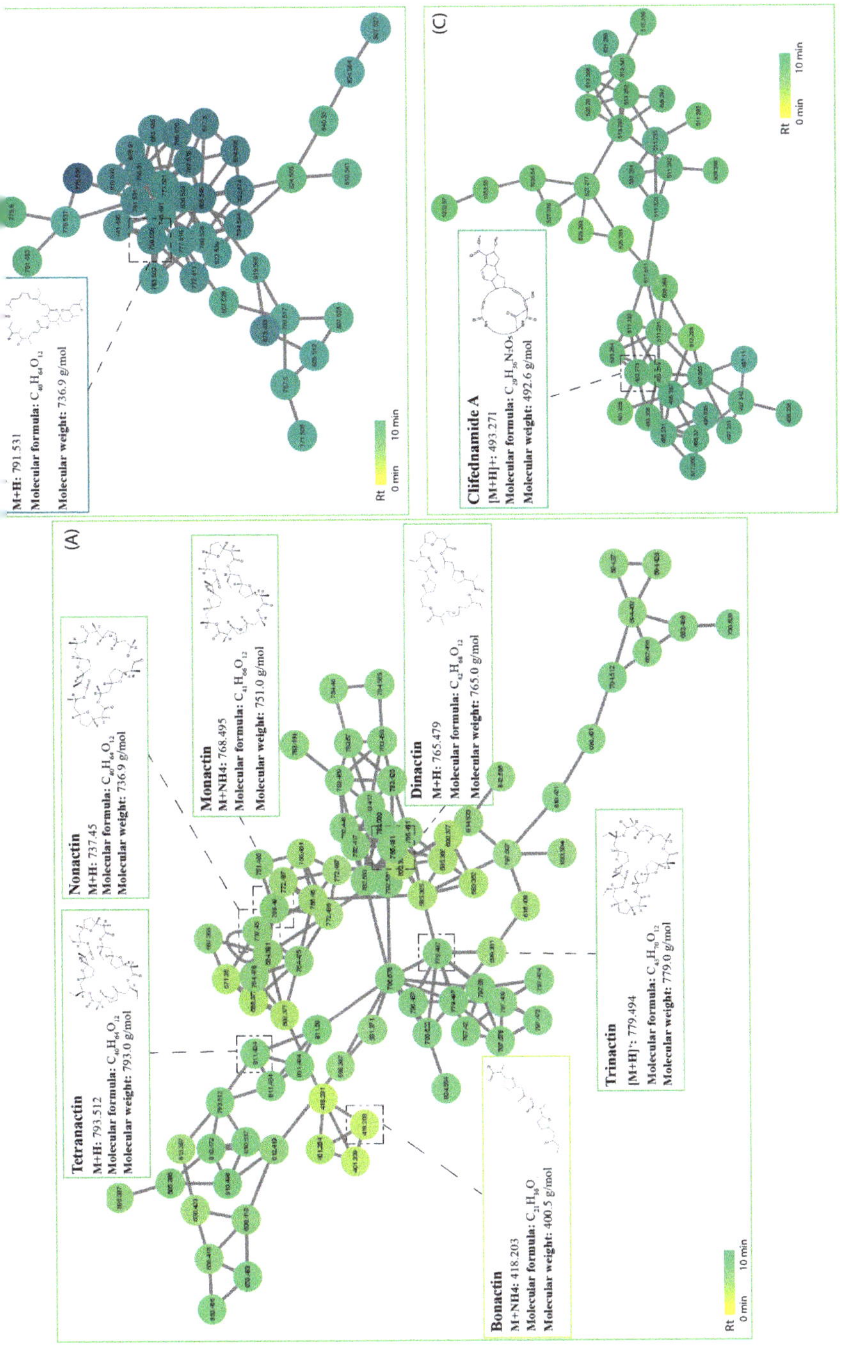

Figure 5. Macrotetrolide cluster identified in the termite gut-associated streptomycetes molecular network. Nodes represent an individual metabolite; each node displays its parent mass and is coloured based on its retention time. **(A)** macrotetrolides **(B)** macrocyclic lactones **(C)** polycyclic tetramate macrolactums.

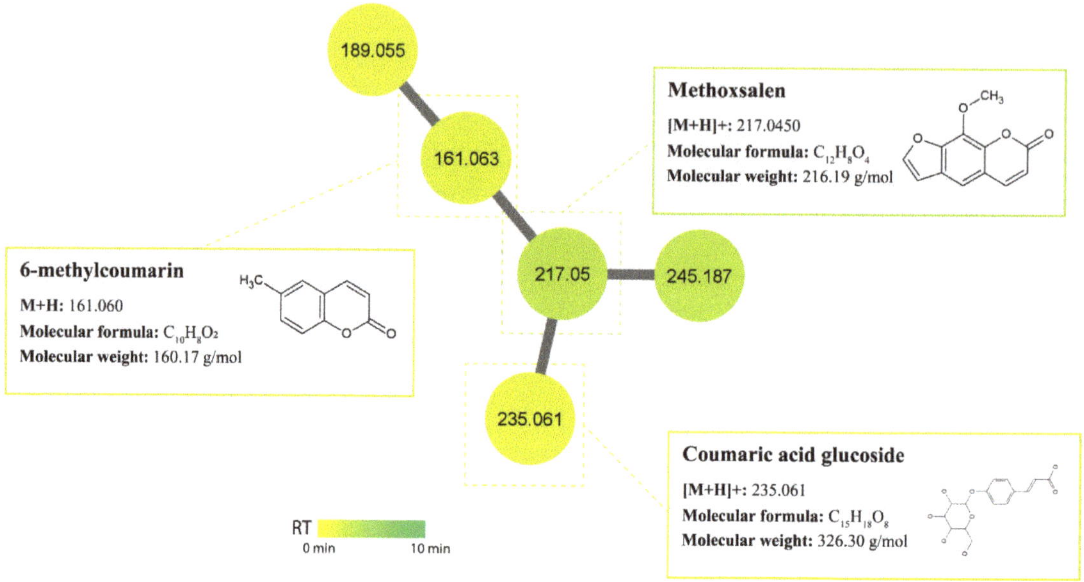

Figure 6. Coumarin cluster identified in the termite gut-associated streptomycetes molecular network. Nodes represent an individual metabolite; each node displays its parent mass and is coloured based on its retention time.

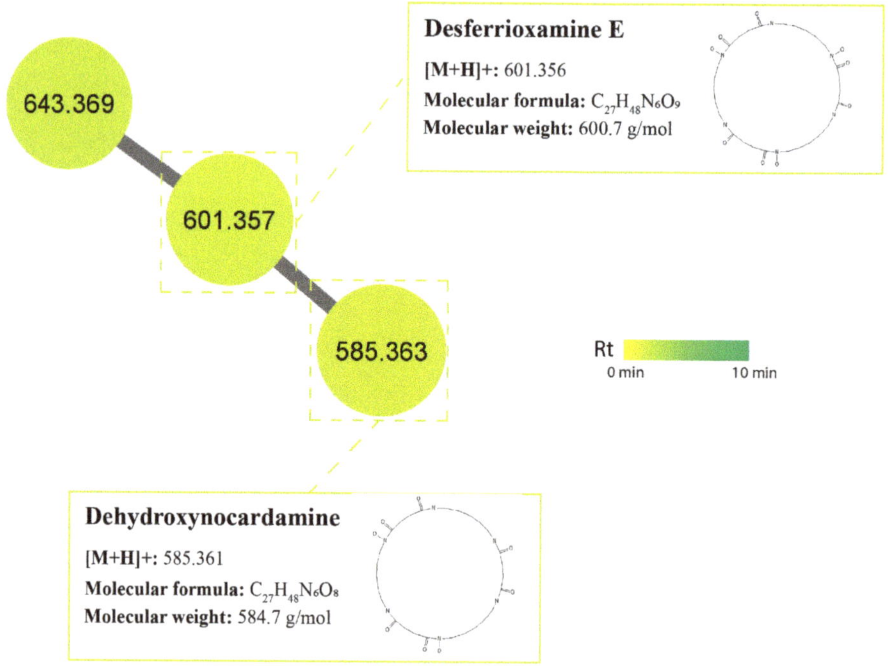

Figure 7. Peptide cluster identified in the termite gut-associated streptomycetes molecular network. Nodes represent an individual metabolite; each node displays its parent mass and is coloured based on its retention time.

2.6. Pathway Enrichment Analysis within the Streptomycetes Fermentation Extracts

Pathway enrichment analysis (PEA) and overrepresentation analysis (ORA) play vital roles in high-dimensional molecular data and are becoming popular methods for analysing metabolic data as these include specific fingerprints of cellular biochemistry [78,79]. These analyses were conducted to investigate the pathways that were amplified across all 15 streptomycete extracts and how they may support the metabolite identification within the molecular networks of this study. The PEA revealed multiple pathways of interest due to their role in antibiotic synthesis. These include streptomycin biosynthesis, polyketide sugar unit biosynthesis, monobactam synthesis, siderophore group nonribosomal peptides synthesis, terpenoid backbone synthesis, biosynthesis of secondary metabolites, streptomycin synthesis, and novobiocin synthesis. Similarly, the ORA demonstrated enriched metabolite sets of interest, including (i) purine metabolism; (ii) phenylalanine, tyrosine, and tryptophan synthesis; and (iii) neomycin, kanamycin, and gentamicin biosynthesis.

Purine metabolism plays a crucial role in supplying energy and essential precursors for antibiotic biosynthesis in *Streptomyces* spp. [80]. Furthermore, it was reported that phenylalanine deaminates to trans-cinnamic acid, which ultimately can be converted into secondary metabolites, such as lignins, flavonoids, and coumarins, all of which were identified within the molecular network in this study [58,81]. One study investigated an endophytic *Streptomyces* isolate and discovered antibiotics kanamycin and gentamicin within their metabolomic analysis [82,83]. The ORA supports the detected metabolites in the molecular network and validates key compounds such as flavonoids and courmarins.

Important pathways responsible for secondary metabolite production include polyketide synthases (PKS) type I, II, and III, as well as nonribosomal peptide synthase (NRPS), both of which were identified within the PEA of this study. PKS genes encode polyketide metabolites such as antifungal antibiotic amphotericin B [84]. Researchers have reported the biosynthesis behind macrocyclic polyketides (macrolide antibiotics) by investigating the encoded PKS pathway. They revealed that these products are synthesised from acetate, propionate, and other short-chain fatty acids within this pathway and, thus, form a hydroxylated macrocyclic lactone ring with three to seven conjugated double bonds and one or more sugars [85,86]. Many of the enriched pathways support the finding presented in the MN and the possibility of the biosynthetic potential of these termite gut-associated *Streptomyces* in producing novel antifungal antibiotics(Figure 8).

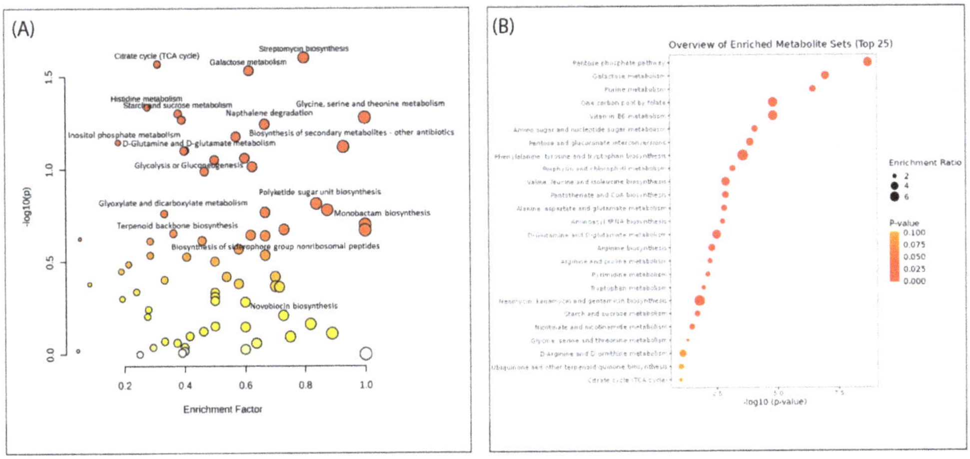

Figure 8. (**A**) Pathway enrichment and (**B**) overrepresentation analyses generated in Metaboanalyst 5.0 of all 15 of the *Streptomyces* isolates from the MS data. The size of the circle is indicative of the pathway impact score, while the colours are based on the *p*-value (the darker the colour, the higher the significance of the metabolites within the relevant pathway).

3. Materials and Methods

3.1. Cultures and Culture Media

The 37 actinomycete isolates screened in this study were previously isolated from the guts of termites, *Coptotermes lacteus* (Froggatt) species located in the Sunshine Coast Region of Queensland, Australia [36,37], and cryogenically preserved in the University of the Sunshine Coast Microbial Library. The actinomycetes were grown onto oatmeal agar [87], and the *P. noxium* strain was previously isolated from infected heritage fig trees within Brisbane, Queensland, and grown on potato dextrose agar (PDA) (OXOID, Australia).

3.2. 16S rDNA Oligonucleotide Sequencing of Actinomycete Isolates

Extraction and amplification of the 16S rDNA were carried out using a commercial QIAGEN HotStarTaq DNA PCR kit as per the manufacturer's instructions and using the eubacterial universal primers 27F 5′ -AGAGTT TGA TCC TGG CTC AG-3′ and 1492R 5′-GGT TAC CTT GTT ACG ACT T-3′. Conventional PCR reactions were performed achieving a final volume of 50 µL; 2× PCR master mix 25 µL, 10× primer mix 5 µL, Q-solution 10 µL, and RNase-free water 10 µL. Amplification was performed using the Bio-Rad T100 Thermocycler, (McHugh Electronic, Australia) with thermocycling conditions beginning with the initial temperature of 95 °C for 15 min, followed by 35 cycles of 94 °C for 1 min, 68 °C for 1 min, and final temperature of 72 °C for 10 min. PCR products were electrophoresed on a 1.5% agarose gel followed by visual confirmation under an IV transilluminator. The PCR products were sent to Macrogen (Seoul, Republic of Korea) (http://dna.macrogen.com/eng/, accessed on 22 November 2021) for sequencing. The sequences were analysed using Geneious Prime 23023.1.2 and putative identifications were made by conducting a BLAST (Basic Local Alignment Search Tool) search using the NCBI (National Centre for Biotechnology Information) database. This database is also where the sequences generated from this study have been deposited. Reference and generated accession numbers can be found in the Supplementary Materials.

3.3. Evaluation of Antifungal Activity of the Actinomycete Isolates

The 37 actinomycete isolates were screened for antifungal activity against the fungal pathogen *P. noxium* using co-culture assays. The isolates were streaked at one end of the agar plate and incubated for 7 days to grow at 28 °C before a fungal plug was placed at the opposing end of the plate. A control plate was also established without the *Streptomyces* isolates. Once the control had reached the edge of the plate, comparisons were made using the following formula:

$$Inhibition\ Zone\ (\%) = \sum \left(\frac{Colony\ diameter\ of\ pathogen - (colony\ diameter\ of\ pathogen + antagonist)}{Colony\ diameter\ of\ pathogen} \right) \times 100$$

3.4. Fermentation and Extraction of Metabolites Produced by Streptomyces Isolates

Fermentation of the *Streptomyces* isolates was conducted using solid-state fermentation. The isolates were inoculated onto oatmeal agar plates and incubated for 7 days at 28 °C. The plates were then inoculated with a fungal plug from the *P. noxium* and grown for three more days using the same conditions. These plates were then twice extracted using ethyl acetate (ChemSupply) (EtOAc) by cutting up the agar and flooding them with the solvent for 24 h. The solvent was filtered and evaporated under a vacuum to obtain the crude extract for further testing.

3.5. Bioassays of the Streptomycetes Fermentation Extracts Using Disk-Diffusion Method

The antifungal activity of the crude extracts was then assessed using the Kirby-Bauer disk-diffusion method [88]. Sterile paper disks were seeded with the crude extract (approx. 100 µL) and placed onto PDA plates (2 cm away from the edge) (left). A blank control disk was also added, containing only DMSO on the opposing side (right). A fungal plug from the fungal pathogen was placed onto the centre of the plate and was incubated at 25 °C for four days. Signs of inhibition of the fungal mycelium were then observed.

3.6. QTOF Chemical Profiling of the Metabolites within the Streptomyces Fermentation Extracts

Liquid chromatography with accurate mass spectroscopy (LC-MS) was performed on a SCIEX ExionLC system coupled to a SCIEX X500 QTOF. The column was a Synergi Fusion RP, 50 mm × 4.6 mm, 2.5 µm. Mobile phase A was 99.9% MilliQ water with 0.1% formic acid, and mobile phase B was 99.9% LC-grade methanol with 0.1% formic acid. The mobile phase program was 1.0 mL/min commencing at 98:2 MPA: MPB for 0.25 min, grading to 100% MPB at 6.25 min, isocratic until 8.25 min, grading back to 98:2 MPA:MPB at 8.5 min, and isocratic until 10.0 min. The mass spectrometer was operated in alternating positive and negative electrospray ionisation modes, providing accurate M+1 and M-1 molecular ions to four decimal places.

3.7. MS Data Analysis and Annotation

The data were initially converted from ".wiff" standard file format to ".mzXML" using MSconvert within the ProteoWizard 3.0 software toolkit. The data were then processed using MZmine 2.53 and exported as a .CSV file where it was analysed using MetaboAnalyst 5.0 (https://new.metaboanalyst.ca, accessed on 22 November 2021), which allowed for the generation of principal component analysis (PCA), pathway enrichment analysis (PEA), and overrepresentation analysis (ORA).

A molecular network was generated using the online workflow (https://ccms-ucsd.github.io/GNPSDocumentation/, accessed on 22 November 2021) on the GNPS website (http://gnps.ucsd.edu, accessed on 22 November 2021) (Wang et al., 2016) [89]. The precursor ion mass tolerance was set to 2.0 Da and a MS/MS fragment ion tolerance of 0.5 Da. A network was then created where edges were filtered to have a cosine score above 0.65 and more than 4 matched peaks. Further, edges between two nodes were kept in the network if and only if each of the nodes appeared in each other's respective top 10 most similar nodes. Finally, the maximum size of a molecular family was set to 100, and the lowest-scoring edges were removed from molecular families until the molecular family size was below this threshold. The resulting network was visualized in Cytoscape. Annotations of the metabolites were first made by matching the spectra against GNPS' spectral libraries. All matches between network spectra and library spectra were required to have a score above 0.7 and at least 4 matched peaks. Generated annotations were validated against the Metabolomics Standards Initiative (MSI) [90] and were in compliance with level 2 annotations corresponding to "putatively annotated compounds" [91].

4. Conclusions

In this study, the phylogeny and antifungal activity of termite gut-associated actinomycetes were investigated. The 16S rRNA gene analysis revealed a dominance of the members of the genus *Streptomyces* in the termite gut. Additionally, 15 of the *Streptomyces* isolates presented with strong antifungal activity against the phytopathogenic fungus *P. noxium*. Fermentation extracts of the selected isolates were analysed through the employment of MS-based metabolomics for the evaluation of their metabolome chemical composition. Known antifungal metabolites of the polyketide chemical class were identified, including isoflavonoids, macrolides, coumarins, and cyclic peptides. Pathway enrichment analyses supported the biosynthetic potential for antibiotic production by the *Streptomyces* isolates and the findings of the molecular networking. The results of this study provide insights into the chemical composition and biopotential of termite gut-associated *Streptomyces* isolates as unexplored sources for novel antifungal compounds for biocontrol applications.

Supplementary Materials: The following supporting information can be downloaded at: https://www.mdpi.com/article/10.3390/antibiotics12091373/s1, Table S1: Termite gut-associated actinomycetes phylogeny; Table S2: Assigned GenBank accession numbers of the actinomycete isolates; Figure S1: Phylogenetic tree of the actinomycetes; Figure S2: Metabolite chemical class distribution; Figure S3: Distribution of polyketide metabolites across the fermentation extracts of streptomycetes.

Author Contributions: Conceptualization, C.A., T.D.T. and D.İ.K.; methodology, C.A.; formal analysis, C.A.; investigation, C.A. and T.D.T., resources, D.İ.K., T.D.T., K.F. and R.T.; data curation, C.A. and T.D.T.; writing—original draft preparation, C.A.; writing—review and editing, D.İ.K. and T.D.T.; supervision, T.D.T. and D.İ.K.; funding acquisition, D.İ.K. All authors have read and agreed to the published version of the manuscript.

Funding: This research was funded by Brisbane City Council, grant number R22_0098.

Institutional Review Board Statement: Not applicable.

Informed Consent Statement: Not applicable.

Data Availability Statement: Data are contained within the article or Supplementary Material.

Acknowledgments: Authors thank Louisa Parkinson and Elizabeth Dann, QAAFI, UQ for the provision of *P. noxium*.

Conflicts of Interest: The authors declare no conflict of interest.

References

1. Fones, H.N.; Bebber, D.P.; Chaloner, T.M.; Kay, W.T.; Steinberg, G.; Gurr, S.J. Threats to global food security from emerging fungal and oomycete crop pathogens. *Nat. Food* **2020**, *1*, 332–342. [CrossRef]
2. Thambugala, K.M.; Daranagama, D.A.; Phillips, A.J.; Kannangara, S.D.; Promputtha, I. Fungi vs. fungi in biocontrol: An overview of fungal antagonists applied against fungal plant pathogens. *Front. Cell. Infect. Microbiol.* **2020**, *10*, 604923. [CrossRef]
3. Stewart, J.; Kim, M.-S.; Ota, Y.; Sahashi, N.; Hanna, J.; Akiba, M.; Ata, J.; Atibalentja, N.; Brooks, F.; Chung, C.-L. Phylogenetic and population genetic analyses reveal three distinct lineages of the invasive brown root-rot pathogen, *Phellinus noxius*, and bioclimatic modeling predicts differences in associated climate niches. *Eur. J. Plant Pathol.* **2020**, *156*, 751–766. [CrossRef]
4. Ann, P.; Lee, H.; Tsai, J. Survey of brown root disease of fruit and ornamental trees caused by *Phellinus noxius* in Taiwan. *Plant Pathol. Bull.* **1999**, *8*, 51–60.
5. Chou, H.; Xiao, Y.-T.; Tsai, J.-N.; Li, T.-T.; Wu, H.-Y.; Liu, L.-y.D.; Tzeng, D.-S.; Chung, C.-L. In Vitro and *in planta* evaluation of *Trichoderma asperellum* TA as a biocontrol agent against *Phellinus noxius*, the cause of brown root rot disease of trees. *Plant Dis.* **2019**, *103*, 2733–2741. [CrossRef] [PubMed]
6. Hsiao, W.-W.; Ting-Hsuan, H.; En-Jang, S. The pathogenicity of basidiospores of *Phellinus noxius* which causes brown root rot disease in Taiwan. *Taiwania* **2019**, *64*, 189–194.
7. Akiba, M.; Ota, Y.; Tsai, I.J.; Hattori, T.; Sahashi, N.; Kikuchi, T. Genetic differentiation and spatial structure of *Phellinus noxius*, the causal agent of brown root rot of woody plants in Japan. *PLoS ONE* **2015**, *10*, e0141792. [CrossRef] [PubMed]
8. Sahashi, N.; Akiba, M.; Ishihara, M.; Ota, Y.; Kanzaki, N. Brown root rot of trees caused by *Phellinus noxius* in the Ryukyu Islands, subtropical areas of Japan. *For. Pathol.* **2012**, *42*, 353–361. [CrossRef]
9. Wang, Y.-F.; Meng, H.; Gu, V.W.; Gu, J.-D. Molecular diagnosis of the brown root rot disease agent *Phellinus noxius* on trees and in soil by rDNA ITS analysis. *Appl. Environ. Biotechnol.* **2016**, *1*, 81–91. [CrossRef]
10. Zhang, H.; Ng, T.K.; Lee, K.C.; Leung, Z.W.; Yau, W.F.; Wong, W.S. Development and evaluation of loop-mediated isothermal amplification (LAMP) as a preliminary diagnostic tool for brown root rot disease caused by *Phellinus noxius* (Corner) GH cunningham in hong kong urban tree management. *Sustainability* **2022**, *14*, 9708. [CrossRef]
11. Adra, C.; Panchalingam, H.; Kurtböke, D.İ. Termite–gut-associated actinomycetes as biological control agents against phytopathogen *Pyrrhoderma noxium*. *Microbiol. Aust.* **2022**, *43*, 190–193. [CrossRef]
12. Dann, K.; Smith, L.; Shuey, L.; Begum, F. Phellinus noxius: A basidiomycete fungus impacting productivity of Australian avocados. In Proceedings of the 7th World Avocado Congress, Cairns, Australia, 5–9 September 2011.
13. Panchalingam, H.; Ashfield-Crook, N.; Naik, V.; Frenken, R.; Foster, K.; Tomlin, R.; Shapcott, A.; Kurtböke, D.İ. Testing the biocontrol ability of a *Trichoderma*-Streptomycetes consortium against *Pyrrhoderma noxium* (Corner) LW Zhou and YC Dai in soil. *J. Fungi* **2022**, *9*, 67. [CrossRef] [PubMed]
14. Schwarze, F.W.; Jauss, F.; Spencer, C.; Hallam, C.; Schubert, M. Evaluation of an antagonistic *Trichoderma* strain for reducing the rate of wood decomposition by the white rot fungus *Phellinus noxius*. *Biol. Control* **2012**, *61*, 160–168. [CrossRef]
15. Caballero, J.R.I.; Ata, J.P.; Leddy, K.; Glenn, T.C.; Kieran, T.J.; Klopfenstein, N.B.; Kim, M.-S.; Stewart, J.E. Genome comparison and transcriptome analysis of the invasive brown root rot pathogen, *Phellinus noxius*, from different geographic regions reveals potential enzymes associated with degradation of different wood substrates. *Fungal Biol.* **2020**, *124*, 144–154. [CrossRef]

6. Liu, T.Y.; Chen, C.H.; Yang, Y.L.; Tsai, I.J.; Ho, Y.N.; Chung, C.L. The brown root rot fungus *Phellinus noxius* affects microbial communities in different root-associated niches of Ficus trees. *Environ. Microbiol.* **2022**, *24*, 276–297. [CrossRef]
7. Wu, Z.-C.; Chang, Y.-Y.; Lai, Q.-J.; Lin, H.-A.; Tzean, S.-S.; Liou, R.-F.; Tsai, I.J.; Chung, C.-L. Soil is not a reservoir for *Phellinus noxius*. *Phytopathology* **2020**, *110*, 362–369. [CrossRef]
8. Tsang, K.S.W.; Cheung, M.K.; Lam, R.Y.C.; Kwan, H.S. A preliminary examination of the bacterial, archaeal, and fungal rhizosphere microbiome in healthy and *Phellinus noxius*-infected trees. *MicrobiologyOpen* **2020**, *9*, e1115. [CrossRef] [PubMed]
9. Chung, C.-L.; Huang, S.-Y.; Huang, Y.-C.; Tzean, S.-S.; Ann, P.-J.; Tsai, J.-N.; Yang, C.-C.; Lee, H.-H.; Huang, T.-W.; Huang, H.-Y. The genetic structure of *Phellinus noxius* and dissemination pattern of brown root rot disease in Taiwan. *PLoS ONE* **2015**, *10*, e0139445. [CrossRef]
0. Tchotet Tchoumi, J.M.; Coetzee, M.P.A.; Vivas, M.; Rajchenberg, M.; Roux, J. Wood-rotting basidiomycetes associated with declining native trees in timber-harvesting compartments of the Garden Route National Park of South Africa. *Austral Ecol.* **2017**, *42*, 947–963. [CrossRef]
1. Poveda, J.; Abril-Urias, P.; Escobar, C. Biological control of plant-parasitic nematodes by filamentous fungi inducers of resistance: *Trichoderma*, mycorrhizal and endophytic fungi. *Front. Microbiol.* **2020**, *11*, 992. [CrossRef]
2. Ghanem, G.A.; Gebily, D.A.; Ragab, M.M.; Ali, A.M.; Soliman, N.E.-D.K.; El-Moity, T.H.A. Efficacy of antifungal substances of three *Streptomyces* spp. against different plant pathogenic fungi. *Egypt. J. Biol. Pest Control* **2022**, *32*, 1–13. [CrossRef]
3. Wang, D.; Li, Y.; Yuan, Y.; Chu, D.; Cao, J.; Sun, G.; Ai, Y.; Cui, Z.; Zhang, Y.; Wang, F. Identification of non-volatile and volatile organic compounds produced by *Bacillus siamensis* LZ88 and their antifungal activity against *Alternaria alternata*. *Biol. Control* **2022**, *169*, 104901. [CrossRef]
4. Donald, L.; Pipite, A.; Subramani, R.; Owen, J.; Keyzers, R.A.; Taufa, T. *Streptomyces*: Still the biggest producer of new natural secondary metabolites, a current perspective. *Microbiol. Res.* **2022**, *13*, 418–465. [CrossRef]
5. Kurtböke, D.İ. Bioactive actinomycetes: Reaching rarity through sound understanding of selective culture and molecular diversity. In *Microbial Resources*; Kurtböke, D.İ., Ed.; Academic Press: Cambridge, MA, USA; Elsevier: Amsterdam, The Netherlands, 2017; pp. 45–76.
6. Ding, T.; Yang, L.-J.; Zhang, W.-D.; Shen, Y.-H. The secondary metabolites of rare actinomycetes: Chemistry and bioactivity. *RSC Adv.* **2019**, *9*, 21964–21988. [CrossRef]
7. Chen, S.; Cheng, D.; Liu, Z.; Hassan, B.; Xu, Y. Community structure and antifungal activity of actinobacteria in a fungus-growing termite. *Ecol. Entomol.* **2023**, *48*, 251–262. [CrossRef]
8. Kurtböke, D.İ.; French, J.R.; Hayes, R.A.; Quinn, R.J. Eco-taxonomic insights into actinomycete symbionts of termites for discovery of novel bioactive compounds. *Biotechnol. Appl. Biodivers.* **2015**, 111–135.
9. Ayu Krishanti, N.P.R.; Zulfiana, D.; Wikantyoso, B.; Zulfitri, A.; Yusuf, S. Antimicrobial production by an actinomycetes isolated from the termite nest. *J. Trop. Life Sci.* **2018**, *8*.
0. Malavika, S.; Kranthi, S.; Shishira Rao, H.; Moily, S.S.; Martin Paul, A. Isolation of gut Actinobacteria from termites. In *Methods in Actinobacteriology*; Dharumadurai, D., Ed.; Springer Protocols Handbooks; Humana: New York, NY, USA, 2022; pp. 51–60. [CrossRef]
1. Zeki, Ö.C.; Eylem, C.C.; Reçber, T.; Kır, S.; Nemutlu, E. Integration of GC–MS and LC–MS for untargeted metabolomics profiling. *J. Pharm. Biomed. Anal.* **2020**, *190*, 113509. [CrossRef]
2. Abdelrahman, S.M.; Dosoky, N.S.; Hanora, A.M.; Lopanik, N.B. Metabolomic profiling and molecular networking of nudibranch-associated *Streptomyces* sp. SCSIO 001680. *Molecules* **2022**, *27*, 4542. [CrossRef]
3. Nothias, L.-F.; Petras, D.; Schmid, R.; Dührkop, K.; Rainer, J.; Sarvepalli, A.; Protsyuk, I.; Ernst, M.; Tsugawa, H.; Fleischauer, M. Feature-based molecular networking in the GNPS analysis environment. *Nat. Methods* **2020**, *17*, 905–908. [CrossRef]
4. Tinte, M.M.; Masike, K.; Steenkamp, P.A.; Huyser, J.; van der Hooft, J.J.; Tugizimana, F. Computational metabolomics tools reveal metabolic reconfigurations underlying the effects of biostimulant seaweed extracts on maize plants under drought stress conditions. *Metabolites* **2022**, *12*, 487. [CrossRef]
5. Kurtböke, D.İ. Actino-Rush on the Sunshine Coast: Prospects for Bioprospecting. In Proceedings of the 10th International Congress on Culture Collections, Tsukuba, Japan, 10–15 October 2004; pp. 319–322.
6. Kurtböke, D.İ.; French, J.R. Use of phage battery to investigate the actinofloral layers of termite gut microflora. *J. Appl. Microbiol.* **2007**, *103*, 722–734. [CrossRef]
7. Kurtböke, D.İ.; French, J.R. Actinobacterial resources from termite guts for regional bioindustries. *Microbiol. Aust.* **2008**, *29*, 42–44. [CrossRef]
8. Berasategui, A.; Shukla, S.; Salem, H.; Kaltenpoth, M. Potential applications of insect symbionts in biotechnology. *Appl. Microbiol. Biotechnol.* **2016**, *100*, 1567–1577. [CrossRef] [PubMed]
9. Kaltenpoth, M. Actinobacteria as mutualists: General healthcare for insects? *Trends Microbiol.* **2009**, *17*, 529–535. [CrossRef] [PubMed]
0. Sujada, N.; Sungthong, R.; Lumyong, S. Termite nests as an abundant source of cultivable actinobacteria for biotechnological purposes. *Microbes Environ.* **2014**, *29*, 211–219. [CrossRef] [PubMed]
1. Watanabe, Y.; Shinzato, N.; Fukatsu, T. Isolation of actinomycetes from termites' guts. *Biosci. Biotechnol. Biochem.* **2003**, *67*, 1797–1801. [CrossRef]

42. Ayed, A.; Kalai-Grami, L.; Ben Slimene, I.; Chaouachi, M.; Mankai, H.; Karkouch, I.; Djebali, N.; Elkahoui, S.; Tabbene, O.; Limam, F. Antifungal activity of volatile organic compounds from *Streptomyces* sp. strain S97 against *Botrytis cinerea*. *Biocontrol Sci. Techno.* **2021**, *31*, 1330–1348. [CrossRef]
43. Lyu, A.; Liu, H.; Che, H.; Yang, L.; Zhang, J.; Wu, M.; Chen, W.; Li, G. Reveromycins A and B from *Streptomyces* sp. 3–10: Antifungal activity against plant pathogenic fungi in vitro and in a strawberry food model system. *Front. Microbiol.* **2017**, *8*, 550. [CrossRef]
44. Wan, M.; Li, G.; Zhang, J.; Jiang, D.; Huang, H.-C. Effect of volatile substances of *Streptomyces platensis* F-1 on control of plant fungal diseases. *Biol. Control* **2008**, *46*, 552–559. [CrossRef]
45. Igarashi, M.; Kinoshita, N.; Ikeda, T.; Kameda, M.; Hamada, M.; Takeuchi, T. Resormycin, a novel herbicidal and antifungal antibiotic produced by a strain of *Streptomyces platensis* I. Taxonomy, production, isolation and biological properties. *J. Antibiot.* **1997**, *50*, 1020–1025. [CrossRef]
46. Chen, Y.-L.; Zhao, J.; Liu, W.; Gao, J.-F.; Tao, L.-M.; Pan, H.-X.; Tang, G.-L. Identification of phoslactomycin biosynthetic gene clusters from *Streptomyces platensis* SAM-0654 and characterization of PnR1 and PnR2 as positive transcriptional regulators. *Gene* **2012**, *509*, 195–200. [CrossRef] [PubMed]
47. Lai, B.; Dai, R.; Zeng, T.; Yao, J. Optimization of Fermentation Conditions and Properties of Secondary Metabolite of Streptomyces catenulae XG40. *Chin. J. Biol. Control* **2022**, *38*, 1242.
48. Budi, M.B.S.; Giyanto, G.; Tondok, E.T. Isolation of actinomycetes from peatland to suppress the growth of *Ganoderma boninense* the causal agent of basal stem rot disease in oil palm. *Biodiversitas J. Biol. Divers.* **2022**, *23*. [CrossRef]
49. Zhang, R.; Han, X.; Xia, Z.; Luo, X.; Wan, C.; Zhang, L. *Streptomyces luozhongensis* sp. nov., a novel actinomycete with antifungal activity and antibacterial activity. *Antonie Van Leeuwenhoek* **2017**, *110*, 195–203. [CrossRef]
50. Marian, M.; Ohno, T.; Suzuki, H.; Kitamura, H.; Kuroda, K.; Shimizu, M. A novel strain of endophytic *Streptomyces* for the biocontrol of strawberry anthracnose caused by *Glomerella cingulata*. *Microbiol. Res.* **2020**, *234*, 126428. [CrossRef]
51. Yang, P.; Li, M.; Zhao, J.; Zhu, M.; Shang, H.; Li, J.; Cui, X.; Huang, R.; Wen, M. Oligomycins A and C, major secondary metabolites isolated from the newly isolated strain *Streptomyces diastaticus*. *Folia Microbiol.* **2010**, *55*, 10–16. [CrossRef]
52. De Simeis, D.; Serra, S. Actinomycetes: A never-ending source of bioactive compounds—An overview on antibiotics production. *Antibiotics* **2021**, *10*, 483. [CrossRef]
53. Wang, Y.; Yang, D.; Bi, Y.; Yu, Z. Macrolides from *Streptomyces* sp. SN5452 and their antifungal activity against *Pyricularia oryzae*. *Microorganisms* **2022**, *10*, 1612. [CrossRef] [PubMed]
54. Risdian, C.; Mozef, T.; Wink, J. Biosynthesis of polyketides in *Streptomyces*. *Microorganisms* **2019**, *7*, 124. [CrossRef] [PubMed]
55. Li, S.; Yang, B.; Tan, G.-Y.; Ouyang, L.-M.; Qiu, S.; Wang, W.; Xiang, W.; Zhang, L. Polyketide pesticides from actinomycetes. *Curr. Opin. Biotechnol.* **2021**, *69*, 299–307. [CrossRef]
56. Wang, J.-F.; Liu, S.-S.; Song, Z.-Q.; Xu, T.-C.; Liu, C.-S.; Hou, Y.-G.; Huang, R.; Wu, S.-H. Naturally occurring flavonoids and isoflavonoids and their microbial transformation: A review. *Molecules* **2020**, *25*, 5112. [CrossRef] [PubMed]
57. Ebada, S.S.; Eze, P.; Okoye, F.B.; Esimone, C.O.; Proksch, P. The fungal endophyte *Nigrospora oryzae* produces quercetin monoglycosides previously known only from plants. *ChemistrySelect* **2016**, *1*, 2767–2771. [CrossRef]
58. Moore, B.S.; Hertweck, C.; Hopke, J.N.; Izumikawa, M.; Kalaitzis, J.A.; Nilsen, G.; O'Hare, T.; Piel, J.; Shipley, P.R.; Xiang, L. Plant-like biosynthetic pathways in bacteria: From benzoic acid to chalcone. *J. Nat. Prod.* **2002**, *65*, 1956–1962. [CrossRef]
59. Matseliukh, B.; Polishchuk, L.; Lutchenko, V.; Bambura, O.; Kopiyko, O. Synthesis of daidzein and genistein by streptomycetes and their effect on production of antibiotics. *Mikrobiolohichnyi Zhurnal* **2005**, *67*, 12–21.
60. Kang, H.R.; Lee, D.; Benndorf, R.; Jung, W.H.; Beemelmanns, C.; Kang, K.S.; Kim, K.H. Termisoflavones A–C, isoflavonoid glycosides from termite-associated *Streptomyces* sp. RB1. *J. Nat. Prod.* **2016**, *79*, 3072–3078. [CrossRef]
61. Savi, D.C.; Shaaban, K.A.; Gos, F.M.; Thorson, J.S.; Glienke, C.; Rohr, J. Secondary metabolites produced by *Microbacterium* sp. LGMB471 with antifungal activity against the phytopathogen *Phyllosticta citricarpa*. *Folia Microbiol.* **2019**, *64*, 453–460. [CrossRef] [PubMed]
62. Pancu, D.F.; Scurtu, A.; Macasoi, I.G.; Marti, D.; Mioc, M.; Soica, C.; Coricovac, D.; Horhat, D.; Poenaru, M.; Dehelean, C. Antibiotics: Conventional therapy and natural compounds with antibacterial activity—A pharmaco-toxicological screening. *Antibiotics* **2021**, *10*, 401. [CrossRef]
63. Bao, Y.; Li, H.; Dong, Y.; Duan, H.; Li, H.; Li, W. Genome-guided discovery of antifungal filipins from a deep-sea-derived *Streptomyces antibioticus*. *J. Nat. Prod.* **2022**, *85*, 365–374. [CrossRef]
64. Jizba, J.; Sedmera, P.; Zima, J.; Beran, M.; Blumauerová, M.; Kandybin, N.; Samoukina, G. Macrotetrolide antibiotics produced by *Streptomyces globisporus*. *Folia Microbiol.* **1991**, *36*, 437–443. [CrossRef]
65. Wang, J.; Tan, H.; Lu, Y.; Cao, L. Determination of ionophore antibiotics nactins produced by fecal *Streptomyces* from sheep. *BioMetals* **2014**, *27*, 403–407. [CrossRef]
66. Shishlyannikova, T.A.; Kuzmin, A.V.; Fedorova, G.A.; Shishlyannikov, S.M.; Lipko, I.A.; Sukhanova, E.V.; Belkova, N.L. Ionofore antibiotic polynactin produced by *Streptomyces* sp. 156A isolated from Lake Baikal. *Nat. Prod. Res.* **2017**, *31*, 639–644. [CrossRef] [PubMed]
67. Zhang, K.; Gu, L.; Zhang, Y.; Liu, Z.; Li, X. Dinactin from a new producer, *Streptomyces badius* gz-8, and its antifungal activity against the rubber anthracnose fungus *Colletotrichum gloeosporioides*. *Microbiol. Res.* **2020**, *240*, 126548. [CrossRef] [PubMed]

. Khebizi, N.; Boudjella, H.; Bijani, C.; Bouras, N.; Klenk, H.-P.; Pont, F.; Mathieu, F.; Sabaou, N. Oligomycins A and E, major bioactive secondary metabolites produced by *Streptomyces* sp. strain HG29 isolated from a Saharan soil. *J. De Mycol. Medicale* **2018**, *28*, 150–160. [CrossRef]

. Chakraborty, M.; Mahmud, N.U.; Muzahid, A.N.M.; Rabby, S.F.; Islam, T. Oligomycins inhibit *Magnaporthe oryzae* Triticum and suppress wheat blast disease. *PLoS ONE* **2020**, *15*, e0233665. [CrossRef]

. Yu, L.; Li, H.; Zhou, Z.; Liu, F.; Du, L. An antifungal polycyclic tetramate macrolactam, heat-stable antifungal factor (HSAF), is a novel oxidative stress modulator in *Lysobacter enzymogenes*. *Appl. Environ. Microbiol.* **2021**, *87*, e03105–e03120. [CrossRef]

. Gray, K.C.; Palacios, D.S.; Dailey, I.; Endo, M.M.; Uno, B.E.; Wilcock, B.C.; Burke, M.D. Amphotericin primarily kills yeast by simply binding ergosterol. *Proc. Natl. Acad. Sci. USA* **2012**, *109*, 2234–2239. [CrossRef]

. Al-Amiery, A.A.; Kadhum, A.A.H.; Mohamad, A.B. Antifungal activities of new coumarins. *Molecules* **2012**, *17*, 5713–5723. [CrossRef]

. Hussain, M.I.; Syed, Q.A.; Khattak, M.N.K.; Hafez, B.; Reigosa, M.J.; El-Keblawy, A. Natural product coumarins: Biological and pharmacological perspectives. *Biologia* **2019**, *74*, 863–888. [CrossRef]

. Chen, Y.-Z.; Wang, S.-R.; Li, T.; Zhang, G.-C.; Yang, J. Antifungal activity of 6-Methylcoumarin against *Valsa mali* and its possible mechanism of action. *J. Fungi* **2022**, *9*, 5. [CrossRef] [PubMed]

. Rodrigues, J.L.; Gomes, D.; Rodrigues, L.R. Challenges in the heterologous production of furanocoumarins in *Escherichia coli*. *Molecules* **2022**, *27*, 7230. [CrossRef]

. Yamanaka, K.; Oikawa, H.; Ogawa, H.-o.; Hosono, K.; Shinmachi, F.; Takano, H.; Sakuda, S.; Beppu, T.; Ueda, K. Desferrioxamine E produced by *Streptomyces griseus* stimulates growth and development of *Streptomyces tanashiensis*. *Microbiology* **2005**, *151*, 2899–2905. [CrossRef]

. Jarmusch, S.A.; Lagos-Susaeta, D.; Diab, E.; Salazar, O.; Asenjo, J.A.; Ebel, R.; Jaspars, M. Iron-meditated fungal starvation by lupine rhizosphere-associated and extremotolerant *Streptomyces* sp. S29 desferrioxamine production. *Mol. Omics* **2021**, *17*, 95–107. [CrossRef] [PubMed]

. Chen, Y.; Li, E.-M.; Xu, L.-Y. Guide to metabolomics analysis: A bioinformatics workflow. *Metabolites* **2022**, *12*, 357. [CrossRef]

. Wieder, C.; Frainay, C.; Poupin, N.; Rodríguez-Mier, P.; Vinson, F.; Cooke, J.; Lai, R.P.; Bundy, J.G.; Jourdan, F.; Ebbels, T. Pathway analysis in metabolomics: Recommendations for the use of over-representation analysis. *PLoS Comput. Biol.* **2021**, *17*, e1009105. [CrossRef]

. Sivapragasam, S.; Grove, A. The link between purine metabolism and production of antibiotics in *Streptomyces*. *Antibiotics* **2019**, *8*, 76. [CrossRef]

. Kumavath, R.N.; Barh, D.; Azevedo, V.; Kumar, A.P. Potential pharmacological applications of enzymes associated with bacterial metabolism of aromatic compounds. *J. Microbiol. Antimicrob.* **2017**, *9*, 1–13.

. Alam, K.; Mazumder, A.; Sikdar, S.; Zhao, Y.-M.; Hao, J.; Song, C.; Wang, Y.; Sarkar, R.; Islam, S.; Zhang, Y. *Streptomyces*: The biofactory of secondary metabolites. *Front. Microbiol.* **2022**, *13*, 968053. [CrossRef] [PubMed]

. Baba, M.S.; Mohamad Zin, N.; Ahmad, S.J.; Mazlan, N.W.; Baharum, S.N.; Ahmad, N.; Azmi, F. Antibiotic biosynthesis pathways from endophytic *Streptomyces* SUK 48 through metabolomics and genomics approaches. *Antibiotics* **2021**, *10*, 969. [CrossRef]

. Zhang, B.; Zhou, Y.-T.; Jiang, S.-X.; Zhang, Y.-H.; Huang, K.; Liu, Z.-Q.; Zheng, Y.-G. Amphotericin B biosynthesis in *Streptomyces nodosus*: Quantitative analysis of metabolism via LC–MS/MS based metabolomics for rational design. *Microb. Cell Factories* **2020**, *19*, 1–12. [CrossRef]

5. Mulks, M.; Nair, M.; Putnam, A. In vitro antibacterial activity of faeriefungin, a new broad-spectrum polyene macrolide antibiotic. *Antimicrob. Agents Chemother.* **1990**, *34*, 1762–1765. [CrossRef] [PubMed]

5. Yao, T.; Liu, Z.; Li, T.; Zhang, H.; Liu, J.; Li, H.; Che, Q.; Zhu, T.; Li, D.; Li, W. Characterization of the biosynthetic gene cluster of the polyene macrolide antibiotic reedsmycins from a marine-derived *Streptomyces* strain. *Microb. Cell Factories* **2018**, *17*, 1–12. [CrossRef] [PubMed]

7. Williams, S.; Wellington, E. Actinomycetes. In *Methods of Soil Analysis: Part 2, Chemical and Microbiological Properties, Agronomy Monograph*; Page, A.L., Ed.; American Society of Agronomy Inc.; Soil Science of America Inc.: Madison, WI, USA, 1983; pp. 969–987.

3. Bauer, A.W.; Perry, D.M.; Kirby, W.M. Single-disk antibiotic-sensitivity testing of staphylococci: An analysis of technique and results. *AMA Arch. Intern. Med.* **1959**, *104*, 208–216. [CrossRef]

9. Wang, M.; Carver, J.J.; Phelan, V.V.; Sanchez, L.M.; Garg, N.; Peng, Y.; Nguyen, D.D.; Watrous, J.; Kapono, C.A.; Luzzatto-Knaan, T.; et al. Sharing and community curation of mass spectrometry data with Global Natural Products Social Molecular Networking. *Nat. Biotechnol.* **2016**, *34*, 828–837. [CrossRef]

). Fiehn, O.; Robertson, D.; Griffin, J.; van der Werf, M.; Nikolau, B.; Morrison, N.; Sumner, L.W.; Goodacre, R.; Hardy, N.W.; Taylor, C. The metabolomics standards initiative (MSI). *Metabolomics* **2007**, *3*, 175–178. [CrossRef]

. Goodacre, R. Water, water, every where, but rarely any drop to drink. *Metabolomics* **2014**, *10*, 5–7. [CrossRef] *antibiotics*

MDPI

Article

Exploring the Multifaceted Potential of a Peptide Fraction Derived from *Saccharomyces cerevisiae* Metabolism: Antimicrobial, Antioxidant, Antidiabetic, and Anti-Inflammatory Properties

Patrícia Branco [1,2,3,*], Elisabete Muchagato Maurício [1,4,5,*], Ana Costa [2], Diogo Ventura [2], Catarina Roma-Rodrigues [6,7], Maria Paula Duarte [8], Alexandra R. Fernandes [6,7] and Catarina Prista [2]

[1] School of Engineering, Lusófona University, Campo Grande 376, 1749-024 Lisboa, Portugal
[2] Linking Landscape, Environment, Agriculture and Food (LEAF), Associated Laboratory TERRA, Instituto Superior de Agronomia, University of Lisbon, Tapada da Ajuda, 1349-017 Lisboa, Portugal
[3] Unit of Bioenergy and Biorefinary, Laboratório Nacional de Energia e Geologia (LNEG), Estrada do Paço do Lumiar, 22, 1649-038 Lisboa, Portugal
[4] CBIOS—Universidade Lusófona's Research Center for Biosciences & Health Technologies, Campo Grande 376, 1749-024 Lisboa, Portugal
[5] Elisa Câmara, Lda, Dermocosmética, Centro Empresarial de Talaíde, n°7 e 8, 2785-723 São Domingos de Rana, Portugal
[6] UCIBIO—Applied Molecular Biosciences Unit, Department Ciências da Vida, NOVA School of Science and Technology, 2829-516 Caparica, Portugal
[7] i4HB, Associate Laboratory—Institute for Health and Bioeconomy, Faculdade de Ciências e Tecnologia, Universidade NOVA de Lisboa, 2829-516 Caparica, Portugal
[8] MEtRICs, Departamento de Química, NOVA School of Science and Technology | FCTNOVA, Campus de Caparica, Universidade Nova de Lisboa, 2829-516 Caparica, Portugal
* Correspondence: patricia.branco@ulusofona.pt (P.B.); elisabete.mauricio@ulusofona.pt (E.M.M.)

Abstract: The rising demand for minimally processed, natural, and healthier food products has led to the search for alternative and multifunctional bioactive food components. Therefore, the present study focuses on the functional proprieties of a peptide fraction derived from *Saccharomyces cerevisiae* metabolism. The antimicrobial activity of the peptide fraction is evaluated against various foodborne pathogens, including *Candida albicans*, *Candida krusei*, *Escherichia coli*, *Listeria monocytogenes*, and *Salmonella* sp. The peptide fraction antioxidant properties are assessed using FRAP and DPPH scavenging capacity assays. Furthermore, the peptide fraction's cytotoxicity is evaluated in colorectal carcinoma and normal colon epithelial cells while its potential as an antidiabetic agent is investigated through α-amylase and α-glucosidase inhibitory assays. The results demonstrate that the 2–10 kDa peptide fraction exhibits antimicrobial effects against all tested microorganisms, except *C. krusei*. The minimal inhibitory concentration for *E. coli*, *L. monocytogenes*, and *Salmonella* sp. remains consistently low, at 0.25 mg/mL, while *C. albicans* requires a higher concentration of 1.0 mg/mL. Furthermore, the peptide fraction displays antioxidant activity, as evidenced by DPPH radical scavenging activity of 81.03%, and FRAP values of 1042.50 ± 32.5 µM TE/mL at 1.0 mg/mL. The peptide fraction exhibits no cytotoxicity in both tumor and non-tumoral human cells at a concentration up to 0.3 mg/mL. Moreover, the peptide fraction presents anti-inflammatory activity, significantly reducing the expression of the TNFα gene by more than 29.7% in non-stimulated colon cells and by 50% in lipopolysaccharide-stimulated colon cells. It also inhibits the activity of the carbohydrate digestive enzymes α-amylase (IC_{50} of 199.3 ± 0.9 µg/mL) and α-glucosidase (IC_{20} of 270.6 ± 6.0 µg/mL). Overall, the findings showed that the peptide fraction exhibits antibacterial, antioxidant, anti-inflammatory, and antidiabetic activity. This study represents a step forward in the evaluation of the functional biological properties of *S. cerevisiae* bioactive peptides.

Keywords: *Saccharomyces cerevisiae*; foodborne pathogens; antimicrobial peptides; bioactive metabolites; biopreservatives; antioxidant activity; antidiabetic activity; anti-inflammatory activity

1. Introduction

Various pathogenic microorganisms, including *E. coli*, *L. monocytogenes*, *Salmonella* spp., and *Candida* species, pose a risk of contaminating food products [1–3]. Foodborne diseases caused by microbial contamination are a significant global concern, leading to adverse health effects and substantial economic impacts such as increased public health expenses, food waste, and limitations on storage and transportation [4].

E. coli, commonly found in the gastrointestinal tract of humans and animals, is widely distributed in the environment. Consumption of contaminated food, such as meat, vegetables, and dairy products, is the primary mode of transmission for these bacteria [5]. While some strains of *E. coli* are harmless, other are pathogenic and can cause severe foodborne illnesses, including diarrhea, kidney failure, and even death [6].

L. monocytogenes is an opportunistic foodborne pathogen responsible for listeriosis, particularly affecting individuals with weakened immune systems [1]. It is commonly found in meat and dairy products, with cheese being an important source. *L. monocytogenes* can contaminate food at any stage of production, and it can survive and grow at refrigeration temperatures, posing a significant public health risk [7].

Salmonella sp. is typically detected in food products such as poultry, pork, milk, and eggs, but it can also originate from other sources, including dairy products, fruits, and vegetables [8,9]. Ingesting *Salmonella* can lead to a range of symptoms, including diarrhea, fever, and abdominal cramps. In severe cases, hospitalization and even death can occur, particularly among vulnerable populations such as young children and the elderly [10].

C. albicans, a yeast commonly present in the human body as a commensal microorganism, can become an opportunistic pathogen causing infections. While candidiasis is typically associated with mucocutaneous infections, recent studies have shown that *C. albicans* is able to survive and grow in several food products, including dairy, meat, and vegetables, posing risks to human health, and causing economic losses due to food spoilage [11–13].

C. krusei, known for its surface-growing film on foods, possesses characteristics that enable it to tolerate the conditions found in food products, resulting in spoilage. It is highly tolerant to low pH and high preservative concentrations, making acid-preserved food susceptible to spoilage, leading to excessive carbon dioxide production, bloating, and packaging ruptures [14,15]. Additionally, *C. krusei* has been reported as a cause of candidiasis, possibly due to its resistance to the antifungal agent fluconazole [16].

To mitigate the growth of these pathogenic microorganisms, prevent physical and chemical changes in food products, and extend their shelf life, the food industry commonly employs chemical preservatives. However, many of these preservatives, such as sulphur dioxide, sodium benzoate, benzoic acid, sodium sorbate, potassium sorbate, and sodium nitrite, pose risks to human health. Sulphur dioxide, for example, can cause headaches, palpitations, and allergies, while benzoates have been associated with allergies, asthma, and skin rashes. Sorbates and sorbic acid, though rarely, have been linked to urticaria and contact dermatitis [17–19].

Therefore, the increasing consumer interest in healthier foods, without chemical preservatives, is pressing the food industry towards less processed and more natural products, the so-called "clean label" products [20]. In response, the application of natural preservatives, such as bioactive metabolites produced by microorganisms, emerges as a compelling alternative strategy to develop healthier, "clean label", food products while maintaining their safety and shelf life.

One alternative approach gaining traction involves the use of natural preservatives, such as bioactive metabolites produced by microorganisms. Numerous bioactive metabolites with antimicrobial activity, including antimicrobial peptides, bacteriocins, and mycocins, have been identified [21–23].

Certain strains of *S. cerevisiae* have been found to produce bioactive peptides, which exhibit antagonistic effects against some wine yeasts and lactic-acid bacteria [24–27]. In a previous work [25], a 2–10 kDa peptide fraction obtained from *S. cerevisiae* strain CCMI 885 supernatant was fractionated by gel-filtration chromatography and four peaks were obtained and collected for antimicrobial assays. Peak II, with an apparent MW of 8 kDa, exhibited antagonistic effects against some wine yeasts and lactic acid bacteria and was found to contain bioactive peptides derived from the glyceraldehyde 3-phosphate dehydrogenase enzyme [25]. Afterward, Branco et al. (2017) [26] screened eight *S. cerevisiae* strains, including the one used in this study (ISA 1028), for their production of peptides. The results of 2–10 kDa fractions gel-filtration chromatography, obtained from each strain supernatant, showed that all *S. cerevisiae* strains presented a similar chromatographic profile, with four peaks ranging from approximately 2 to 10 kDa [26]. A correlation between the peak II area and the antagonistic effect of each *S. cerevisiae* strain against one wine yeast was detected, revealing that *S. cerevisiae* CCMI 885 and *S. cerevisiae* ISA 1028 were the strains with the higher antagonistic effect against *Hanseniaspora guilliermondii*. Similarly, Al-Sahlany et al. (2020) [28] fractioned, by gel-filtration chromatography, a 2–10 kDa peptide fraction obtained from *S. cerevisiae* strain ATCC 36858 supernatant and found a peptide with a molecular weight of 9.77 kDa that inhibits the growth of *E. coli* and *Staphylococcus aureus*.

Bioactive peptides have a wide range of functional properties, including antimicrobial, anticancer, antihypertensive, antidiabetics, antioxidative, and anti-inflammatory activity [29]. For instance, Mudgil et al. (2018) [30] discovered that peptides derived from camel milk protein hydrolysates demonstrated antidiabetic and anti-obesity properties. Likewise, a peptide derived from mushroom *Ganoderma lucidum* exhibited antioxidant activity and exerted antimicrobial effects against *E. coli* and *Salmonella typhi* [31–33].

Antioxidants are particularly important in preventing the oxidation of compounds and are found abundantly in natural foods such as fruits and vegetables; they include polyphenols, such as flavonoids [34–37], and bioactive peptides [38]. In addition, they have been found in yeast peptide fractions and yeast extracts [39,40]. The consumption of antioxidant-rich foods has been associated with numerous health benefits [41]. Antioxidants play a vital role in safeguarding cells against oxidative stress caused by free radicals, which have been linked to various chronic diseases such as cancer, heart disease, and Alzheimer's disease [41–43].

Taking into account the previous findings on the 2–10 kDa fraction derived from *S. cerevisiae* strains metabolism and all the possible functional properties previously detected on bioactive peptides, the aim of this study is to assess the functional properties of the 2–10 kDa peptide fraction derived from *S. cerevisiae* strain (ISA 1028) metabolism, namely its (i) antimicrobial activity against foodborne pathogens, (ii) potential as natural antioxidants, (iii) anti-inflammatory activity on human colon epithelial cells, and (iv) antidiabetic activity.

2. Results

2.1. Antimicrobial Activity of the 2–10 kDa Fraction against FOODBORNE Pathogens

2.1.1. Minimal Inhibitory Concentration (MIC)

The antimicrobial activity of the peptide fraction (2–10 kDa) derived from the fermentation supernatant of *S. cerevisiae* strain ISA 1028 was evaluated by determining the MIC against three bacterial strains (*E. coli* ATCC 25922, *L. monocytogenes* ISA 4008, and *Salmonella* sp. ISA 4348) and two yeast strains (*C. albicans* ISA 2289 and *C. krusei* ISA 2290).

The analysis of the peptide fraction by High-Performance Liquid Chromatography (HPLC) revealed the presence of ethanol (8% (*v*/*v*)). Hence, the maximum concentration of ethanol in the MIC assay is 4% (*v*/*v*), since the 2–10 kDa fraction is mixed with the growth medium in a proportion of 1:1 and serial dilutions of them were made in the microplate according to the assay methodology.

The results confirmed that all tested microorganisms exhibited normal growth in the absence of the peptide fraction and in the presence of ethanol (negative control). In addition, the microorganisms under trial were also tested against suitable antibiotics (positive controls) to ensure the proper sensitivity of each strain.

The MIC of the 2–10 kDa fraction against the three bacterial strains was 0.25 mg/mL, indicating that the antimicrobial peptide effectively inhibited the growth of these bacteria at a relatively low concentration (Table 1).

Table 1. Minimum inhibitory concentration (MIC) of 2–10 kDa fraction against the tested microorganisms.

	MIC (mg/mL)			
Tested Microorganisms	**2–10 kDa Fraction**	**Chloramphenicol [a]**	**Ketoconazole [a]**	**Ethanol [b]**
C. albicans	1.0	-	0.012	n.d
C. krusei	>1.0	-	0.012	n.d
E. coli	0.25	0.012	-	n.d
L. monocytogenes	0.25	0.050	-	n.d
Salmonella sp.	0.25	0.025	-	n.d

[a]—Positive control; [b]—Negative control; - not tested; n.d.—not detected (MIC > 4% (*v*/*v*) of ethanol).

On the other hand, the MIC of the 2–10 kDa fraction against *C. albicans* was 1.0 mg/mL, indicating that a higher concentration is necessary to inhibit the growth of this pathogenic yeast when compared to the tested bacteria. (Table 1). However, the MIC of the 2–10 kDa fraction against *C. krusei* was higher than 1.0 mg/mL, indicating that this particular pathogenic yeast resisted the 2–10 kDa fraction at the tested concentrations (Table 1). Considering that several authors classify the antimicrobial activity of natural substances as strong inhibitors (MIC up to 0.5 mg/mL), moderate inhibitors (MIC between 0.6 and 1.5 mg/mL), and weak inhibitors (MIC above 1.6 mg/mL) [44–46], it is possible to confirm that the peptide fraction under study presents a strong inhibitory activity against bacteria and moderate against the *C. albicans*.

2.1.2. Fungicidal and Bactericidal Activity

To assess the bactericidal and fungicidal activity of the 2–10 kDa fraction against the microorganism showing higher sensitivity to these peptides, i.e., *C. albicans*, *E. coli*, *L. monocytogenes* and *Salmonella* sp., we studied the viability of the microorganisms by counting colony-forming units per milliliter (CFU/mL). Samples from the negative control and test assays at the MIC value (0.25 mg/mL for bacteria and 1.0 mg/mL for *C. albicans*) and at concentrations higher than the MIC were taken after 24 h and colony-forming units were counted.

The results demonstrated that in the absence of the 2–10 kDa fraction (negative control), all microorganisms reached a viability of 10^7 CFU/mL to 10^8 CFU/mL within 24 h (Figure 1). When the 2–10 kDa fraction was present at the MIC value (0.25 mg/mL), the viability of bacteria decreased by 2 to 3 orders of magnitude when compared to the negative control without the peptide fraction. Specifically, the CFU/mL values were 1.5×10^5 CFU/mL (*E. coli*), 2.5×10^5 CFU/mL (*Salmonella* sp.), and 3.2×10^5 CFU/mL (*L. monocytogenes*) (Figure 1).

As expected, the peptide fraction exhibited increased antimicrobial activity at a higher concentration of 0.5 mg/mL, resulting in a decrease of 4 to 5 orders of magnitude in CFU/mL when compared to the control-assay values. The CFU/mL values at this concentration were 2.9×10^3 CFU/mL (*E. coli*), 2.3×10^3 CFU/mL (*Salmonella* sp.) and 4.7×10^3 CFU/mL (*L. monocytogenes*) (Figure 1). Interestingly, at the highest concentration tested (1.0 mg/mL), no growth was observed for *E. coli* and *Salmonella* sp., while *L. monocytogenes* showed a partial inhibition with a viability of 70 CFU/mL (Figure 1). Once again, *C. albicans* demonstrated greater resistance to the peptide fraction compared to bacteria, as a viability of 1.0×10^6 CFU/mL was detected when cells were exposed to 1.0 mg/mL of the 2–10 kDa fraction (Figure 1).

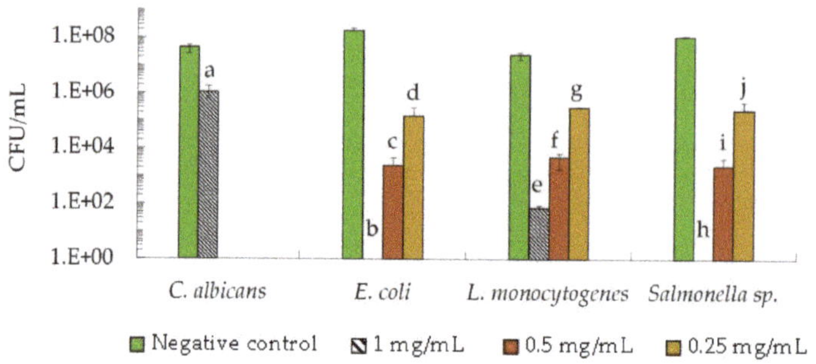

Figure 1. Viability (CFU/mL), of *C. albicans* ISA 2289, E. coli ATCC 25922, *L. monocytogenes* ISA 4008 and *Salmonella* sp. ISA in the presence of fraction 2–10 kDa at a final concentration of 1.0, 0.5 and 0.25 mg/mL and in the absence of fraction 2–10 kDa (Negative control). Data represented correspond to means ± SD (error bars) of three independent biological assays. Different letters (a–j) represent p-value < 0.05 relative to respective control.

2.2. Antioxidant Activity of 2–10 kDa Fraction

The antioxidant capacity of the 2–10 kDa peptide fraction at 1.0, 0.5, and 0.25 mg/mL was determined by two methods, one scavenging based assay (DPPH radical scavenging activity (RSA%)) and one measuring reducing power (Ferric Reducing Antioxidant Power (FRAP)). Trolox was used as the standard for the DPPH and FRAP assays (Figure S1 depicts the calibration curve for each assay), and the results are represented in Figure 2. In the case of FRAP assay (Figure 2A), the fraction at 1.0 mg/mL showed the highest antioxidant capacity (1042.50 ± 32.5 µM TE/mL). With the fraction concentration reduction to 0.5 mg/mL, a reduction to 355.00 ± 42.5 µM TE/mL is also observed (Figure 2A). The same phenomenon occurs with the 2–10 kDa fraction at 0.25 mg/mL, in which the presented value also decreases around 50% (173.75 ± 3.75 µM TE/mL) (Figure 2A). Among the concentrations tested, the 2–10 kDa fraction at 1.0 mg/mL concentration had the highest capacity to inhibit DPPH radical with a %RSA of 81.03%, while at 0.5 mg/mL it was 60.99% and at 0.25 mg/mL was 56.22% (Figure 2B).

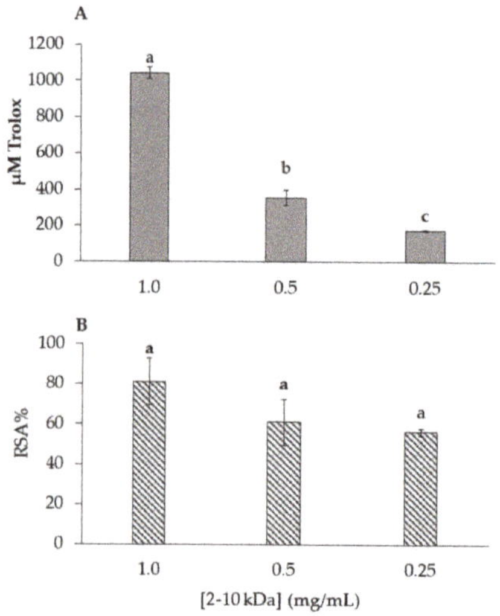

Figure 2. Values obtained for FRAP expressed in µM Trolox (**A**) and for DPPH expressed in RSA% (**B**), for the 2–10 kDa fraction at 1.0, 0.5 and 0.25 mg/mL. Different letters (a–c) represent significant differences (p < 0.05).

2.3. Cytotoxicity of 2–10 kDa Fraction

To assess if the 2–10 kDa fraction presents cytotoxicity in human cells, the viability of normal colon epithelial cells (coN) and colorectal carcinoma cells (HCT116) was evaluated after 48 h incubation with increasing concentrations of the peptide fraction. Results showed that the fraction presented no cytotoxicity for concentrations up to 0.25–0.30 mg/mL. Moreover, independently of the tumorigenic characteristic of the cells, an IC_{50} of 0.4 mg/mL was observed in colon epithelial cells (Table 2, Figure 3). Importantly, at the MIC concentration for antibacterial activity (0.25 mg/mL, Table 1), there is no significant impact on the viability of the human cells (100% activity).

Table 2. Relative IC_{50} values (mg/mL) of peptide fraction 2–10 kDa in colorectal carcinoma (HCT116) and normal colon (coN) cell lines.

Cell Line	HCT116	coN
IC_{50} (mg/mL)	0.44 ± 0.10	0.42 ± 0.08

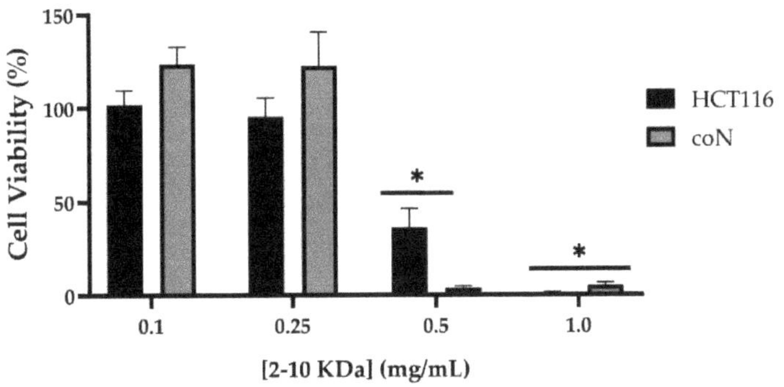

Figure 3. Cytotoxicity of peptide fraction 2–10 kDa in colorectal carcinoma cells (HCT116) and normal colon cells (coN). Cells were incubated for 48 h with crescent concentrations of the peptide fraction 2–10 kDa and cell viability was measured through the MTS assay. Bars represent the average ± SEM of four experiments. * p-value < 0.05 relative to respective control.

2.4. Analysis of Anti-Inflammatory Effect of Peptide Fraction 2–10 kDa

To understand if the peptide fraction has some effect on the inflammatory process, coN cells were exposed to an inflammatory stimulus (lipopolysaccharides (LPS)) and then incubated for 3 h with 0.25 mg/mL of the peptide fraction 2–10 kDa or 1% (*v*/*v*) ethanol (as the vector control). Cells that were not exposed to the inflammatory stimulus were used for comparison.

The inflammatory potential was evaluated by RT-qPCR, measuring the expression levels of the gene encoding the pro-inflammatory cytokine tumor necrosis factor alpha (TNF-α) (Figure 4). The analysis of TNF-α expression on coN cells after 2 h incubation in the presence of LPS confirmed the induction of the cytokine expression when cells are exposed to an inflammatory stimulus (Figure 4A). After additional 3 h of incubation in the presence of LPS, TNF-α expression remained high (Figure 4B). However, a 52% reduction of TNF-α expression was observed when LPS-treated colon cells were incubated in the presence of the peptide fraction 2–10 kDa when compared to the cells only treated with LPS (Figure 4B,C (+LPS, orange bars)), demonstrating the anti-inflammatory capability of the peptide fraction. Moreover, this anti-inflammatory potential is also observed even in the absence of the LPS stimulus (−LPS) but to a lower extent (only a 29% reduction of TNF-α expression was observed) (Figure 4C,D, blue bars).

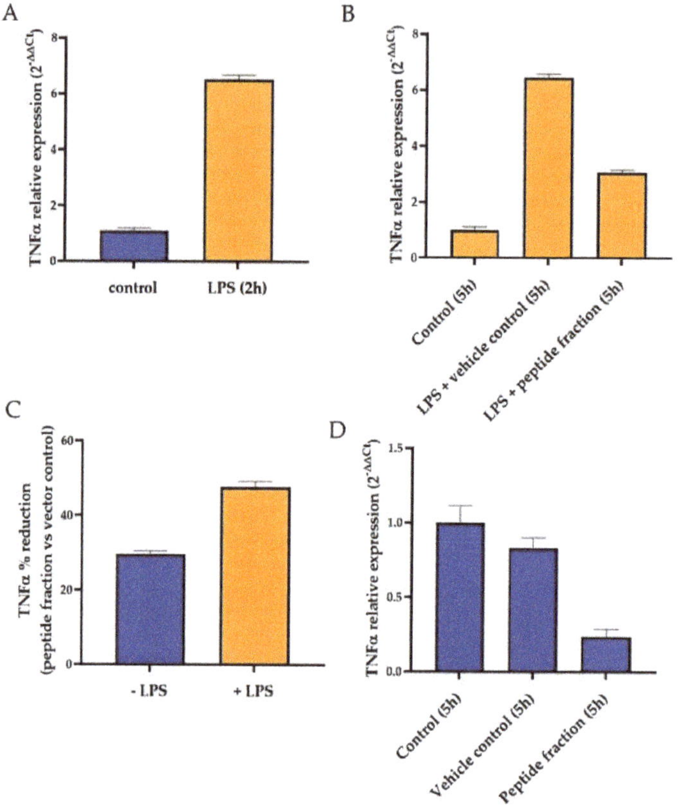

Figure 4. Effect of peptide fraction 2–10 kDa in TNF-α expression in colon epithelial cells (coN). Cells were incubated for 2 h with 7 µg/mL lipopolysaccharide (+LPS) and then for 3 h with ethanol 1% (*v*/*v*) or 0.25 mg/mL peptide fraction 2–10 kDa. In parallel, cells were submitted to the same treatment but without LPS (-LPS, (**A**)) Gene expression after 2 h in +LPS samples (orange bars), calculated by $2^{-\Delta\Delta Ct}$, using as reference RNA 18S gene and—LPS samples (blue bars, (**B**)). Gene expression in LPS treated samples and exposed to vehicle control and peptide fraction calculated by $2^{-\Delta\Delta Ct}$, using as reference RNA 18S gene and samples only treated with LPS for 5 h, (**C**). Percentage of TNF-α reduction, after incubation with LPS (orange bars) or without LPS exposure (blue bars, (**D**)). Gene expression in LPS untreated samples (-LPS) and exposed to vehicle control and peptide fraction calculated by $2^{-\Delta\Delta Ct}$, using as reference RNA 18S gene and untreated samples collected after 5 h. Error bars represent SEM of at least three independent experiments. *p*-value < 0.05 relative to respective control sample (treated with ethanol).

2.5. *Antidiabetic Effect of Peptide Fraction 2–10 kDa*

The antidiabetic potential of the peptide fraction was evaluated by its in vitro α-amylase and α-glucosidase inhibitory activities through colorimetric methods. Acarbose, an antidiabetic drug that acts by inhibiting α-amylase and α-glucosidase activities, was used as a positive control (Table 3).

The peptide fraction inhibited both enzymes in a dose-dependent way. However, an inhibitory activity lower than the positive control (acarbose) was observed. The inhibitory activity of the peptide fraction was more pronounced against α-amylase. In fact, under the experimental conditions used, the peptide fraction totally inhibited the α-amylase activity, making it possible to determine the IC_{50} value i.e., the concentration that inhibits 50% of the enzyme activity under the specific set of assay conditions. In what concerns α-glucosidase, under the experimental conditions used, only a maximum inhibition of about 20% was verified, whereby the IC_{20} value was determined.

Table 3. α-Amylase and α-glucosidase inhibitory activity of peptide fraction 2–10 kDa.

	Peptide Fraction	Acarbose
α-Amylase (IC_{50} (μg/mL))	199.3 ± 0.9	11.6 ± 1.0
α-Glucosidase (IC_{20} (μg/mL))	270.6 ± 6.0	134.1 ± 4.0

3. Discussion

Microbiological contamination in food products is a global concern due to the increasing incidence of foodborne diseases and their significant economic impact, including limitations on storage and transport time, increased public health expenses, and food waste. Therefore, it is crucial to prevent and control the growth of spoilage microorganisms. Foodborne pathogens such as *L. monocytogenes*, *E. coli*, *Salmonella* sp., and *Candida* sp. pose a significant food safety risk and can cause illnesses [3,4]. The use of chemical preservatives to control the growth of these microorganisms is a common practice. However, these preservatives can adversely affect human health [47]. Hence, there is a need to explore alternative preservatives that are both less toxic and can effectively eliminate and prevent the proliferation of foodborne pathogens in food products. In fact, some antimicrobial peptides (AMPs) produced by bacteria that are classified as Generally Recognized As Safe (GRAS) [48] were applied as food preservatives, such is the case of nisin [49–51].

Nisin, a bacteriocin derived from the bacterium *Lactococcus lactis*, has emerged as a promising natural preservative in the food industry. Its antimicrobial activity has been extensively studied, demonstrating efficacy against a wide range of spoilage and pathogenic microorganisms, including Gram-positive bacteria and some Gram-negative bacteria [52,53]. Nisin has been successfully utilized in various food products, including dairy, meat, poultry, bakery, and beverages, to inhibit the growth of spoilage and pathogenic microorganisms, thus extending the shelf life and enhancing food safety [54].

In addition to bacteriocins, yeasts produce extracellular proteins known as mycocins, which exhibit antimicrobial activity [55]. Research conducted by Izgü et al. (2007) [56] demonstrated that *Wickerhamomyces anomalus* NCYC 434 secretes a mycocin that effectively inhibits *C. krusei*. Similarly, Al-Qaysi et al. (2017) [57] found that *Debaromyces hansenii* DSMZ 70238 secretes a mycocin with antagonistic effects against *Candida* spp., *E. coli*, and *S. aureus*. Furthermore, mycocins secreted by *Tetrapisispora phaffii* DBVPG 6706 and *W. anomalus* DBVPG 3003 have demonstrated activity against various wine yeasts, including strains of *Brettanomyces bruxellensis*, which is a major cause of wine spoilage worldwide [58–60].

Another yeast well known for its ability to produce several bioactive compounds and metabolites, including peptides, carbohydrates such β-glucans and mannans, vitamins and antioxidant enzymes, is *S. cerevisiae* [61,62]. Previous studies [24–28] have identified the presence of antimicrobial activity within a peptide fraction ranging from 2–10 kDa derived from *S. cerevisiae* metabolism. This fraction contains antimicrobial peptides that play a crucial role in inhibiting the growth of some yeast, i.e., *Hanseniaspora guilliermondii*, *Lachancea thermotolerans*, *Torulaspora delbrueckii*, *Kluyveromyces marxianus*, *B. bruxellensis*, and bacteria such as lactic acid bacteria [24–27], *E. coli*, and *S. aureus* [28].

In this study, for the first time, the antimicrobial activity of the 2–10 kDa fraction derived from the metabolic processes of *S. cerevisiae* was tested against *L. monocytogenes*, *Salmonella* sp., and *Candida* species as well as its functional proprieties, i.e., antioxidant, anti-inflammatory, and antidiabetic effects.

The antimicrobial assays demonstrated that the 2–10 kDa fraction had an inhibitory effect on all the tested microorganisms, except for the yeast *C. krusei* (Table 1, Figure 1), suggesting an inherent resistance of this pathogenic yeast to the peptide fraction.

It can be observed that the fraction exhibits a significant antibacterial activity against the Gram-positive (*L. monocytogenes*) and Gram-negative (*Salmonella* sp. and *E. coli*) bacteria used in this study (Table 1, Figure 1). These findings show that the fraction under study has a broad spectrum of action and thus demonstrates potential to control foodborne pathogens and food spoilage. Moreover, the results highlight the advantage of these peptides over other natural antimicrobials compounds, namely some natural extracts rich in polyphenols that are typically more specific towards Gram-positive bacteria [37,63,64].

Several factors influence the antioxidant capacity of peptides, including their amino acid composition, sequence, and hydrophobic amino acid content [65–67]. As can be seen from Figure 2, the 2–10 kDa fraction at the different concentrations hold antioxidant capacity. The differences between DPPH and FRAP results may result from to their underlying principles. The DPPH assay is based on the measurement of the scavenging capacity of antioxidants towards it [68] and the FRAP assay is based on the compounds capability to reduce Fe^{3+} to Fe^{2+} [69], which may justify the lowest results in the case of FRAP. However, while for the DPPH assay there were no significant differences between the concentrations, in the case of the FRAP assay, a significant correlation with the peptide concentration was observed, which may indicate that, for DPPH, other factors besides peptides can be disturbing the results. Particularly, the concentration of 1.0 mg/mL stood out, while the concentrations of 0.5 and 0.25 mg/mL showed very similar values.

Mirzaei et al. (2015) [61] conducted a study to assess the antioxidant activity of different peptide fractions derived from *S. cerevisiae* using the DPPH inhibition assay. The study examined peptide fractions with varying molecular weights (>3 kDa, 3–5 kDa, 5–10 kDa, and <10 kDa). The results showed that the <3 kDa fraction obtained from trypsin isolate exhibited the highest antioxidant capacity, measuring 489.12 $\pm$ 0.001 µM TE/mg protein. In a more recent study by Mirzaei et al. (2021) [67], the researchers performed the FRAP assay in order to evaluate the antioxidant potential of the peptide fractions. Interestingly, the <3 kDa fraction showed a lower result than the 3–5 kDa fraction. It is worth mentioning that the antioxidant activity observed in this current study, using both the DPPH and FRAP assays, was higher than the values reported by the author in their previous work.

While the use of bioactive metabolites as food preservatives shows to be great promise, it is crucial to ensure that these metabolites are safe for consumption and do not have any adverse effects on human health. Studies investigating the cytotoxicity of bioactive metabolites such as antimicrobial peptides on gastrointestinal tract cell lines have shown minimal or no cytotoxic effects at concentrations effective for controlling microbial growth [70]. Therefore, and for the first time, the biological effect of the 2–10 kDa fraction derived from *S. cerevisiae* metabolism was evaluated in human cell cultures, namely in colorectal carcinoma cells (HCT116) and normal colon cells (coN). Regarding the biological effect of the 2–10 kDa fraction in human cells, it was observed that the two tested cell cultures presented an IC_{50} higher than 0.4 mg/mL (Table 2), with 100% viability at 0.3 mg/mL. This result suggests the 2–10 kDa fraction possesses antibacterial activity (MIC 0.25 mg/mL) in concentrations lower than the IC_{50} obtained in this study. These findings emphasize the potential application of the peptide fraction as an antibacterial or as co-adjuvant preservative in food or in health care. Moreover, the peptide fraction showed an anti-inflammatory capability as evidenced by its impact on the expression of TNF-α, regardless of the presence or absence of the inflammatory stimulus, LPS. However, this anti-inflammatory potential was higher when coN cells were previously stimulated with LPS (Figure 4). Overall, these results suggest that the 2–10 kDa fraction possesses an anti-inflammatory potential in colon cells, independent of their previous stimulation with an inflammatory agent, although this response is more pronounced when cells were previously exposed to the inflammatory stimulus.

Type 2 diabetes is a major public health concern, characterized by persistent hyperglycemia coupled with several metabolic dysfunctions. Inhibiting the activity of the carbohydrate digestive enzymes to avoid the degradation of polysaccharides into glucose is one approach to prevent postprandial hyperglycemia. Therefore, α-amylase and α-glucosidase inhibitors have been considered first-line drugs to control blood sugar levels, thus helping to prevent and control type 2 diabetes [71]. Some currently available antidiabetic drugs, such as acarbose, act by inhibiting the activity of these two enzymes. However, these drugs are associated with gastrointestinal disturbance, abdominal pain, and flatulence [72]. Thus, many studies have focused on identifying alternative α-amylase and α-glucosidase inhibitors, such as plant extracts [73] or food derived bioactive peptides [30,71,72,74–76], that may simultaneously have inhibitory activity and a low incidence of undesirable side effects.

The peptide fraction was able to inhibit the activity of α-amylase and α-glucosidase, the effect being more pronounced towards α-amylase. Other peptides have been shown to inhibit α-amylase and α-glucosidase, with IC_{50} values within the range of 0.027–23.30 mg/mL and 1.45–10 mg/mL, respectively [30,71,74–76], with these inhibitory activities related to their amino acid composition. Some authors suggested that branched chain, such as isoleucine, aromatic, such as tyrosine and tryptophan, and positively charged, such as arginine, residues are preferably bound to α-amylase and could be key amino acids associated with α-amylase inhibitory activity [72,74]. On the other hand, the presence of basic amino acids at the N-terminal and of proline within the chain and alanine or methionine at the C-terminal seems to be important for the α-glucosidase inhibitory activity [72].

4. Materials and Methods

4.1. Production of the 2–10 kDa Fraction

The peptide fraction was obtained as described in [25]. Briefly, *S. cerevisiae* ISA 1028 cells were grown at 25 °C, without agitation in Synthetic Grape Juice at pH 4.5 (glucose 110 g/L, fructose 110 g/L; acids solution: tartaric acid 6 g/L, malic acid 3 g/L, citric acid 0.5 g/L; amino acids solution: yeast nitrogen base without aa 1.7 g/L, casamino acids 2 g/L, calcium chloride 0.2 g/L, arginine 0.8 g/L, proline 1 g/L, tryptophan 0.1 g/L and 2.5 g/L yeast extract). All solutions were autoclaved, except the amino acid's solution, which was sterilized by filtration (0.22 μm). After seven days, cells were removed by centrifugation and the supernatant was first filtered through a 0.45 μm membrane and then through a 0.22 μm membrane (Merck Millipore, EUA, Burlington, MA, USA). Then, the obtained cell-free supernatant was first ultrafiltrate through centrifugal filter units (Vivaspin 15R, Sartorius, Göttingen, Germany) equipped with 10 kDa membranes and then concentrated (40-fold) with 2 kDa membranes, obtaining a 2–10 kDa fraction. Ethanol and organic acids present in the 2–10 kDa fraction were analysed in duplicate using a High-Performance Liquid Chromatography (HPLC) VWR Hitachi Chromaster system (VWR, Radnor, PA, USA) equipped with a refractive index detector (Chromaster HPLC 5310) and UV-Vis detector (Chromaster HPLC 5420) (Waters, Dublin, Ireland) equipped with a refractive index detector (2414Waters). The 2–10 kDa fraction was first filtrated by 0.22 μm Millipore membranes (Merck Millipore, EUA, Burlington, MA, USA) and then injected in a Rezex™ ROA Organic Acid H+ (8%) column (300 mm 7.8 mm, Phenomenex, Torrance, CA, USA) and eluted with sulfuric acid (5 mmol/L) at 65 °C with a flow rate of 0.5 mL/min.

4.2. Antimicrobial-Assays

The antimicrobial activity was determined using the broth microdilution method, according Clinical and Laboratory Standards Institute guidelines [75]. All assays were performed in triplicate, and positive (chloramphenicol for bacteria or ketoconazole for yeasts) and negative (ethanol) controls were included in all assays.

4.2.1. Microorganisms and Media

In this work, the following microorganisms from Culture collection of Instituto Superior de Agronomia, Portugal, were used: *C. albicans* ISA 2289; *C. krusei* ISA 2290; *L. monocytogenes* ISA 4008; *S. cerevisiae* ISA 1028, *Salmonella* sp. ISA 4348, and *E. coli* ATCC

25922 from the American Type Culture Collection. The antimicrobial activity was assessed against the microorganisms described above. In order to prepare the working culture of the microorganisms, a subculture was prepared from the stock culture in plates with recommended media and incubation conditions for each microorganism, Luria-Bertani (LB) agar plates (10 g/L tryptone, 10 g/L NaCl, 5 g/L yeast extract, 20 g/L agar, pH was adjusted to 7.0 with 5 N NaOH) for *E. coli* ATCC 25922 and *Salmonella* sp. ISA 4348 and Brain Heart Infusion (BHI) (Biokar Diagnostics, Allonne, France) for *L. monocytogenes* ISA 4008 at 35 ± 2 °C, or Sabouraud Dextrose Agar (SDA) (Biokar Diagnostics, Allonne, France) for yeasts at 30 ± 2 °C [37].

4.2.2. Determination of Minimum Inhibitory Concentrations

The conventional minimal inhibitory concentration (MIC) was determined by the broth microdilution method in 96-well microtiter plates, according to Pereira et al. (2022) [46]. The peptide fraction was prepared at 2 mg/mL and a dilution of 1:2 was introduced in the first line, followed by a series of 2-fold dilutions in Sabouraud Dextrose broth (Biokar Diagnostics, Allonne, France) for yeasts (*C. albicans* and *C. krusei*) and in Mueller-Hinton broth (Biokar Diagnostics, Allonne, France) for the remaining bacteria. Ethanol at 8% (*v*/*v*) was used as negative control and positive controls, 100 mg/L of chloramphenicol (Merck KGaA, Darmstadt, Germany) for bacteria or 0.1 mg/mL of ketoconazole (Merck KGaA, Darmstadt, Germany) for yeasts, were used in the same conditions and dilutions.

To inoculate the 96-well microtiter plates, a standardized saline suspension (NaCl, 0.85% *w*/*v*) was prepared, and turbidity adjusted to 0.5 McFarland scale (DEN-1, McFarland Densitometer, Biosan) corresponding to 1–2 × 10^8 CFU/mL for bacteria and 1–5 × 10^6 CFU/mL for yeasts. The inoculation was performed according to the CLSI document M07-A9, 2012 [77]. The microplates were then incubated at 35 °C ± 2 for all microorganisms, for 24 h, except for yeasts, which were incubated for 24 h at 30 °C ± 2. After the incubation period, microplates were visually observed to determine the MIC (defined as the lowest peptide fraction concentration at which no visible growth could be detected) and samples were taken to evaluate the fungicidal and bactericidal activity.

4.2.3. Evaluation of Fungicidal and Bactericidal Activity

After the incubation period (24 h) the target microorganisms' culturability (CFU/mL) was determined by the classical plating method. Briefly, 100 μL of cells were plated onto LB agar plates (*E. coli* and *Salmonella* sp.), onto BHI agar (*L. monocytogenes*) or SDA plates (*C. albicans*) after appropriate dilution (decimal serial dilution method). Plates were incubated at 35 ± 2 °C (bacteria) and 30 ± 2 °C (yeasts) and the number of CFU enumerated after 1–2 days.

4.3. Antioxidant Activity

4.3.1. Ferric Reducing Antioxidant Power (FRAP)

The FRAP assay is based on the compound's capability to reduce Fe^{3+} to Fe^{2+}, which will translate into antioxidant activity. A version of the Benzie and Strain (1996) [69] assay was used with modifications. The FRAP reagent utilized was composed of 25 mL of 0.3 M of acetate buffer ($C_2H_3NaO_2 \cdot 3H_2O$, Merck KGaA, Darmstadt, Germany) pH 3.6, 2.5 mL of TPTZ 10 mM (2,4,6-tripyridyl-S-triazine, Alfa Aesar, MA, EUA) in 40 mM HCl and 2.5 mL of 20 mM ferric (III) chloride hexahydrate ($FeCl_3 \cdot 6H_2O$, Sigma, St. Louis, MO, USA) at room temperature. In a 96 well microplate, 10.3 μL of sample diluted 1:10, 30.9 μL of ultrapure water (Merck KGaA, Darmstadt, Germany) and 309 μL of the FRAP reagent were added. It was then incubated in a MOBI microplate reader (μ2 MicroDigital Co., Ltd., Seoul, Republic of Korea) at 37 °C for 45 min. The absorbance was read at 595 nm before and after incubation. The blank was made with ultrapure water (Merck kGaA, Darmstadt, Germany) to substitute the sample. A calibration curve was made with TROLOX 2000 μM in concentrations of 0, 100, 200, 400, 500, 600, 750, 100, 1500, and 2000 μM. The results were expressed in μM.

4.3.2. DPPH Radical Scavenging Assay (2,2-Diphenyl-1-picrylhydrazyl)

DPPH (2,2-Diphenyl-1-picrylhydrazyl, TCI, Tokyo, Japan) is a free radical that absorbs at 515 nm in its radical form but not in its reduced form. The protocol used was a modified version of Brand–Williams (1995) [68]. The DPPH 0.24 mg/mL solution was made with 0.6 mg of DPPH diluted in ethanol 100% (Aga, Lisbon, Portugal) at room temperature. In a 96 well microplate, 8.75 µL of sample and 341.25 µL of DPPH solution were added and incubated in a MOBI microplate reader (µ2 MicroDigital Co., Ltd., Republic of Korea) at 30 °C for 45 min. The absorbance was read at 515 nm before and after incubation, and the blank was made with ethanol 100% (Aga, Lisbon, Portugal) to replace the sample. The absorbance of the resulting solution was measured at 515 nm and radical-scavenging activity (%) was calculated as follows: Radical-scavenging activity (%) = $(\text{Absorbance}_{\text{control}} - \text{Absorbance}_{\text{sample}} \times 100)/\text{Absorbance}_{\text{control}})$. A calibration curve was made with TROLOX 2000 µM in concentrations of 0, 100, 200, 400, 500, 600, 750, 100, 1500 and 2000 µM. The results were expressed in µM.

4.4. Cytotoxicity Assay

The cytotoxic effect of peptide fraction 2–10 kDa was determined in colorectal carcinoma cells (HCT116, CCL-247 ATCC, Manassas, VA, USA) and colon epithelial cells (coN, CRL-1790, ATCC) using Dulbecco's modified Eagle medium (DMEM, ThermoFisher Scientific, Waltham, MA, USA) supplemented with 10% (*v*/*v*) fetal bovine serum (FBS, ThermoFisher Scientific) and a mixture of 100 U/mL Penicilin and 100 µg/mL Streptomycin (ThermoFisher Scientific). Cells were seeded in a density of 7500 cells/well in a 96-well plate and after 24 h, submitted to crescent concentrations of peptide fraction 2–10 kDa or crescent concentrations of ethanol (vector control). After 48 h incubation at 37 °C, 5% (*v*/*v*) CO_2 and saturated humidity, the medium was removed and replaced by a mixture of medium with Cell Titer 96® Aqueous One solution cell proliferation assay (Promega, Madison, WI, USA) and protocol followed the manufacturer's instructions. The viability percentage of cells exposed to each concentration of peptide fraction was normalized to the viability percentage of cells exposed to respective to the concentration of ethanol (vector control).

4.5. Inflammatory Effect

The anti-inflammatory potential of the peptide fraction 2–10 kDa was examined in coN cells with few alterations from what was previously published [78]. Fibroblasts and coN cells were seeded in 25 cm^2 T-flasks and then incubated for 24 h at 37 °C, 5% (*v*/*v*) CO_2 and 99% (*v*/*v*) relative humidity. Cells were incubated for 2 h with 7 µg/mL lipopolysaccharide (LPS, Sigma Aldrich, St. Louis, MO, USA), then 0.25 mg/mL peptide fraction 2–10 kDa or 1% (*v*/*v*) ethanol were added, and cells incubated for further 3 h. Similar samples without LPS were also prepared in parallel. Samples treated and untreated with LPS (+LPS and −LPS, respectively) were collected after 2 h incubation with LPS, and +LPS and −LPS samples incubated with peptide fraction or respective negative control were collected after 3 h (5 h since the beginning of the experiment) by cell detachment with Tryple Express (ThermoFisher Scientific), centrifugation at 500× *g* and pellet resuspension in NZYol (NZYtech, Lisbon, Portugal). RNA was extracted according to manufacturer's instructions, cDNA synthesized using the NZY M-MULV First strand cDNA synthesis kit (NZYtech), and TNF-α and RNA 18S genes were amplified using the NZY Supreme qPCR Green Master Mix (NZYtech) in a Corbett-Rotor Gene thermal cycler (Qiagen, Hilden, Germany). The expression of TNF-α in samples was determined with the $2^{-\Delta\Delta Ct}$ method [79].

4.6. α-Amylase Inhibitory Assay

The inhibition of the α-amylase was determined according to the procedure described by Romeiras et al. (2023) [78]. Reaction mixtures with 100 µL of Type VI-B porcine pancreatic α-amylase (0.5 mg/mL in 100 mM sodium phosphate buffer, containing 6.7 mM sodium chloride, pH 6.7) and 100 µL of the peptide fraction 2–10 kDa (concentrations up to 1.53 mg/mL) were incubated at 37 °C for 10 min. Then, 100 μL of 1% starch (*w*/*v*), in 100 mM sodium phosphate buffer, containing 6.7 mM sodium chloride, pH 6.7) was added. After 10 min at 37 °C, 200 μL of DNS reagent (20 mL of 96 mM DNS, 8 mL of 5.315 M sodium potassium tartrate tetrahydrate in 2 M NaOH and water up to 40 mL) was added. The reaction mixtures were boiled at 100 °C for 15 min, cooled until room temperature, diluted with 2 mL of water and the absorbance (Abs) was measured at 520 nm (SPEKOL 1500, Analytik, Jena, Germany). The reaction mixture without peptide fraction was used as negative control (NC) and reaction mixtures prepared with inactive α-amylase were used as samples' blanks (SB). Acarbose was used as positive control. The enzyme inhibitory rate was calculated as follows:

$$\text{Inhibition (\%)} = [(\text{Abs}_{\text{NC}} - (\text{Abs}_{\text{sample}} - \text{Abs}_{\text{SB}}) \times 100/(\text{Abs}_{\text{NC}})]$$

Values were assessed in triplicate and the results were expressed as the final concentration (μg/mL), in the reaction mixture, which reduces the enzyme activity by 50% (IC_{50}). Data are presented as means ± standard deviations.

4.7. α-Glucosidase Inhibitory Assay

The inhibition of the α-glucosidase enzyme was determined according to the procedure described by Romeiras et al. (2023) [78]. Reaction mixtures containing 5 μL of α-glucosidase from *S. cerevisiae* (6.25 U/mL in phosphate buffer (pH 6.9, 0.1 M)), 125 μL of phosphate buffer (pH 6.9, 0.1 M), and 20 μL of the peptide fraction 2–10 kDa (concentrations up to 3.6 mg/mL) were prepared in a 96 wells microplate (Greiner Bio-One, Rainbach im Mühlkreis, Austria) and incubated for 15 min at 37 °C. Then, 20 μL of substrate solution (p-Nitrophenyl-α-D-glucopyranoside, 2.75 mM in phosphate buffer (pH 6.9, 0.1 M)) was added and the plates were incubated for an additional 15 min, at 37 °C. The reaction was stopped with 80 μL of 0.2 M Na_2CO_3. The absorbance of the wells was measured in a microplate reader (FLUOstar® Omega Plate Reader, BMG Labtech, Ortenberg, Germany) at 405 nm. Samples' blanks were prepared (SB) without α-glucosidase and the reaction mixture without peptide fraction was used as negative control (NC). The enzyme inhibitory rate was calculated as follows:

$$\text{Inhibition (\%)} = [(\text{Abs}_{\text{NC}} - (\text{Abs}_{\text{sample}} - \text{Abs}_{\text{SB}}) \times 100/(\text{Abs}_{\text{NC}})]$$

Acarbose was used as positive control. Values were assessed in quadruplicate, and the results were expressed as the final concentration (μg/mL) in the reaction mixture, which reduces the enzyme activity by 20% (IC_{20}). Data are presented as means ± standard deviations.

4.8. Statistical Analysis

The experimental data were analysed in GraphPad Prism v8.2.1. Data were expressed as mean ± standard deviation (SD), except when specified otherwise, of at least three independent experiments. When the *p*-value between two results was lower than 0.05, it was considered statistically significant.

5. Conclusions

In conclusion, this study demonstrates that the 2–10 kDa fraction has an antimicrobial, antioxidant, anti-inflammatory, and antidiabetic activity. Although the present study constitutes the first attempt at investigating the 2–10 kDa fraction activities, further studies are needed to confirm its efficacy and to evaluate its safety in food and health care. Overall, this research represents a significant advance in the evaluation of the functional biological properties of *S. cerevisiae* bioactive peptides.

Supplementary Materials: The following supporting information can be downloaded at: https://www.mdpi.com/article/10.3390/antibiotics12081332/s1, Figure S1: Calibration curve and equation of DPPH test (A) and FRAP method (B). Relative antioxidant activity was expressed as percentage of absorbance decrease and subsequently calculated to equivalent content of Trolox.

Author Contributions: E.M.M., C.P. and P.B. conceived and designed the research. A.C., C.R.-R., E.M.M., D.V., M.P.D. and P.B. conducted experiments. A.R.F., C.P., E.M.M., M.P.D. and P.B. analyzed data. A.R.F. gave a special contribution in the conception of the cytotoxicity and anti-inflammatory studies in human cells, C.R.-R. conducted these experiments, both analysed the data and wrote, revised, and edited this part of the manuscript. A.C., C.P., E.M.M., M.P.D. and P.B. wrote and edited the manuscript. E.M.M., C.P., M.P.D. and P.B. revised the manuscript. Funding acquisition: A.R.F., C.P., M.P.D. and P.B. All authors have read and agreed to the published version of the manuscript.

Funding: The present work was financed by Research Unit, Linking Landscape, Environment, Agriculture and Food (LEAF), in the scope of the project LEAF, UIDB/04129/2020 and by national funds from FCT—Fundação para a Ciência e a Tecnologia, I.P., in the scope of the project UIDP/04378/2020 and UIDB/04378/2020 of the Research Unit on Applied Molecular Biosciences—UCIBIO and the project LA/P/0140/2020 of the Associate Laboratory Institute for Health and Bioeconomy—i4HB. This work was also supported by the Mechanical Engineering and Resource Sustainability Center- MEtRICs which is funded by national funds from FCT/MCTES (UIDB/04077/2020 and UIDP/04077/2020).

Data Availability Statement: The data presented in this study are available in the manuscript and supplementary material.

Acknowledgments: The authors thank to LEAF and to FCT for the financial support of this work.

Conflicts of Interest: The authors declare they have no conflict of interest.

References

1. Bhunia, A.K. *Foodborne Microbial Pathogens: Mechanisms and Pathogenesis*, 2nd ed.; Springer: New York, NY, USA, 2018; pp. 1–23.
2. Brenier-Pinchart, M.P.; Faure, O.; Garban, F.; Fricker-Hidalgo, H.; Mallaret, M.R.; Trens, A.; Lebeau, B.; Pelloux, H.; Grillot, R. Ten-year surveillance of fungal contamination of food within a protected haematological unit. *Mycoses* **2006**, *49*, 421–425. [CrossRef] [PubMed]
3. Pitt, I.J.; Hocking, A.D. *Fungi and Food Spoilage*, 3rd ed.; Springer: New York, NY, USA, 2009; p. 519.
4. Hussain, M.A.; Dawson, C.O. Economic Impact of Food Safety Outbreaks on Food Businesses. *Foods* **2013**, *2*, 585–589. [CrossRef] [PubMed]
5. Rasko, D.A.; Webster, D.R.; Sahl, J.W.; Bashir, A.; Boisen, N.; Scheutz, F.; Nataro, J.P. Origins of the *E. coli* strain causing an outbreak of hemolytic–uremic syndrome in Germany. *N. Engl. J. Med.* **2011**, *365*, 709–717. [CrossRef]
6. Majowicz, S.E.; Scallan, E.; Jones-Bitton, A.; Sargeant, J.M.; Stapleton, J.; Angulo, F.J.; O'Brien, S.J. Global incidence of human Shiga toxin–producing *Escherichia coli* infections and deaths: A systematic review and knowledge synthesis. *Foodborne Pathog. Dis.* **2014**, *11*, 447–455. [CrossRef]
7. Farber, J.M.; Peterkin, P.I. *Listeria monocytogenes*, a food-borne pathogen. *Microbiol. Rev.* **1991**, *55*, 476–511. [CrossRef]
8. Todd, E.C. Epidemiology of foodborne diseases: A worldwide review. *World Health Stat. Q.* **1997**, *50*, 30–50. [PubMed]
9. Munck, N.; Leekitcharoenphon, P.; Litrup, E.; Kaas, R.; Meinen, A.; Guillier, L.; Tang, Y.; Malorny, B.; Palma, F.; Borowiak, M.; et al. Four European *Salmonella typhimurium* datasets collected to develop WGS-based source attribution methods. *Sci. Data* **2020**, *7*, 75. [CrossRef]
10. Scallan, E.; Hoekstra, R.M.; Angulo, F.J.; Tauxe, R.V.; Widdowson, M.A.; Roy, S.L.; Jones, J.L.; Griffin, P.M. Foodborne illness acquired in the United States—Major pathogens. *Emerg. Infect. Dis.* **2011**, *17*, 7–15. [CrossRef] [PubMed]
11. Zeng, Q.; Liu, X.; Liu, Y.; Wang, Q.; Zhang, Z. Occurrence and characteristics of *Candida* spp. in vegetables and fresh-cut fruits. *J. Food Saf.* **2020**, *40*, e12717. [CrossRef]
12. Diba, K.; Ehsani, A.; Khodadadi, E. Prevalence of *Candida albicans* in raw meat products in Tehran. *Iran. J. Food Saf.* **2020**, *40*, e12760.
13. Khodavaisy, S.; Nabili, M.; Davari, B.; Vahedi, M.; Aslani, N. The prevalence of *Candida* species in vegetables and characterization of isolates by ITS-RFLP analysis. *Curr. Med. Mycol.* **2017**, *3*, 17–22. [CrossRef]
14. Deak, T.; Beuchat, L.R. *Handbook of Food Spoilage Yeasts*; CRS Press: New York, NY, USA, 1996; pp. 111–154.
15. Casey, G.D.; Dobson, A.D. Molecular detection of *Candida krusei* contamination in fruit juice using the citrate synthase gene cs1 and a potential role for this gene in the adaptive response to acetic acid. *J. Appl. Microbiol.* **2003**, *95*, 13–22. [CrossRef] [PubMed]
16. Pelletier, R.; Alarie, I.; Lagacé, R.; Walsh, T.J. Emergence of disseminated candidiasis caused by *Candida krusei* during treatment with caspofungin: Case report and review of literature. *Med. Mycol. J.* **2005**, *43*, 559–564. [CrossRef]

17. Freedman, B.J. Sulphur dioxide in foods and beverages: Its use as a preservative and its effect on asthma. *Br. J. Dis. Chest* **1980**, *74*, 128–134. [CrossRef]
18. Piper, J.D.; Piper, P.W. Benzoate and Sorbate Salts: A Systematic Review of the Potential Hazards of These Invaluable Preservatives and the Expanding Spectrum of Clinical Uses for Sodium. *Compr. Rev. Food Sci. Food Saf.* **2017**, *16*, 868–880. [CrossRef]
19. Vally, H.; Misso, N.L.; Madan, V. Clinical effects of sulphite additives. Clinical and experimental allergy. *Clin. Exp. Allergy* **2009**, *39*, 1643–1651. [CrossRef]
20. Asioli, D.; Aschemann-Witzel, J.; Caputo, V.; Vecchio, R.; Annunziata, A.; Næs, T.; Varela, P. Making sense of the "clean label" trends: A review of consumer food choice behavior and discussion of industry implications. *Food Res. Int.* **2017**, *99 Pt 1*, 58–71. [CrossRef]
21. Hancock, R.E.; Sahl, H.G. Antimicrobial and host-defense peptides as new anti-infective therapeutic strategies. *Nat. Biotechnol.* **2006**, *24*, 1551–1557. [CrossRef]
22. Cotter, P.D.; Hill, C.; Ross, R.P. Bacteriocins: Developing innate immunity for food. *Nat. Rev. Microbiol.* **2005**, *3*, 777–788. [CrossRef]
23. Nascimento, B.L.; Delabeneta, M.F.; Rosseto, L.R.B.; Junges, D.S.B.; Paris, A.P.; Persel, C.; Gandra, R.F. Yeast Mycocins: A great potential for application in health. *FEMS Yeast Res.* **2020**, *20*, foaa016. [CrossRef]
24. Albergaria, H.; Francisco, D.; Gori, K.; Arneborg, N.; Gírio, F. *Saccharomyces cerevisiae* CCMI 885 secretes peptides that inhibit the growth of some non-*Saccharomyces* wine-related strains. *Appl. Microbiol. Biotechnol.* **2010**, *86*, 965–972. [CrossRef] [PubMed]
25. Branco, P.; Francisco, D.; Chambon, C.; Hébraud, M.; Arneborg, N.; Almeida, M.G.; Caldeira, J.; Albergaria, H. Identification of novel GAPDH-derived antimicrobial peptides secreted by *Saccharomyces cerevisiae* and involved in wine microbial interactions. *Appl. Microbiol. Biotechnol.* **2014**, *98*, 843–853. [CrossRef] [PubMed]
26. Branco, P.; Francisco, D.; Monteiro, M.; Almeida, M.G.; Caldeira, J.; Arneborg, N.; Prista, C.; Albergaria, H. Antimicrobial properties and death-inducing mechanisms of Saccharomycin, a biocide secreted by *Saccharomyces cerevisiae*. *Appl. Microbiol. Biotechnol.* **2017**, *101*, 159–171. [CrossRef]
27. Branco, P.; Coutinho, R.; Malfeito-Ferreira, M.; Prista, C.; Albergaria, H. Wine spoilage control: Impact of Saccharomycin on *Brettanomyces bruxellensis* and its conjugated effect with sulfur dioxide. *Microorganisms* **2021**, *9*, 2528. [CrossRef]
28. Al-Sahlany, S.T.G.; Altemimi, A.B.; Al-Manhel, A.J.A.; Niamah, A.K.; Lakhssassi, N.; Ibrahim, S.A. Purification of Bioactive Peptide with Antimicrobial Properties Produced by *Saccharomyces cerevisiae*. *Foods* **2020**, *9*, 324. [CrossRef]
29. Akbarian, M.; Khani, A.; Eghbalpour, S.; Uversky, V.N. Bioactive Peptides: Synthesis, Sources, Applications, and Proposed Mechanisms of Action. *Int. J. Mol. Sci.* **2022**, *23*, 1445. [CrossRef]
30. Mudgil, P.; Kamal, H.; Yuen, G.C.; Maqsood, S. Characterization and identification of novel antidiabetic and anti-obesity peptides from camel milk protein hydrolysates. *Food Chem.* **2018**, *259*, 46–54. [CrossRef]
31. Sun, J.; He, H.; Xie, B.J. Novel Antioxidant Peptides from Fermented Mushroom *Ganoderma lucidum*. *J. Agric. Food Chem.* **2004**, *52*, 6646–6652. [CrossRef] [PubMed]
32. Heleno, S.A.; Barros, L.; Martins, A.; Queiroz, M.J.R.P.; Santos-Buelga, C.; Ferreira, I.C.F.R. Fruiting body, spores and in vitro produced mycelium of *Ganoderma lucidum* from Northeast Portugal: A comparative study of the antioxidant potential of phenolic and polysaccharidic extracts. *Food Res. Int.* **2012**, *46*, 135–140. [CrossRef]
33. Mishra, J.; Rajput, R.; Singh, K. Antibacterial Natural Peptide Fractions from Indian *Ganoderma lucidum*. *Int. J. Pept. Res. Ther.* **2018**, *24*, 543–554. [CrossRef]
34. Kelebek, H.; Selli, S.; Canbas, A.; Cabaroglu, T. HPLC determination of organic acids, sugars, phenolic compositions and antioxidant capacity of orange juice and orange wine made from a Turkish cv. *Kozan. Microchem. J.* **2009**, *91*, 187–192. [CrossRef]
35. Stinco, C.M.; Fernández-Vázquez, R.; Hernanz, D.; Heredia, F.J.; Meléndez-Martínez, A.J.; Vicario, I.M. Industrial orange juice debittering: Impact on bioactive compounds and nutritional value. *J. Food Eng.* **2013**, *116*, 155–161. [CrossRef]
36. Muchagato Maurício, E.; Rosado, C.; Duarte, M.P.; Fernando, A.L.; Díaz-Lanza, A.M. Evaluation of industrial sour cherry liquor wastes as an ecofriendly source of added value chemical compounds and energy. *Waste Biomass Valorization* **2020**, *11*, 201–210. [CrossRef]
37. Correia, R.; Duarte, M.P.; Maurício, E.M.; Brinco, J.; Quintela, J.C.; da Silva, M.G.; Gonçalves, M. Chemical and Functional Characterization of Extracts from Leaves and Twigs of Acacia dealbata. *Processes* **2022**, *10*, 2429. [CrossRef]
38. Guo, Q.; Chen, P.; Chen, X. Bioactive Peptides Derived from Fermented Foods: Preparation and Biological Activities. *J. Funct. Foods* **2023**, *101*, 105422. [CrossRef]
39. Oliveira, A.S.; Pereira, J.O.; Ferreira, C.; Faustino, M.; Durão, J.; Pintado, M.E.; Carvalho, A.P. Peptide-rich extracts from spent yeast waste streams as a source of bioactive compounds for the nutraceutical market. *Innov. Food Sci. Emerg. Technol.* **2022**, *81*, 103148. [CrossRef]
40. Lopes, A.; Azevedo-Silva, J.; Carsanba, E.; Pintado, M.; Oliveira, A.S.; Ferreira, C.; Oliveira, C. Peptide extract from spent yeast improves resistance of *Saccharomyces cerevisiae* to oxidative stress. *Appl. Microbiol. Biotechnol.* **2023**, *107*, 3405–3417. [CrossRef]
41. Diplock, A.T. Antioxidants and disease prevention. *Mol. Aspects Med.* **1994**, *15*, 293–376. [CrossRef]
42. Halliwell, B. Free radicals and antioxidants: Updating a personal view. *Nutr. Rev.* **2012**, *70*, 257–265. [CrossRef]
43. Morén, C.; deSouza, R.M.; Giraldo, D.M.; Uff, C. Antioxidant Therapeutic Strategies in Neurodegenerative Diseases. *Int. J. Mol. Sci.* **2022**, *23*, 9328. [CrossRef] [PubMed]

. Duarte, M.C.T.; Leme, E.E.; Delarmelina, C.; Soares, A.A.; Figueira, G.M.; Sartoratto, A. Activity of essential oils from Brazilian medicinal plants on *Escherichia coli*. *J. Ethnopharmacol.* **2007**, *111*, 197–201. [CrossRef] [PubMed]

5. Vieitez, I.; Maceiras, L.; Jachmanián, I.; Alborés, S. Antioxidant and antibacterial activity of different extracts from herbs obtained by maceration or supercritical technology. *J. Supercrit. Fluids* **2018**, *133*, 58–64. [CrossRef]

5. Pereira, P.; Mauricio, E.M.; Duarte, M.P.; Lima, K.; Fernandes, A.S.; Bernardo-Gil, G.; Cebola, M.J. Potential of supercritical fluid myrtle extracts as an active ingredient and co-preservative for cosmetic and topical pharmaceutical applications. *Sustain. Chem. Pharm.* **2022**, *28*, 100739. [CrossRef]

7. Sharma, S. Food Preservatives and their harmful effects. *Int. J. Sci. Res. Publ.* **2015**, *5*, 1–2.

3. EFSA Panel on Additives and Products or Substances used in Animal Feed. Scientific opinion on the safety and efficacy of nisin (E 234) as a feed additive for all animal species. *EFSA J.* **2017**, *15*, 4749. [CrossRef]

9. Bodley, M.D. Application of Bacteriocins in the Preservation of Fruit Juice. Ph.D. Thesis, Faculty of Science at the Nelson Mandela Metropolitan University, Gqeberha, South Africa, January 2015.

). Gharsallaoui, A.; Oulahal, N.; Joly, C.; Degraeve, P. Nisin as a Food Preservative: Part 1: Physicochemical Properties, Antimicrobial Activity, and Main Uses. *Crit. Rev. Food Sci. Nutr.* **2016**, *56*, 1262–1274. [CrossRef] [PubMed]

1. Campion, A.; Casey, P.G.; Field, D.; Cotter, P.D.; Hill, C.; Ross, R.P. In vivo activity of nisin A and nisin V against *Listeria monocytogenes* in mice. *BMC Microbiol.* **2013**, *13*, 23. [CrossRef]

2. Madrazo, A.L.; Segura Campos, M.R. Review of antimicrobial peptides as promoters of food safety: Limitations and possibilities within the food industry. *J. Food Saf.* **2020**, *40*, e12854. [CrossRef]

3. Maurício, E.; Rosado, C.; Duarte, M.P.; Verissimo, J.; Bom, S.; Vasconcelos, L. Efficiency of Nisin as Preservative in Cosmetics and Topical Products. *Cosmetics* **2017**, *4*, 41. [CrossRef]

4. Delves-Broughton, J.P.; Blackburn, P.; Evans, R.J.; Hugenholtz, J. Application of the bacteriocin, nisin. *Anton Leeuw* **1996**, *69*, 193–202. [CrossRef]

5. Golubev, W.I. Mycocins (killer toxins). In *The Yeasts*, 4th ed.; Kurtzman, C.P., Fell, J.W., Eds.; Elsevier: Amsterdam, The Netherlands, 1998; pp. 55–62.

5. Izgü, F.; Altinbay, D.; Türeli, A.E. In vitro susceptibilities of *Candida* spp. to Panomycocin, a novel exo-beta-1,3-glucanase isolated from *Pichia anomala* NCYC 434. *Microbiol. Immunol.* **2017**, *51*, 797–803. [CrossRef] [PubMed]

7. Al-Qaysi, S.; Al-Haideri, H.; Thabit, Z.A.; Al-Kubaisy, W.; Ibrahim, J. Production, Characterization, and Antimicrobial Activity of Mycocin Produced by *Debaryomyces hansenii* DSMZ70238. *Int. J. Microbiol.* **2017**, *2017*, 2605382. [CrossRef] [PubMed]

3. Comitini, F.; Pietro, N.D.; Zacchi, L.; Mannazzu, I.; Ciani, M. *Kluyveromyces phaffii* killer toxin active against wine spoilage yeasts: Purification and characterization. *Microbiology* **2004**, *150*, 2535–2541. [CrossRef]

9. Ciani, M.; Fatichenti, F. Killer toxin of *Kluyveromyces phaffii* DBVPG 6076 as a biopreservative agent to control apiculate wine Yeasts. *Appl. Environ. Microbiol.* **2001**, *67*, 3058–3063. [CrossRef]

). De Ingeniis, J.; Raffaelli, N.; Ciani, M.; Mannazzu, I. *Pichia anomala* DBVPG 3003 secretes a ubiquitin-like protein that has antimicrobial activity. *Appl. Environ. Microbiol.* **2009**, *75*, 1129–1134. [CrossRef]

1. Mirzaei, M.; Mirdamadi, S.; Ehsani, M.R.; Aminlari, M.; Hosseini, E. Purification and Identification of Antioxidant and ACE-Inhibitory Peptide from *Saccharomyces cerevisiae* Protein Hydrolysate. *J. Funct. Foods* **2015**, *19*, 259–268. [CrossRef]

2. Pereira, P.R.; Freitas, C.S.; Paschoalin, V.M.F. *Saccharomyces cerevisiae* Biomass as a Source of Next-generation Food Preservatives: Evaluating Potential Proteins as a Source of Antimicrobial Peptides. *Compr. Rev. Food Sci. Food Saf.* **2021**, *20*, 4450–4479. [CrossRef]

3. Pereira, P.; Bernardo-Gil, M.G.; Cebola, M.J.; Mauricio, E.; Romano, A. Supercritical fluid extracts with antioxidant and antimicrobial activities from myrtle (*Myrtus communis* L.) leaves. Response surface optimization. *J. Supercrit. Fluids* **2013**, *83*, 57–64. [CrossRef]

4. Loebler, M.; Sánchez, C.; Muchagato Maurício, E.; Diogo, E.; Santos, M.; Vasilenko, P.; Duarte, M.P. Potential application of propolis extracts to control the growth of *Stemphylium vesicarium* in "Rocha" pear. *Appl. Sci.* **2020**, *10*, 1990. [CrossRef]

5. Zou, T.B.; He, T.P.; Li, H.B.; Tang, H.W.; Xia, E.Q. The Structure-Activity Relationship of the Antioxidant Peptides from Natural Proteins. *Molecules* **2016**, *21*, 72. [CrossRef]

5. Hur, S.J.; Lee, S.Y.; Kim, Y.C.; Choi, I.; Kim, G.B. Effect of Fermentation on the Antioxidant Activity in Plant-Based Foods. *Food Chem.* **2014**, *160*, 346–356. [CrossRef] [PubMed]

7. Mirzaei, M.; Shavandi, A.; Mirdamadi, S.; Soleymanzadeh, N.; Motahari, P.; Mirdamadi, N.; Moser, M.; Subra, G.; Alimoradi, H.; Goriely, S. Bioactive Peptides from Yeast: A Comparative Review on Production Methods, Bioactivity, Structure-Function Relationship, and Stability. *Trends Food Sci. Technol.* **2021**, *118*, 297–315. [CrossRef]

8. Brand-Williams, W.; Cuvelier, M.E.; Berset, C. Use of a Free Radical Method to Evaluate Antioxidant Activity. *LWT Food Sci. Technol.* **1995**, *28*, 25–30. [CrossRef]

9. Benzie, I.F.F.; Strain, J.J. The Ferric Reducing Ability of Plasma (FRAP) as a Measure of "Antioxidant Power": The FRAP Assay. *Anal. Biochem.* **1996**, *239*, 70–76. [CrossRef]

0. Maher, S.; McClean, S. Investigation of the cytotoxicity of eukaryotic and prokaryotic antimicrobial peptides in intestinal epithelial cells in vitro. *Biochem. Pharmacol.* **2006**, *71*, 1289–1298. [CrossRef]

1. Antony, P.; Vijayan, R. Bioactive peptides as potential nutraceuticals for diabetes therapy: A comprehensive review. *Int. J. Mol. Sci.* **2021**, *22*, 9059. [CrossRef] [PubMed]

72. Rivero-Pino, F.; Espejo-Carpio, F.J.; Guadix, E.M. Antidiabetic food-derived peptides for functional feeding: Production, functionality and in vivo evidences. *Foods* **2020**, *9*, 983. [CrossRef] [PubMed]
73. Romeiras, M.M.; Essoh, A.P.; Catarino, S.; Silva, J.; Lima, K.; Varela, E.; Moura, M.; Gomes, I.; Duarte, M.C.; Duarte, M.P. Diversity and biological activities of medicinal plants of Santiago island (Cabo Verde). *Heliyon* **2023**, *9*, e14651. [CrossRef]
74. Zhou, H.; Safdar, B.; Li, H.; Yang, L.; Ying, Z.; Liu, X. Identification of a novel α-amylase inhibitory activity peptide from quinoa protein hydrolysate. *Food Chem.* **2023**, *403*, 134434. [CrossRef]
75. González-Montoya, M.; Hernández-Ledesma, B.; Mora-Escobedo, R.; Martínez-Villaluenga, C. Bioactive peptides from germinated soybean with anti-diabetic potential by inhibition of dipeptidyl peptidase-IV, α-amylase, and α-glucosidase enzymes. *Int. J. Mol. Sci.* **2018**, *19*, 2883. [CrossRef]
76. Wang, R.; Zhao, H.; Pan, X.; Orfila, C.; Lu, W.; Ma, Y. Preparation of bioactive peptides with antidiabetic, antihypertensive, and antioxidant activities and identification of α-glucosidase inhibitory peptides from soy protein. *Food Sci. Nutr.* **2019**, *7*, 1848–1856. [CrossRef] [PubMed]
77. *CLSI Document M07-A9*; Methods for Dilution Antimicrobial Susceptibility Tests for Bacteria that Grow Aerobically. Approved Standard—Ninth Edition. Clinical and Laboratory Standards Institute: Wayne, PA, USA, 2012.
78. Fernandes, A.R.; Mendonça-Martins, I.; Santos, M.F.A.; Raposo, L.R.; Mendes, R.; Marques, J.; Romão, C.C.; Romão, M.J.; Santos-Silva, T.; Baptista, P.V. Improving the Anti-inflammatory Response via Gold Nanoparticle Vectorization of CO-Releasing Molecules. *ACS Biomater. Sci. Eng.* **2020**, *6*, 1090–1101. [CrossRef] [PubMed]
79. Livak, K.J.; Schmittgen, T.D. Analysis of relative gene expression data using real-time quantitative PCR and the 2(-Delta Delta C(T)) Method. *Methods* **2001**, *25*, 4. [CrossRef] [PubMed]

antibiotics

MDPI

rticle

Rhamnolipid Nano-Micelles Inhibit SARS-CoV-2 Infection and Have No Dermal or Eye Toxic Effects in Rabbits

laa M. Ali [1], Harriet J. Hill [2], Gehad E. Elkhouly [3,4], Marwa Reda Bakkar [5], Nermeen R. Raya [3,4], ania Stamataki [2] and Yasmin Abo-zeid [3,4,*]

1 Department of Pathology, Faculty of Veterinary Medicine, Cairo University, Giza 12211, Egypt
2 Institute of Immunology and Immunotherapy, University of Birmingham, Birmingham B15 2TT, UK
3 Department of Pharmaceutics and Industrial Pharmacy, Faculty of Pharmacy, Helwan University, Cairo 11795, Egypt
4 Helwan Nanotechnology Center, Helwan University, Cairo 11792, Egypt
5 Botany and Microbiology Department, Faculty of Science, Helwan University, Cairo 11795, Egypt
* Correspondence: yasmin.abozeid@pharm.helwan.edu.eg; Tel.: +20-10-927-928-46

Abstract: Hand hygiene is considered to be the key factor in controlling and preventing infection, either in hospital care settings or in the community. Alcohol-based hand sanitizers are commonly used due to their rapid action and broad spectrum of microbicidal activity, offering protection against bacteria and viruses. However, their frequent administration during COVID-19 pandemic was associated with serious hazards, such as skin toxicity, including irritation, skin dermatitis, skin dryness or cracking, along with peeling redness or itching, with the higher possibility of getting infections. Thus, there is a need to find alternative and novel approaches for hand sanitation. In our previous publications, we reported that rhamnolipids nano-micelles had a comparable antibacterial activity to alcohol-based hand sanitizer and a lower cytotoxicity against human dermal fibroblast cells. In the current study, we investigated the antiviral activity of rhamnolipids nano-micelles against SARS-CoV-2. There was no cytotoxic effect on Vero cells noted at the tested concentrations of rhamnolipids nano-micelles. The rhamnolipids nano-micelles solution at 20, 78, and 312 μg/mL all demonstrated a significant ($p < 0.05$) decrease of virus infectivity compared to the virus only and the blank vehicle sample. In addition, an acute irritation test was performed on rabbits to further ascertain the biosafety of rhamnolipids nano-micelles. In the eye and skin irritation tests, no degree of irritation was recorded after topical application of rhamnolipids nano-micelles. In addition, histopathological, biomarker, and hematological analyses from animals treated with rhamnolipids nano-micelles were identical to those recorded for untreated animal. From the above, we can conclude that rhamnolipids nano-micelles are a good candidate to be used as a hand sanitizer instead of alcohol-based hand sanitizers. However, they must still be tested in the future among healthcare workers (HCW) in a health care setting to ascertain their antimicrobial efficacy and safety compared to alcohol-based hand sanitizers.

Keywords: SARS-CoV-2; rhamnolipids nano-micelles; hospital acquired infections; hand sanitation

1. Introduction

Hospital-acquired infections (HAIs) are considered to be a major cause of morbidity and mortality, and are the second most prevalent cause of death, globally [1]. Various multicentric studies presented that around 3.5 to 12% of hospitalized patients acquired at least one HAIs [2,3], with a possibility of 10 million deaths by 2050 due to HAIs [4]. The emergence and re-emergence of micro-organisms, including viruses, pose a serious threat to human health at a global level [5]. This was exemplified by the COVID-19 pandemic, due to SARS-CoV-2 and its variants, which resulted in a global death toll of 6,515,964 persons, as reported by the John Hopkin dashboard on 13 September 2022. The absence of standard treatment and the appearance of mutated strains of SARS-CoV-2 might render the developed vaccine inactive, therefore, a quick spread of infection among patients admitted to hospitals is expected [6]. This puts healthcare workers (HCWs) at a higher risk of getting the infection due to shedding of the infectious agents, from either infected patient or carrier patients to HCWs, who further spread the infection among others in the healthcare setting, as has already been reported [7–9].

The World Health Organization (WHO) advised people to adopt a healthy lifestyle, in addition to following prevention and control measures, to control HAIs including SARS-CoV-2 [6]. The latter involved keeping social distance, wearing masks, and keeping hand hygiene. Hand hygiene was reported to be a critical factor for controlling the spread of infections among HCWs and patients in hospital care settings [6,10,11]. Although washing hands with soap and water are effective for keeping hand hygiene, alcohol-based hand sanitizers are commonly used for hand disinfection in hospital settings due to their broad-spectrum antimicrobial activity against bacteria, viruses, parasites, and fungi [12,13]. However, frequent administration of alcohol-based hand sanitizers has been reported to be associated with several side effects, such as redness and cracking of skin, with the possibility of infection, as well as developing microbial resistance and virus outbreaks [6,11,14–19]. Based on this, it is highly recommended to find alternative approaches to replace alcohol-based hand sanitizers. Rhamnolipids are biosurfactants produced by *Pseudomonas aeruginosa* and could be produced efficiently and economically at industrial scale [20–22]. We previously reported the antibacterial activity of rhamnolipids nano-micelles against selected resistant Gram-negative and Gram-positive bacteria, and a docking study showed rhamnolipids nano-micelles to be effective against SARS-CoV-2 [6,23]. In addition, an in-vitro study confirmed the superior safety of rhamnolipids nano-micelles against human dermal fibroblast cells over alcohol-based hand sanitizer [23]. Thus, we previously recommended rhamnolipids nano-micelles to replace alcohol-based hand sanitizer for hand sanitization [23].

The antiviral activity of rhamnolipids mixtures was previously reported against the Severe Acute Respiratory Syndrome (SARS-CoV-2) strain VR PV10734 clinical isolate, CoV-229E (ATCC VR-740), HCoV-OC43 (ATCC VR-1558), HSV-1 strain SC16, fluorescent HSV-1 (GFP-HSV-1), HSV-2 strain 333, Poliovirus Type 1 (PV-1) strain Chat (ATCC VR-1562) [5]. However, the antiviral activity of rhamnolipids nano-micelles against SARS-CoV-2 has not yet been studied. Herein, we investigated the antiviral activity of rhamnolipids nano-micelles against replicating SARS-CoV-2. To expand on the previous experiments addressing the safety of rhamnolipids nano-micelles in-vitro on human dermal fibroblast cells [23], we also performed skin and eye irritation tests for rhamnolipids nano-micelles on rabbits.

2. Results

2.1. Production of Rhamnolipids

The successful production of rhamnolipids was confirmed using an ESI-MS spectrometer coupled with UPLC (LC/ESI-MS) in our previous publication [6,23]. The obtained rhamnolipids were composed of a higher proportion of mono-rhamnolipids to di-rhamnolipids (Table S1, Supplementary Materials).

2.2. Preparation and Characterization of Rhamnolipids Nano-Micelles

Rhamnolipid nano-micelles solution was prepared as described in Section 4.2.2. The particle size and zeta potential were recorded using a Malvern Zeta-sizer instrument (Malvern Instruments Ltd., Malvern, Worcestershire, UK), and are presented as an average diameter (D, nm) $\pm$ SD and an average zeta potential (mv) $\pm$ SD, respectively. The particle size of the nano-micelles solution prepared at 20, 78, and 312 µg/mL were 191 $\pm$ 22.16, 265 $\pm$ 33.56, and 188 $\pm$ 46.98, respectively. The polydispersity index (PDI) values were 0.33, 0.30, and 0.36 for samples prepared at concentration, 20, 78, and 312 µg/mL, respectively, and this is indicative of a monodisperse sample. The concentration of rhamnolipids had a non-significant ($p > 0.05$) effect on particle size. Zeta potential values were -47.93 ± 1.81, -35.23 ± 4.32, and -41.57 ± 13.70 for samples prepared at 20, 78, and 312 µg/mL, respectively, indicating highly stable samples.

Transmission electron microscopy (TEM) images of rhamnolipid nano-micelles solution prepared at 20, 78, and 312 µg/mL are shown in Figure 1. The images showed spherical nano-micelles with no sign of aggregation, and the particle size ranged from 84.5 to 95.5 nm, from 80.8 to 105, and from 76.5 to 114 nm for nano-micelles prepared at 20, 78, and 312 µg/mL, respectively. However, samples prepared at 20 µg/mL showed some particles of lower size. The particle size identified with TEM was smaller than that recorded with the Malvern Zeta-sizer instrument.

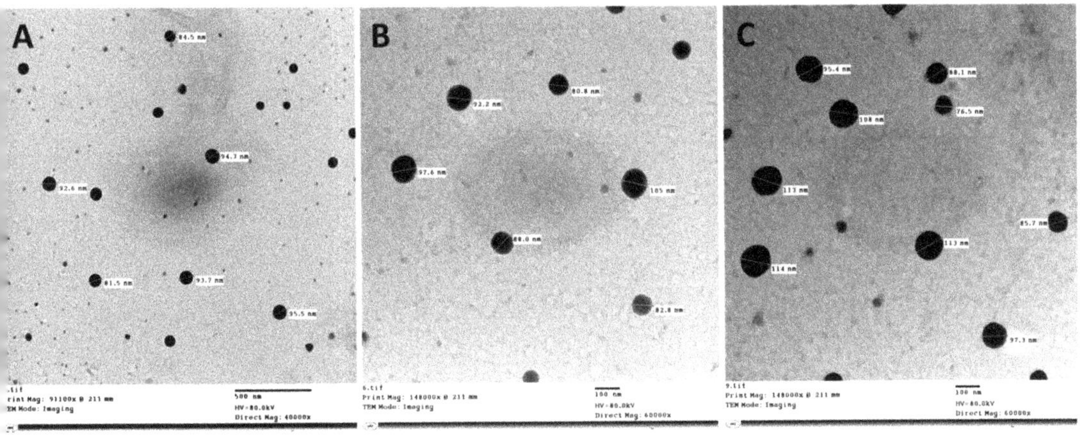

Figure 1. TEM images of rhamnolipid nano-micelles prepared at rhamnolipids concentration (**A**) 20, (**B**) 78, and (**C**) 312 µg/mL, respectively.

The pH for rhamnolipid nano-micelles prepared at different concentration of rhamnolipids was determined as described in Section 4.2.3 where pH value of all solution was very close to each other and was equivalent to 6.32 ± 0.02, 6.22 ± 0.05, and 6.43 ± 0.1 for samples prepared at 20, 78, and 312 µg/mL, respectively.

2.3. *Antiviral Activity of Rhamnolipids Nano-Micelles*

The antiviral activity of rhamnolipids nano-micelles against SARS-CoV-2 are presented in Figure 2. Vero cells are the gold standard for coronavirus infection in vitro and support the full virus life cycle for SARS-CoV-2. To assess antiviral activity of rhamnolipids nano-micelles, we infected Vero cells with SARS-CoV-2 Wuhan (Public Health England), as described in Section 4.2.4. The virus was pre-treated for 1 h with blank PBS (10 mM, pH 7.4) solution, a vehicle was used to disperse rhamnolipids nano-micelles or with the indicated concentrations of nano-micelles. Representative images of infected cells are shown in Figure 2A, and quantification of the results in Figure 2B. There was no cytotoxic effect noted at the tested concentrations. Rhamnolipids nano-micelles solution at 20, 78, and 312 µg/mL all demonstrated a significant ($p < 0.05$) decrease of virus infectivity compared to virus only and the blank solution sample.

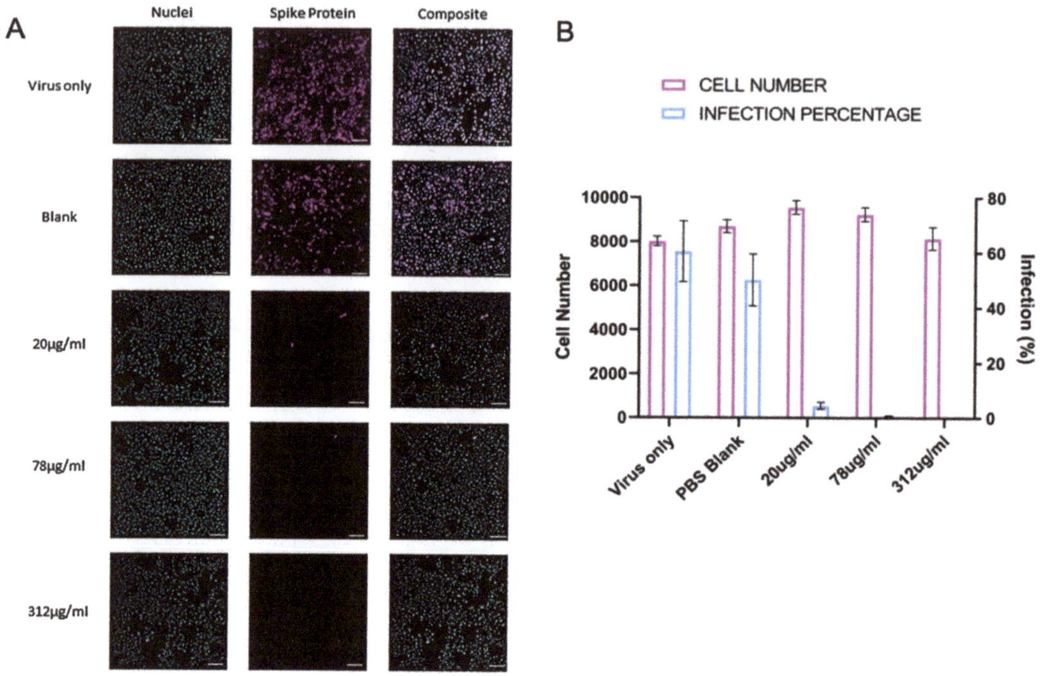

Figure 2. Antiviral activity of rhamnolipids nano-micelles solution prepared at 20, 78, and 312 μg/mL against SARS-CoV-2. (**A**) Representative immunofluorescence images for infected Vero cells where nuclei were detected by Hoechst 33342 (Cyan) and SARS-CoV-2 spike protein by rabbit anti-spike antibody detected with anti-rabbit-Alexa-555 secondary antibody (magenta). Scale bars represent 100 μm. (**B**) Automated quantification of cell number per field of view and % infected cells 48 h after infection. Results are the average of three independent experiments, with three replicates in each. Error Bars represent standard deviation.

2.4. Irritation and Skin Sensitization Test

2.4.1. Skin Irritation Test

In the skin irritation test, the skin was observed for signs of irritation, e.g., erythema/edema based on the Magnusson and Kligman scale [24]. The items of skin irritation on the scoring charts consisted of erythema, eschar, and edema. The degree of irritation on the skin was checked twice over three days, once at 24 h and once at 72 h. The obtained results, presented in Table 1, demonstrate the absence of any differences between untreated (control) and treated sites for both abraded and intact skin regarding all items checked: erythema, eschar, and edema. Additional data concerning skin irritation was provided in Supplementary Materials (Table S2).

During the test, the control sites for the abraded skin were more reddened compared to intact skin due to being scratched with the needle of the syringe, however, after 45 min, the redness disappeared, and the skin returned to its original state. As presented in Figure 3, control (untreated) and treated sites for both abraded and intact skin were lacking any abnormal signs, where eschar, edema, and erythema were not observed on the back of the rabbits. The photos presented in Figure 3 were taken at the end of the test (after 72 h) for the intact and abraded sites of untreated site (control) and sites treated with rhamnolipids nano-micelles (0.625 mg/mL), and PBS (10 mM, pH 7.4). Additional images presenting skin irritation was provided in Supplementary Materials (Figure S1).

Table 1. Skin irritation evaluation after treatment with rhamnolipids nano-micelles solution and PBS (10 mM, pH 7.4), vehicle used to disperse rhamnolipids nano-micelles *.

		Control Site								Treated Site							
		Erythema and Eschar				Edema				Erythema and Eschar				Edema			
		Intact		Abraded		Intact		Abraded		Intact		Abraded		Intact		Abraded	
Tested sample/tested time (h)	Animal Number	24	72	24	72	24	72	24	72	24	72	24	72	24	72	24	72
Rhamnolipids nano-micelles (0.625 μg/mL)	1	0	0	0	0	0	0	0	0	0	0	0	0	0	0	0	0
	2	0	0	0	0	0	0	0	0	0	0	0	0	0	0	0	0
	3	0	0	0	0	0	0	0	0	0	0	0	0	0	0	0	0
PBS (10 mM, pH 7.4)	4	0	0	0	0	0	0	0	0	0	0	0	0	0	0	0	0
	5	0	0	0	0	0	0	0	0	0	0	0	0	0	0	0	0
	6	0	0	0	0	0	0	0	0	0	0	0	0	0	0	0	0

* All skin irritation ratings were 0 after 24 and 72 h, both intact and abraded skin showed no symptoms (erythema, eschar, and oedema) indicating incidence of irritation.

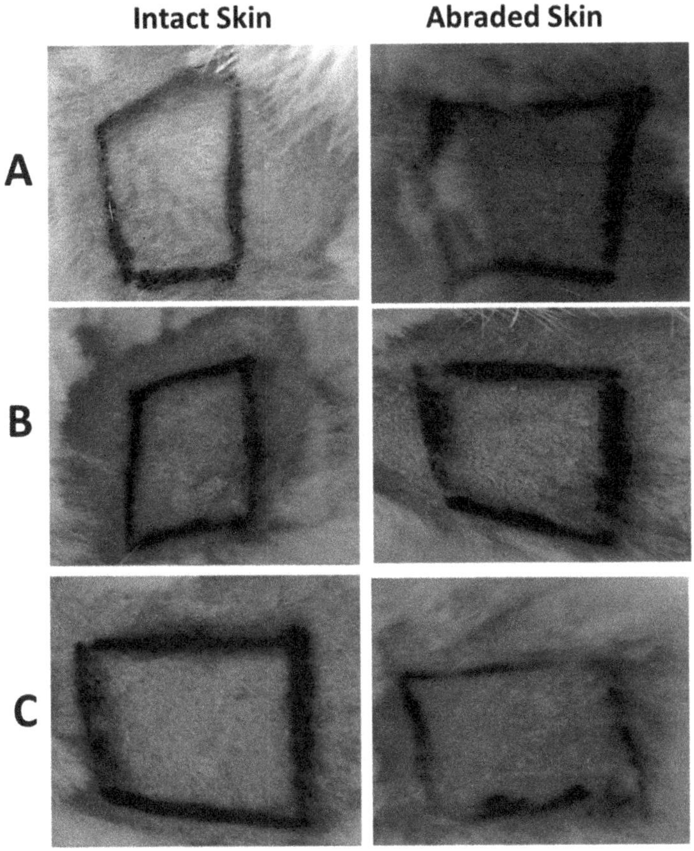

Figure 3. Representative photos of the skin irritation test for one representative rabbit for (**A**) Group I; untreated group, act as a control, (**B**) Group II; rhamnolipids nano-micelles solution (0.625 mg/mL) dispersed in PBS (10 mM, pH 7.4) treated group, and (**C**) Group III; PBS (10 mM, pH 7.4) treated group. Black lines were drawn with a non-irritating pen. No differences in score were observed after 72 h between the test and control site for both abraded and intact skin.

2.4.2. Eye Irritation Test

The observations of eye cytotoxicity were conducted at 24 and 72 h after initiating the test for both the untreated animal and for the animal treated with tested solutions of rhamnolipids nano-micelles (0.625 mg/mL), PBS (10 mM, pH 7.4), the vehicle used to disperse nano-micelles. The assessment of eye irritation was conducted according to the histological grading system for eye irritation [25]. Different grading from 0 to 3 was carried out on the basis of the absence or presence, and severity of, symptoms, including inflammation, redness, and tearing. The scoring chart of the guideline involved checking the following: corneal opacity, reactivity of iris, chemosis, and discharge of eyes after the sample instillation into the eye (treated sites). The obtained data are presented in Table 2. The instillation of samples into the eye was not associated with any irritation or abnormality in the treated eye after 24 and 72 h of sample application compared to the untreated eye. Additional data concerning eye irritation was provided in Supplementary Materials (Table S3).

Table 2. Evaluation of eye irritations following treatment with rhamnolipids nano-micelles solution and PBS, vehicle used to disperse rhamnolipids nano-micelles *.

Tested Solution	Tissues Examined in the Eye	Number of Rabbits					
		1		2		3	
		RT. Untreated	LT. Treated	RT. Untreated	LT. Treated	RT. Untreated	LT. Treated
Rhamnolipids nano-micelles (0.625 mg/mL)	Cornea	0	0	0	0	0	0
	Iris	0	0	0	0	0	0
	Conjunctiva	0	0	0	0	0	0
PBS (10 mM, pH 7.4)	Cornea	0	0	0	0	0	0
	Iris	0	0	0	0	0	0
	Conjunctiva	0	0	0	0	0	0

* All eye irritation scores were 0. The observations were concerned with corneal opacity, reactivity of iris, conjunctival edema, and ocular discharge. No symptoms indicated irritations were noted.

The untreated (negative control) and treated eyes were imaged after 72 h and the images are presented in Figure 4. The pupil sizes differ due to the varying light levels while taking photos. As can be seen, the images of the treated eyes, in comparison to the untreated eyes, demonstrated no differences between them. This indicates the absence of any abnormality that might occur to the cornea, iris, and conjunctiva due to treatment with tested solutions, rhamnolipids nano-micelles (0.625 mg/mL) and PBS (10 mM, pH 7.4). Additional images presenting eye irritation was provided in Supplementary Materials (Figure S2).

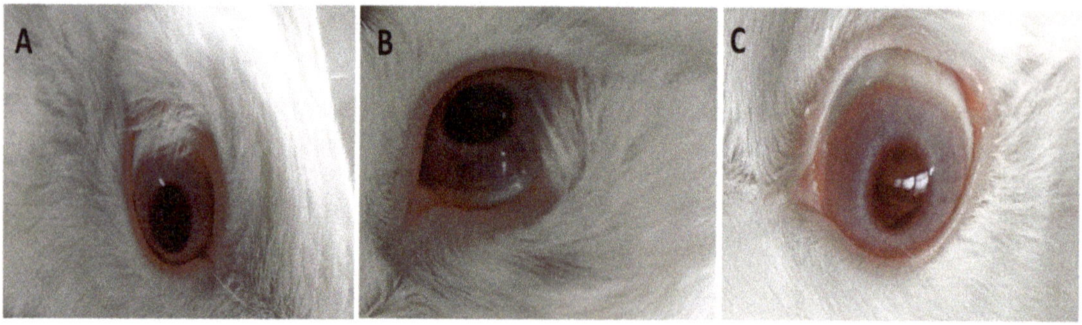

Figure 4. Representative photos of the eye irritation test for one representative rabbit for (**A**) Group I; untreated group, act as a control, (**B**) Group II; PBS (10 mM, pH 7.4) treated group, and (**C**) Group III; rhamnolipids nano-micelles solution (0.625 mg/mL) dispersed in PBS (10 mM, pH 7.4) treated group. The cornea, iris, and conjunctiva were observed after 72 h. In this study, no differences of cornea, iris and conjunctiva were observed among different groups.

2.4.3. Skin Sensitization Test

In the skin sensitization test, the skin was observed for signs of irritation, e.g., erythema/edema based on the scoring system of the Magnusson and Kligman grading score [24]. These parameters were checked once at 24, 48, and 72 h after the intradermal injection of tested solutions, and the results are presented in Table 3 and Figure 5. The obtained results revealed that animals treated with rhamnolipids nano-micelles (0.625 mg/mL) showed no clinical signs of sensitization and were identical to the negative control, contrary to marked signs of skin sensitization identified with positive control. This included redness, edema and erythema.

Table 3. * Skin sensitization scores recorded for guinea pigs injected intradermally with rhamnolipids nano-micelles solution versus positive and negative control **.

Hours/ Number of Animals	Positive Control Erythema/Edema			Rhamnolipids Nano-Micelles (0.625 mg/mL) Erythema/Edema			Negative Control Erythema/Edema		
	24	48	72	24	48	72	24	48	72
1	0/0	2/1	2/2	0/0	0/0	0/0	0/0	0/0	0/0
2	0/0	2/1	2/2	0/0	0/0	0/0	0/0	0/0	0/0
3	0/0	2/1	2/3	0/0	0/0	0/0	0/0	0/0	0/0
4	0/0	3/1	2/3	0/0	0/0	0/0	0/0	0/0	0/0
5	0/0	3/1	3/3	0/0	0/0	0/0	0/0	0/0	0/0
6				0/0	0/0				
7				0/0	0/0				
8				0/0	0/0				
9				0/0	0/0				
10				0/0	0/0				

* All skin sensitization ratings were 0 after 24, 48 and 72 h for all animals treated with rhamnolipids nano-micelles solution and they were identical to negative control. This indicated the absence of any signs of sensitization and the safety of rhamnolipids nano-micelles. ** Positive control; animals injected intradermally with skin sensitizing agent, mixture of Freunds's Complete Adjuvant (FCA), and physiological buffered saline (PBS) (1:1 mixture, 0.1 mL); Negative control; animals injected intradermally with physiological buffered saline (PBS, 0.1 mL).

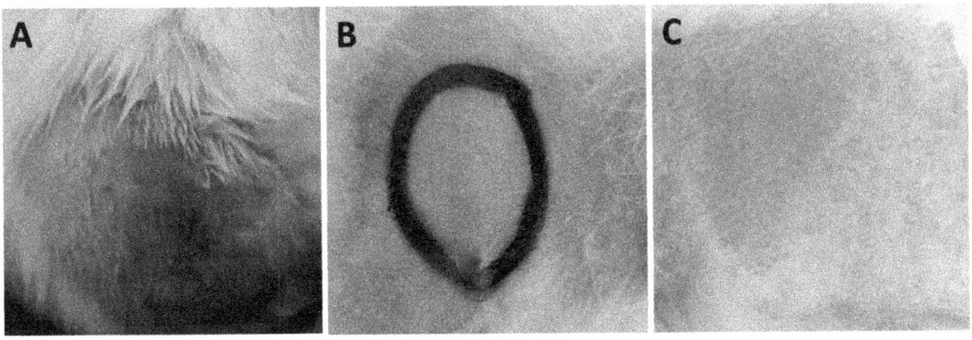

Figure 5. Representative photos of skin sensitization test for one representative guinea pigs for (**A**) Positive control; animals injected intradermally with skin sensitizing agent, a mixture of Freund's Complete Adjuvant (FCA), and physiological buffered saline (PBS) (1:1 mixture, 0.1 mL) (**B**) Rhamnolipids nano-micelles solution (0.625 mg/mL) dispersed in PBS (10 mM, pH 7.4), and (**C**) Negative control; animals injected intradermally with physiological buffered saline (PBS, 0.1 mL). Black lines were drawn with a non-sensitizing or irritating pen. Rhamnolipids nano-micelles showed no clinical signs of skin sensitization (no erythema nor edema) that was identical to negative control contrary to positive control where marked signs of sensitization (erythema, edema, and redness) were identified.

2.4.4. Histopathological and Immunohistochemical Evaluations

The skin, eyes, lungs, and liver of the animals treated with rhamnolipids nano-micelles (0.625 mg/mL), and PBS (10 mM, pH 7.4) were investigated for any histopathological changes in comparison to the untreated animal. The histopathological features for skin are presented in Figures 6 and 7.

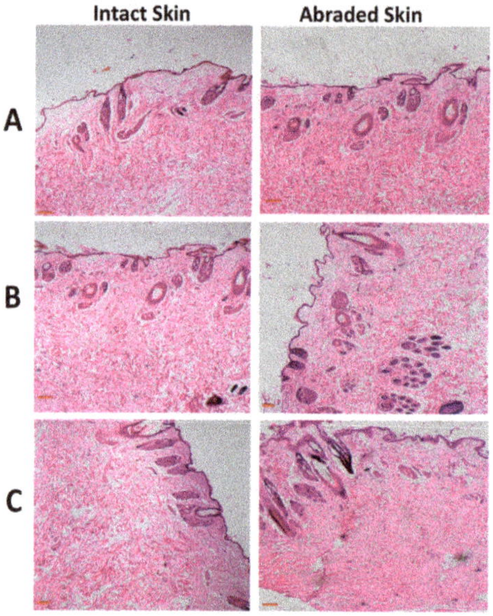

Figure 6. Representative images of H and E stained photomicrographs of skin for one representative rabbit after skin irritation test for (**A**) Group I; untreated group, act as a control, (**B**) Group II; Rhamnolipids nano-micelles solution group, and (**C**) Group III; PBS (10 mM, pH 7.4) treated group. All groups treated with samples showed normal histological tissue identical to untreated group after 72 h of treatment with samples. All skin layers appear normal and similar to control with the absence of any sign of erosion, ulcers, necrotic cells, or inflammatory cells. Scale bar, 100 μm.

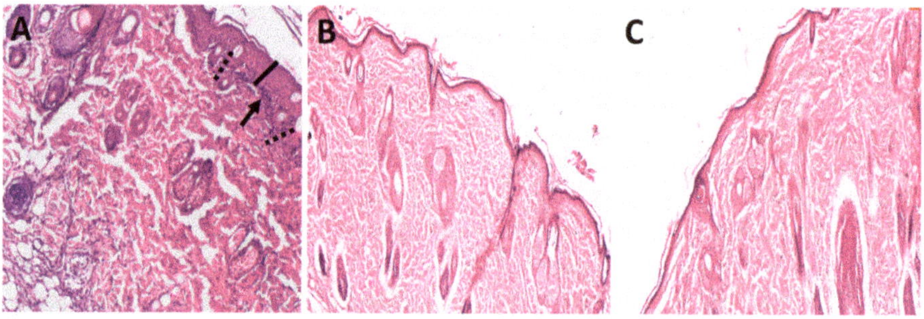

Figure 7. Representative images of H and E stained photomicrographs of skin for one representative guinea pigs after skin sensitization test for (**A**) Positive control; animals injected intradermally with skin sensitizing agent, a mixture of Freund's Complete Adjuvant (FCA), and physiological buffered saline (PBS) (1:1 mixture, 0.1 mL) (**B**) Rhamnolipids nano-micelles solution (0.625 mg/mL) dispersed in PBS (10 mM, pH 7.4), and (**C**) Negative control; animals injected intradermally with physiological buffered saline (PBS, 0.1 mL). Positive control revealed thickening of epiderma (black line), inflammatory cells infiltration (black head arrow) and spongiosis (dashed line). (**B**,**C**) group revealed normal skin layers.

The skin histopathology conducted after the skin irritation test are presented in Figure 6 (additional images were also presented in Figure S3 in Supplementary Materials), and revealed that both the treated and untreated animals showed normal skin histology where skin layers appeared normal with the absence of any sign of erosion, ulcers, necrotic cells, or inflammatory cells. The histopathological images of the skin sensitization test are presented in Figure 7. As revealed, positive control (Figure 7A) showed that the structure of dermis and epidermis was thickened and disrupted, with some areas of necrosis with infiltration of pleomorphic inflammatory cells. This is in contrast to the animals injected intradermally with rhamnolipids nano-micelles (0.625 mg/mL) (Figure 7B), where skin epidermis and dermis appeared completely normal, and showed no histopathological alterations compared to the negative control group (Figure 7C).

The histopathological features for the eyes, lungs and liver after the irritation test are presented in Figures 8–10.

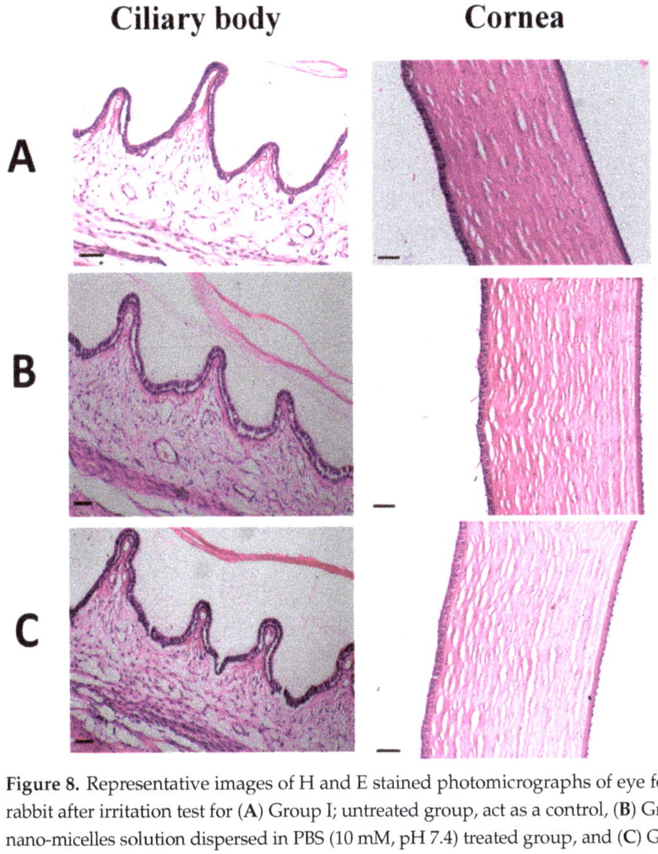

Figure 8. Representative images of H and E stained photomicrographs of eye for one representative rabbit after irritation test for (**A**) Group I; untreated group, act as a control, (**B**) Group II; rhamnolipids nano-micelles solution dispersed in PBS (10 mM, pH 7.4) treated group, and (**C**) Group III; PBS (10 mM, pH 7.4) treated group. All groups treated with samples showed normal histological tissue identical to untreated group after 72 h of treatment with samples. Cornea, fibrous connective tissue appeared normal with no signs of inflammation, erosion, ulcers or necrobiotic changes. Ciliary body appeared also normal with no cutting in the filament of ciliary body with no oedema. Scale bar, 50 μm.

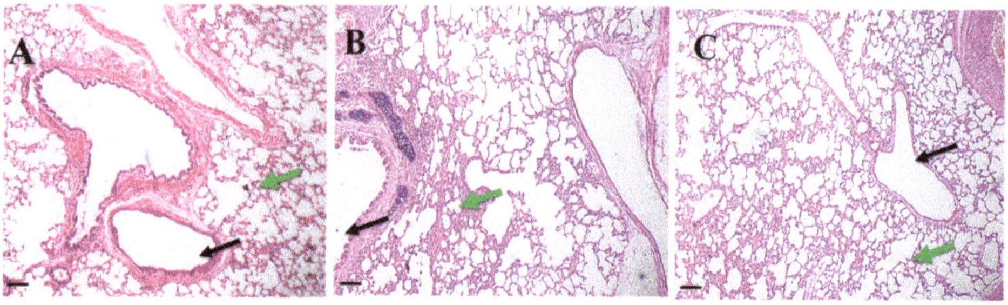

Figure 9. Representative photos of H and E stained photomicrographs of lung for one representative rabbit after irritation test for (**A**) Group I; untreated group, act as a control, (**B**) Group II; rhamnolipids nano-micelles solution dispersed in PBS (10 mM, pH 7.4) treated group, and (**C**) Group III; PBS (10 mM, pH 7.4) treated group. All groups treated with samples showed normal histological tissue identical to untreated group after 72 h of treatment with samples. Black arrows pointed to bronchioles while green arrows pointed to alveolar cells. Scale bar, 50 μm.

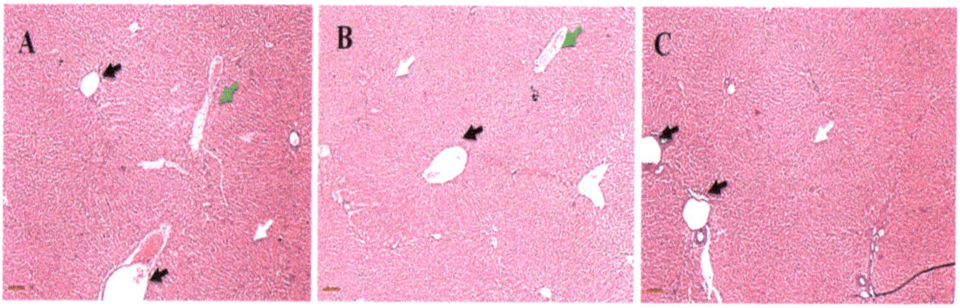

Figure 10. Representative images of H and E stained photomicrographs of liver for one representative rabbit after irritation test for (**A**) Group I; untreated group, act as a control, (**B**) Group II; rhamnolipids nano-micelles solution dispersed in PBS (10 mM, pH 7.4) treated group, and (**C**) Group III; PBS (10 mM, pH 7.4) treated group. All groups treated with samples showed normal histological tissue identical to untreated group after 72 h of treatment with samples. The black arrow points to hepatic central vein, the green arrow points to portal triads, and the white arrows point to hepatocytes. Scale bar, 100 μm.

For the eye histopathological examination (Figure 8), the cornea and fibrous connective tissue of the treated and untreated animals appeared normal, with no signs of inflammation, erosion, ulcers or necrobiotic changes. The ciliary body also appeared normal, with no cutting in the filament of the ciliary body nor oedema. Additional images concerned with eye histopathology were also presented in Figure S4 in Supplementary Materials.

In terms of the lung histopathological examination of both the treated and untreated animals (Figure 9), both showed normal bronchioles and alveolar structure. Liver histopathology images of both treated and untreated animals (Figure 10) revealed normal periportal regions, normal hepatocyte morphology, with no evidence of inflammation or necrosis.

2.5. Biochemical Analysis

Liver function biomarkers were analyzed in serum samples of both the treated and untreated animals, and the data are presented in Table 4. The levels of alanine aminotransferase (ALT) and aspartate aminotransferase (AST) measured for the animals treated with rhamnolipids nano-micelles (0.625 mg/mL) and PBS (10 mM, pH 7.4), were non-significantly ($p > 0.05$) different than the levels of ALT and AST measured in the untreated animals.

Table 4. * Effect of tested solutions, rhamnolipids nano-micelles and PBS (10 mM, pH 7.4) on serum biochemical markers. Data are presented as an average ± SE, results are average of three replicates.

Animal Group	AST (U/L)	ALT (U/L)
Untreated animal (Negative Control)	30.00 ± 1.53	27.33 ± 1.20
PBS	32.70 ± 3.20	28.20 ± 2.80
Rhamnolipids nano-micelles (0.625 mg/mL)	35.00 ± 4.35	28.67 ± 3.71

* Data were analyzed by two-way ANOVA, there was non-significant differences between untreated (control) and treated animals at $p < 0.05$. (U/L) = unit/liter.

2.6. Hematological Examination

The hematological examinations for the animals treated with rhamnolipids nano-micelles (0.625 mg/mL), PBS (10 mM, pH 7.4) and the untreated animals are presented in Table 5. The mean values of Erythrogram for the treated animal were non-significantly ($p > 0.05$) different than those determined for the untreated animal.

Table 5. * Effect of tested solutions, rhamnolipids nano-micelles and PBS (10 mM, pH 7.4) on the animal Erythrogram. Data are presented as an average ± SE, results are average of three replicates.

Animal Groups	PCV (%)	Hb (g/dL)	RBCs × 10^6/µL	MCV (fL)	MCHC (%)
Untreated animal (Control)	38.53 ± 0.56	12.00 ± 0.64	8.84 ± 0.11	48.87±0.22	37.26 ± 0.46
PBS	37.90 ± 0.61	11.51 ± 0.32	7.80 ± 0.13	48.20 ± 0.19	37.3 ± 0.32
Rhamnolipids nano-micelles (0.625 mg/mL)	39.41± 0.26	11.34 ± 0.20	7.61±0.140	47.72 ± 0.26	37.28 ± 0.34

* Data were analyzed by two-way ANOVA, there was non-significant differences between untreated (control) and treated animals at $p < 0.05$. PCV, Packed cell volume, Hb = Hemoglobin (g/dL = gram/deciliter), RBCs = Red blood cells count, MCV = Main corpuscular volume (fL, femto-liter), MCHC = Main corpuscular hemoglobin concentrations.

3. Discussion

Nanotechnology is a cutting-edge science that is known to change the physicochemical properties of materials due to their small size (1–1000 nm) and thus potentiate materials' activity and overcome their side effects [26–35]. Nanoparticles were reported to be applied for antimicrobial purposes against viruses [6,36–39] and bacteria [6,23,40].

Rhamnolipids mixtures are biosurfactants produced by *Pseudomonas aeruginosa* and were reported to have antibacterial and antiviral activity [41–44]. We applied nanotechnology to formulate rhamnolipids nano-micelles and demonstrated their antibacterial activity against selected resistant bacterial strains: *Staphylococcus aureus*, *Streptococcus pneumoniae*, *Salmonella* Montevideo, *Salmonella* Typhimurium, and *Acinetobacter baumannii* [6,23]. The MICs recorded for rhamnolipids nano-micelles against tested bacteria implied superior antibacterial activity when compared to MICs of rhamnolipids mixtures reported in the literature [41,45]. For example, the MIC recorded for rhamnolipids mixtures with different resistant strains of *Staphylococcus aureus* ranged from 128 to 650 µg/mL [41,45], while the MIC recorded for the resistant strain of *Staphylococcus aureus* treated with rhamnolipids nano-micelles ranged from 31 to 39 µg/mL [6,23]. Similarly, the MIC recorded for rhamnolipids mixtures with *Streptococcus pneumonia* (clinically isolated) was 128 µg/mL [41], compared to 31 µg/mL, recorded for rhamnolipids nano-micelles [6]. Furthermore, we reported the superior in-vitro biosafety of rhamnolipids nano-micelles compared to alcohol-based hand sanitizer on human dermal fibroblast cells [23]. Although rhamnolipids mixtures demonstrated an antiviral activity against several viruses, including SARS-CoV2 [5,44], according to the author's knowledge, the antiviral activity of rhamnolipids nano-micelles has not yet been investigated.

In the current study, rhamnolipids were produced from the *P. aeruginosa* strain LeS3, and then rhamnolipid nano-micelles were prepared following our previously reported protocol [6,23]. In our previous studies [6,23], we reported that the maximum MIC value recorded against the resistant strain was 312 μg/mL, thus, this concentration should be used to ascertain the complete eradication of resistant bacterial strain. As viruses have a different nature and cell structure than bacteria, we prepared rhamnolipids nano-micelles at different concentrations of rhamnolipids, 20, 78, and 312 μg/mL, and investigated their antiviral activity against SARS-CoV-2. The different concentrations did not affect the particle size of nano-micelles produced, as revealed from the Malvern Instrument analysis and TEM images. However, the TEM images showed a lower size than that recorded by the Malvern Instrument analysis, and this could be attributed to the different techniques applied, and is consistent with what is reported in the literature [6,23,46]. All samples were homogenously distributed as the recorded PDI values were around 0.3. This is consistent with our previous publications [6,23], and with the literature [47–49]. All samples were stable and had a lower tendency to aggregate, as revealed from the zeta potential values, which were >−30 mv [6,23]. The pH value for all of the nano-micelles samples ranged from 6.22 to 6.43; this is consistent with the acceptable pH range (4–7) recommended for skin application [50,51] and is not likely to induce any skin irritation, as previously reported [6,23,52–54].

The rhamnolipids nano-micelles, at all investigated concentrations (20, 78, and 312 μg/mL), successfully inactivated the virus, and this could be explained by the significant ($p < 0.05$) reduction of viral infectivity compared to the positive control. SARS-CoV-2 infectivity was reduced by 93%, 99.4%, and 100% at rhamnolipids nano-micelles concentration of 20, 78, and 312 μg/mL, respectively. Thus, 312 μg/mL is highly recommended to be used for hand sanitation due to its complete eradication of the virus, and hence limits the spread of infection among patients and HCW.

The rhamnolipids mixtures were reported [5] to inactivate SARS-CoV-2 by 70% at 25 μg/mL. Therefore, the rhamnolipids nano-micelles demonstrated superior antiviral activity compared to the rhamnolipids mixtures. A docking study was performed in our previous publication [6] to explore the possible mechanisms of SARS-CoV-2 deactivation by rhamnolipids nano-micelles or as a singlet molecule. SARS-CoV-2 is an enveloped virus, the envelope is composed of a lipid bilayer anchored with viral spike glycoproteins that are essential for virus binding and entry into host cells [55]. The docking study suggested that rhamnolipids interaction with spike glycoproteins resulted in irreversible changes to their structures, and thus, virus deactivation. Additionally, rhamnolipids were suggested to interact with the lipid membranes (lipid envelope) of SARS-CoV-2, and this is associated with the disruption of membrane permeability, similar to that previously observed for HSV1 and HSV2 [44,56–59], and therefore, virus deactivation. Furthermore, above rhamnolipids critical micelle concentration (CMC), the lipid bilayer of SARS-CoV-2 is expected to be completely solubilized by the surfactants, and only the micellar aggregates remain in the solution [60]. Thus, the complete solubilization of the protective lipid bilayer leads to the potential disintegration of the virus into fragments, neutralizing its infectivity. Alternatively, nano-micelles are able to completely entrap the viral particle internally via hydrophobic—hydrophobic interactions [61].

Based on the previous study we performed [23], the MIC values recorded for rhamnolipids nano-micelles solution against tested multidrug resistant Gram-positive and Gram-negative bacteria ranged from 31 to 39 and 312 to 512 μg/mL, respectively [6,23]. The cytotoxic concentration of the rhamnolipid nano-micelles solution responsible for the death of 50% of human dermal fibroblast cells (CC50) was 604 μg/mL [23]. Additionally, rhamnolipids nano-micelles were able to deactivate SARS-CoV-2 at 312 μg/mL. Taken together, rhamnolipids nano-micells at a concentration $\leq$600 μg/mL were demonstrated to have an effective antibacterial and antiviral activity, as well as being compatible with biological system [6,23].

The acute irritation test, conducted on rabbits, was performed to further ascertain the biosafety of rhamnolipids nano-micelles. For the eye and skin irritation tests, no degree of irritation was recorded after topical application of rhamnolipids nano-micelles (0.625 mg/mL), or after its instillation into the eye of rabbits. Additionally, the skin sensitization test performed confirmed further the compatibility of rhamnolipids nano-micelles with animal skin.

The results obtained from the histopathological, biomarker, and hematological analyses from animals treated with rhamnolipids nano-micelles (0.625 mg/mL) demonstrated an absence of any signs of irritation or sensitization and, thus, assured the high safety of rhamnolipids nano-micelles solution being applied topically for hand sanitation. The obtained data is consistent with work performed by Bharali and colleagues [62], in which an acute dermal irritation study of rhamnolipids mixtures showed no dermal reactions, such as erythema or edema, compared to their negative control at 72 h after the application of rhamnolipids mixtures on the shaved skin of rabbits. Furthermore, the concentration of biosurfactants above their CMC (5–200 mg/L) [63] was also nontoxic to the skin of a rabbit. The acute dermal irritation study with the isolated biosurfactants also showed no adverse effect on the hematological parameters for the treated rabbits compared to untreated rabbits [62].

These results suggested that rhamnolipids nano-micelles (0.625 mg/mL) had no toxicity and could be applied safely for hand sanitation. However, a pilot study in a healthcare setting, to assess the efficacy and safety of rhamnolipids nano-micelles application as a hand sanitizer, should be performed in the future to ascertain its antimicrobial efficacy and safety compared to alcohol-based hand sanitizers.

4. Materials and Methods

4.1. Materials

Microbiological media (tryptone soy broth; TSB, and tryptone soy agar; TSA) were purchased from Hi-Media, Mumbai, India. Peptone and sodium chloride were purchased from Oxoid, UK. Hydrochloric acid, ethyl acetate, and sulfuric acid were purchased from Honeywell™, Charlotte, NC, USA. L-rhamnose was purchased from Sigma-Aldrich, Darmstadt, Germany. Orcinol was obtained from SDFCL, Chennai, India. Carbopol gel, phosphate-buffered saline (PBS) tablets, and absolute ethyl alcohol were purchased from Merck, Darmstadt, Germany. All chemicals and reagents were of analytical grade.

The African green monkey kidney epithelial cell line (Vero) was purchased from the American Type Culture Collection (ATCC). Vero cells were cultured in complete Dulbeccos modified eagle medium (cDMEM), containing 10%, fetal bovine serum (FBS), 1% non-essential amino acids (NEAAs), 1% penicillin, and streptomycin and 1% l-glutamine, purchased from Thermo Fisher Scientific, Waltham, MA, USA.

4.2. Methodology

4.2.1. Production of Rhamnolipids

The production of rhamnolipids was carried out following our previously reported protocol, using the shake-flask technique [6], and then they were further purified from the production medium by acid precipitation and organic solvent extraction [64]. The *Pseudomonas aeruginosa* strain LeS3 was grown in TSB to obtain an OD_{600} of 0.8, corresponding to a density of 8 log cfu/mL. A 250 mL Erlenmeyer flask containing 100 mL of a sterilized production medium was formulated from chicken carcass soup (CCS), containing 5% chicken fat and 0.5% NaCl. The sterilized CCS was inoculated with 1% of the overnight bacterial culture. Inoculated flasks were then incubated in an orbital shaker (Vision Scientific Co., Ltd., Bucheon, Korea. VS-8480SR) at 30 °C and 150 rpm for 5 days. At the end of the incubation period, bacterial cells were removed from the culture broth by centrifugation at 10,000 rpm and 5 °C for 10 min (Sigma, 3–6PK) to obtain cell-free supernatant (CFS). The CFS was acidified to pH 2.0 using 1N HCl and stored overnight at 5 °C. Rhamnolipids were then extracted using an equal volume of ethyl acetate. A yellow–brown viscous paste of rhamnolipids was obtained and then stored in the fridge until further use.

4.2.2. Preparation of Rhamnolipids Nano-Micelles

Rhamnolipid nano-micelles were prepared following our previously published protocol [6]. An aqueous solution of rhamnolipids, at a concentration of 10 mg/mL, was sonicated in phosphate-buffered saline (PBS, 10 mM, pH 7.4), using a probe sonicator (Dr. Hielscher Sonicator, Teltow, Germany) to form rhamnolipid nano-micelles.

4.2.3. Characterization of Rhamnolipids Nano-Micelles

Particle Size and Zeta Potential

The particle size and zeta potential of the rhamnolipid nano-micelle solution were determined using a Malvern Zeta-sizer Nano ZS (Malvern Instruments, Ltd., Malvern, Worcestershire, UK) at 25 °C ± 0.1. Samples were diluted in PBS to give a count rate ranging from 50 to 300 KCPs.

Transmission Electron Microscopy

The rhamnolipid nano-micelles hand sanitizer solution was imaged by TEM (H-700, Hitachi, Ltd., Tokyo, Japan), at an accelerated voltage of 80 kV, using the negative staining method. The rhamnolipid nano-micelles hand sanitizer solution was diluted (1:50) with double-distilled water, and then a drop of the diluted solution was spread on a mesh copper grid coated with carbon film and kept for 5 min to dry. Then, a drop of phosphotungstic acid (2% w/v) was added to the grid for 50 s, and the excess liquid was removed using filter paper.

Determination of pH

The pH of the rhamnolipid nano-micelles were determined at room temperature. The pH was determined using an Ohaus Economical pH bench meter (starter 3100, New Haven, CT, USA) that was previously calibrated with three standard buffer solutions (pH of 4, 7, and 10).

4.2.4. SARS-CoV-2 Infection Experiments (HCoV-19/England/2/202 Strain)

The Vero cells (ATCC® CCL-81) were washed with PBS, dislodged with 0.25% Trypsin-EDTA (Sigma Life Sciences) and seeded into 96-well imaging plates (Greiner, Gloucestershire, UK) at a density of 8×10^3/well in culture media (DMEM containing 10% FBS, 1% Penicillin and Streptomycin, 1% l-Glutamine and 1% non-essential amino acids). The next day, cells were infected with SARS-CoV-2 strain hCOV-19/England/2/2020. Virus stock 10^6 IU/mL (kind gift from Christine Bruce, Public Health England) was diluted 1/150 in culture media allowing 25 μL per well. The virus was then diluted further, with 25 μL per well media containing treatments of interest prepared at 2×concentration to give a final at 3333 IU/mL for infection. The virus-nano-micelles mixture was incubated for 1 h at room temperature before being added to Vero cells for 24- or 48-h infections. The cells were then blocked in PBS containing 10% FBS and stained with rabbit anti-SARS-CoV-2 spike protein, subunit 1 (The Native Antigen Company, Oxford, UK), followed by Alexa Fluor 555-conjugated goat anti-rabbit IgG secondary antibody (Invitrogen, Thermo Fisher Scientific, Waltham, USA). The cell nuclei were stained with Hoechst 33342 (Thermo Fisher Scientific). After being washed with PBS, the cells were imaged and analyzed using a Thermo Scientific CellInsight CX5 High-Content Screening (HCS) platform. Viable cells were counted, and infected cells were scored by spike perinuclear fluorescence above a set threshold determined by positive (untreated) and negative (uninfected) controls. A minimum of nine fields of view and 5,000 nuclei per well in triplicate wells per treatment were scored in each experiment. All experiments were repeated 2–4 times.

4.2.5. Irritation and Skin Sensitization Study

Animal

New Zealand White rabbits (CLAVCAP-VACSER, Cairo, Egypt), weighing 2.5–3 kg, were used to investigate the irritation of rhamnolipids nano-micelles solution on the eyes and skin. Rabbits were maintained under managed conditions: 12 light-dark cycle, 25 ± 2 °C and 50 ± 20% relative humidity. They had free access to food, a standard commercial pellet diet (containing at the very least; 5% fiber, 20% protein, 3.5% fat, 6.5% vitamins and ash mixture) and were offered water and libitum. The rabbits were kept in hygienic conditions throughout the experimental period, and they were left for one week to acclimatize to the lab conditions, before the onset of the experiment. All the procedures in this research were approved by the Faculty of Veterinary Medicine, Cairo University (Vet CU20092022530). All efforts were made to minimize the animals suffering.

Eye Irritation Test

The Draize modified test was carried out to determine ocular irritation in the eyes of ten white Zealand rabbits after the application of rhamnolipids nano-micelles and PBS (10 mM, pH 7.4), where five rabbits were used to test each solution [65]. The rhamnolipids nano micelles solutions and PBS (50 µL) were instilled into the lower cul-de-sac of the rabbits' eye, to be exposed to cornea. The rhamnolipids nano-micelles were administrated five times in the left eye, with each dose separated by 5 min time intervals. The right eye served as a negative control (untreated eye). Eyelids were gently kept together for around 10 s to prevent the loss of the instilled preparation. After dose instilment, the rabbit's eyes were observed for any possible ocular reactions, including redness of the eyes, conjunctival chemosis, discharge, and corneal and iris lesions. These observations were conducted at regular time intervals of 24 and 72 h. The assessment was conducted according to the histological grading system for eye irritation evaluation [25]. Different grading, from 0 to 3, was carried out on the basis of the absence or presence, and severity, of symptoms, including inflammation, redness, and tearing [66]. Photos were taken for all groups after 72 h of treatment.

Skin Irritation Test

Six rabbits were anesthetized to remove the hair on an area of their backs. Six squares, with a dimension of 2.5 × 2.5 cm^2, were marked in the glabrous area on each rabbit. The three squares on the left were scratched with a syringe needle (abraded skin); the three squares on the right side were not scratched (intact skin). The upper two squares on each side act as negative control for abraded and intact skin. On the middle squares of each side, rhamnolipids nano-micelles solution (0.5 mL) was applied. PBS (10 mM, pH 7.4) was the vehicle used to disperse the nano-micelles, thus, on the lower two squares, PBS (10 mM, pH 7.4, 0.5 mL) was applied. All of the squares were covered with sterile gauze of the same size as the square, and nonirritant tape was used to fix the gauze in place. After 24 and 72 h, the skin was observed for signs of irritation, e.g., erythema/edema, based on the Magnusson and Kligman scale [24]. Photos were taken for all groups after 72 h of treatment.

Skin Sensitization Test

The test was carried out on healthy albino guinea-pigs, in accordance with the OECD guideline [67]. Twenty healthy albino guinea-pigs were considered for a sensitization test, and were kept in individual cages for 4 days under 20 °C ± 2 °C, 50 ± 10% relative humidity, natural illumination, with conventional diet and water, with sufficient quantity of ascorbic acid to adjust themselves to the environment. The acclimatized albino guinea-pigs were divided into three groups: the positive control (5 animals) animals received intradermal injections (0.1 mL) of 1:1 mixture (v/v) Freund's Complete Adjuvant (FCA)/physiological saline as skin sensitizing agent; the tested group (10 animals) animals received intradermal injections (0.1 mL) of Rhamnolipids nano-micelles solution (0.625 mg/mL); and the negative control (5 animals) received intradermal injections of PBS (0.1 mL).

Challenge reactions were assessed at 24, 48 and 72 h after intradermal injection of the samples. The intensity of all skin reactions was graded following the sensitization Magnusson and Kligman grading score [24].

Histopathological and Immunohistochemical Evaluations

After 72 h of the treatment protocol, the animals were sacrificed by ether anesthesia and the treated dorsal area was excised and rinsed with ice cold phosphate buffer saline. Next, the skin, eyes, liver, and lungs were fixed in 10% buffered formalin. In a routine manner, formalin fixed skin, eye, liver and lung specimens were dehydrated in different grades of alcohol, followed by clearance in xylol and embedding in paraffin. Serially, sections of 4–5 µm thickness were obtained from the prepared paraffin blocks, followed by their staining with Hematoxylin and Eosin (H & E) [68]. Histopathological findings in the skin were graded semi-quantitatively, according to [69], with some modifications (0 = no abnormality, 1 = slight, 2 = mild, 3 = moderate) for each of the seven findings: hypertrophy, hyperkeratosis, parakeratosis, erosion, inflammatory cells infiltration, extracellular edema and ulcer. Histopathology of skin samples was carried out at the Faculty of Veterinary Medicine, Cairo University, Egypt.

Biochemical Analysis

Liver function biomarkers were analyzed in serum samples of rabbits of all groups, including alanine aminotransferase (ALT) and aspartate aminotransferase (AST) using kits of EIAab® (Wuhan, China).

Hematological Examination

Complete blood pictures (CBC) were obtained using ABC animal blood count apparatus (vet 907 AB 6012).

4.2.6. Statistical Analysis

Statistical analysis was performed using two-way ANOVA. Analyses were carried out using GraphPad Prism 9.0 software at a confidence level of 95%.

5. Conclusions

Hospital- and community-acquired infections are escalating and pose a serious public health problem worldwide. Hands are considered to be an important route for transmitting microbes and infections between individuals. Thus, keeping good hand hygiene is a key factor to control or prevent the spread of infection. Due to reports of skin cytotoxicity caused by the frequent use of alcohol-based hand sanitizer during the COVID-19 pandemic, it was essential to develop alternative and novel approaches for hand sanitation. In our previous publications, we demonstrated the antibacterial activity of rhamnolipids nano-micelles against selected multidrug resistant Gram-positive and Gram-negative bacteria and reported the maximum MIC value to be 312 µg/mL. In the current work, rhamnolipids nano-micelles solution, at 20, 78, and 312 µg/mL, all demonstrated a significant ($p < 0.05$) decrease of SARS-CoV-2 virus infectivity compared to virus only and the blank sample. However, 100 % virus eradication was only recorded with 312 µg/mL. The acute irritation test and skin sensitization test revealed that rhamnolipids nano-micelles were biocompatible with the skin. Furthermore, histopathological studies on skin after the irritation test revealed normal skin histopathology, where skin layers appeared normal with the absence of any signs of erosion, ulcers, necrotic cells, or inflammatory cells. The compatibility of rhamnolipids nano-micelles with the skin was further confirmed by the skin sensitization test, where layers of epidermis and dermis appeared completely normal, with no histopathological alterations in comparison to negative control group. Thus, rhamnolipids nano-micelles are recommended to be used as a safe and effective hand sanitizer. However, a future study in hospitals is still to be performed to ascertain their broad-spectrum antimicrobial efficacy and compatibility with skin.

Supplementary Materials: The following are available online at https://www.mdpi.com/article/10.3390/antibiotics11111556/s1.

Author Contributions: Conceptualization, Y.A.-z. and Z.S.; data curation, Y.A.-z., Z.S., H.J.H., A.M.A., G.E.E. and N.R.R.; formal analysis, Y.A.-z., Z.S., H.J.H. and A.M.A.; investigation, Y.A.-z., A.M.A., Z.S., H.J.H., M.R.B., G.E.E. and N.R.R.; methodology, Y.A.-z., H.J.H., Z.S., A.M.A., M.R.B., G.E.E. and N.R.R.; visualization, Y.A.-z. and Z.S.; writing—original draft, Y.A.-z., Z.S., H.J.H. and A.M.A.; project administration, Y.A.-z.; funding acquisition, Y.A.-z. All authors have read and agreed to the published version of the manuscript.

Funding: This research was funded by the Science, Technology, and Innovation Funding Authority (STIFA), Emergency Targeted Program COVID-19 Emergency Call 2020, grant number 43766, Cairo, Egypt; and Helwan University, Cairo, Egypt. Z.S. is supported by a Medical Research Foundation intermediate career fellowship in viral and autoimmune hepatitis MRF-169-0001-F-STAM-C0826 (UK Research and Innovation).

Institutional Review Board Statement: Not applicable.

Informed Consent Statement: Not applicable.

Data Availability Statement: All authors are happy to share all data (including Supplementary Data).

Acknowledgments: We would like to thank Ebtehal Hassan Mohammed of the TEM unit, Faculty of Agriculture Research Park, for her patience when taking TEM images.

Conflicts of Interest: The authors declare no conflict of interest.

:eferences

1. Haque, M.; McKimm, J.; Sartelli, M.; Dhingra, S.; Labricciosa, F.M.; Islam, S.; Jahan, D.; Nusrat, T.; Chowdhury, T.S.; Coccolini, F.; et al. Strategies to Prevent Healthcare-Associated Infections: A Narrative Overview. *Risk Manag. Healthc. Policy* **2020**, *13*, 1765–1780. [CrossRef] [PubMed]
2. Garcell, H.G.; Arias, A.V.; Pancorbo Sandoval, C.A.; García, E.G.; Valle Gamboa, M.E.; Sado, A.B.; Alfonso Serrano, R.N. Incidence and Etiology of Surgical Site Infections in Appendectomies: A 3-Year Prospective Study. *Oman Med. J.* **2017**, *32*, 31–35. [CrossRef] [PubMed]
3. Al-Mousa, H.H.; Omar, A.A.; Rosenthal, V.D.; Salama, M.F.; Aly, N.Y.; El-Dossoky Noweir, M.; Rebello, F.M.; Narciso, D.M.; Sayed, A.F.; Kurian, A.; et al. Device-Associated Infection Rates, Bacterial Resistance, Length of Stay, and Mortality in Kuwait: International Nosocomial Infection Consortium Findings. *Am. J. Infect. Control* **2016**, *44*, 444–449. [CrossRef] [PubMed]
4. Ananda, T.; Modi, A.; Chakraborty, I.; Managuli, V.; Mukhopadhyay, C.; Mazumder, N. Nosocomial Infections and Role of Nanotechnology. *Bioengineering* **2022**, *9*, 51. [CrossRef] [PubMed]
5. Giugliano, R.; Buonocore, C.; Zannella, C.; Chianese, A.; Esposito, F.P.; Tedesco, P.; De Filippis, A.; Galdiero, M.; Franci, G.; de Pascale, D. Antiviral Activity of the Rhamnolipids Mixture from the Antarctic Bacterium *Pseudomonas gessardii* M15 against Herpes Simplex Viruses and Coronaviruses. *Pharmaceutics* **2021**, *13*, 2121. [CrossRef]
6. Bakkar, M.R.; Faraag, A.H.I.; Soliman, E.R.S.; Fouda, M.S.; Sarguos, A.M.M.; McLean, G.R.; Hebishy, A.M.S.; Elkhouly, G.E.; Raya, N.R.; Abo-zeid, Y. Rhamnolipids Nano-Micelles as a Potential Hand Sanitizer. *Antibiotics* **2021**, *10*, 751. [CrossRef]
7. Kratzel, A.; Kratzel, A.; Todt, D.; V'kovski, P.; Steiner, S.; Steiner, S.; Gultom, M.; Gultom, M.; Thao, T.T.N.; Thao, T.T.N.; et al. Inactivation of Severe Acute Respiratory Syndrome Coronavirus 2 by WHO-Recommended Hand Rub Formulations and Alcohols. *Emerg. Infect. Dis.* **2020**, *26*, 1592–1595. [CrossRef]
8. Kampf, G.; Todt, D.; Pfaender, S.; Steinmann, E. Persistence of Coronaviruses on Inanimate Surfaces and Their Inactivation with Biocidal Agents. *J. Hosp. Infect.* **2020**, *104*, 246–251. [CrossRef]
9. Otter, J.A.; Donskey, C.; Yezli, S.; Douthwaite, S.; Goldenberg, S.D.; Weber, D.J. Transmission of SARS and MERS Coronaviruses and Influenza Virus in Healthcare Settings: The Possible Role of Dry Surface Contamination. *J. Hosp. Infect.* **2016**, *92*, 235–250. [CrossRef]
10. Pittet, D.; Allegranzi, B.; Sax, H.; Dharan, S.; Pessoa-Silva, C.L.; Donaldson, L.; Boyce, J.M. Evidence-Based Model for Hand Transmission during Patient Care and the Role of Improved Practices. *Lancet Infect. Dis.* **2006**, *6*, 641–652. [CrossRef]
11. Emami, A.; Javanmardi, F.; Keshavarzi, A.; Pirbonyeh, N. Hidden Threat Lurking behind the Alcohol Sanitizers in COVID-19 Outbreak. *Dermatol. Ther.* **2020**, *33*, 17–19. [CrossRef]
12. Onyedibe, K.I.; Shehu, N.Y.; Pires, D.; Isa, S.E.; Okolo, M.O.; Gomerep, S.S.; Ibrahim, C.; Igbanugo, S.J.; Odesanya, R.U.; Olayinka, A.; et al. Assessment of Hand Hygiene Facilities and Staff Compliance in a Large Tertiary Health Care Facility in Northern Nigeria: A Cross Sectional Study. *Antimicrob. Resist. Infect. Control* **2020**, *9*, 30. [CrossRef]
13. Chiang, W.L.; Chen, T.W.; Liu, M.Y.; Hsu, C.J. Application and Robust H Control of PDC Fuzzy Controller for Nonlinear Systems with External Disturbance. *J. Mar. Sci. Technol.* **2001**, *9*, 84–90. [CrossRef]
14. Slaughter, R.J.; Mason, R.W.; Beasley, D.M.G.; Vale, J.A.; Schep, L.J. Isopropanol Poisoning. *Clin. Toxicol.* **2014**, *52*, 470–478. [CrossRef]
15. Vogel, L. Hand Sanitizers May Increase Norovirus Risk. *CMAJ* **2011**, *183*, 799–800. [CrossRef]

16. Blaney, D.D.; Daly, E.R.; Kirkland, K.B.; Tongren, J.E.; Kelso, P.T.; Talbot, E.A. Use of Alcohol-Based Hand Sanitizers as a Risk Factor for Norovirus Outbreaks in Long-Term Care Facilities in Northern New England: December 2006 to March 2007. *Am. J. Infect. Control* **2011**, *39*, 296–301. [CrossRef]
17. Pidot, S.J.; Gao, W.; Buultjens, A.H.; Monk, I.R.; Guerillot, R.; Carter, G.P.; Lee, J.Y.H.; Lam, M.M.C.; Grayson, M.L.; Ballard, S.A.; et al. Increasing Tolerance of Hospital *Enterococcus faecium* to Handwash Alcohols. *Sci. Transl. Med.* **2018**, *10*, eaar6115. [CrossRef]
18. Hayat, A.; Munnawar, F. Antibacterial Effectiveness of Commercially Available Hand Sanitizers. *Int. J. Biol. Biotech* **2016**, *13*, 427–431.
19. Mahmood, A.; Eqan, M.; Pervez, S.; Ahmed, H.; Bari, A. COVID-19 and Frequent Use of Hand Sanitizers; Human Health and Environmental Hazards by Exposure Pathways. *Sci. Total Environ.* **2020**, *742*, 140561. [CrossRef]
20. Müller, M.M.; Hörmann, B.; Syldatk, C.; Hausmann, R. *Pseudomonas aeruginosa* PAO1 as a Model for Rhamnolipid Production in Bioreactor Systems. *Appl. Microbiol. Biotechnol.* **2010**, *87*, 167–174. [CrossRef]
21. Chong, H.; Li, Q. Microbial Production of Rhamnolipids: Opportunities, Challenges and Strategies. *Microb. Cell Fact.* **2017**, *16*, 137. [CrossRef]
22. Henkel, M.; Geissler, M.; Weggenmann, F.; Hausmann, R. Production of Microbial Biosurfactants: Status Quo of Rhamnolipid and Surfactin towards Large-Scale Production. *Biotechnol. J.* **2017**, *12*. [CrossRef]
23. Abo-zeid, Y.; Bakkar, M.R.; Elkhouly, G.E.; Raya, N.R.; Zaafar, D. Rhamnolipid Nano-Micelles Versus Alcohol-Based Hand Sanitizer: A Comparative Study for Antibacterial Activity against Hospital-Acquired Infections and Toxicity Concerns. *Antibiotics* **2022**, *11*, 605.
24. Magnusson, B.; Kligman, A.M. The Identification of Contact Allergens by Animal Assay. The Guinea Pig Maximization Test. *J. Invest. Dermatol.* **1969**, *52*, 268–276. [CrossRef]
25. Li, J.; Li, Z.; Liang, Z.; Han, L.; Feng, H.; He, S.; Zhang, J. Fabrication of a Drug Delivery System That Enhances Antifungal Drug Corneal Penetration. *Drug Deliv.* **2018**, *25*, 938–949. [CrossRef]
26. Szunerits, S.; Barras, A.; Khanal, M.; Pagneux, Q.; Boukherroub, R. Nanostructures for the Inhibition of Viral Infections. *Molecules* **2015**, *20*, 14051–14081. [CrossRef]
27. Abo-zeid, Y.; Williams, G.R.; Touabi, L.; Mclean, G.R. An Investigation of Rhinovirus Infection on Cellular Uptake of Poly (Glycerol-Adipate) Nanoparticles. *Int. J. Pharm.* **2020**, *589*, 119826. [CrossRef]
28. Hillaireau, H.; Le Doan, T.; Appel, M.; Couvreur, P. Hybrid Polymer Nanocapsules Enhance in Vitro Delivery of Azidothymidine Triphosphate to Macrophages. *J. Control. Release* **2006**, *116*, 346–352. [CrossRef]
29. Abo-zeid, Y.; Mantovani, G.; Irving, W.L.; Garnett, M.C. Synthesis of Nucleoside-Boronic Esters Hydrophobic pro-Drugs: A Possible Route to Improve Hydrophilic Nucleoside Drug Loading into Polymer Nanoparticles. *J. Drug Deliv. Sci. Technol.* **2018**, *46*, 354–364. [CrossRef]
30. Hashim, F.; El-Ridy, M.; Nasr, M.; Abdallah, Y. Preparation and Characterization of Niosomes Containing Ribavirin for Liver Targeting. *Drug Deliv.* **2010**, *17*, 282–287. [CrossRef]
31. Lembo, D.; Swaminathan, S.; Donalisio, M.; Civra, A.; Pastero, L.; Aquilano, D.; Vavia, P.; Trotta, F.; Cavalli, R. Encapsulation of Acyclovir in New Carboxylated Cyclodextrin-Based Nanosponges Improves the Agent's Antiviral Efficacy. *Int. J. Pharm.* **2013**, *443*, 262–272. [CrossRef] [PubMed]
32. Abo-zeid, Y.; Garnett, M.C. Polymer Nanoparticle as a Delivery System for Ribavirin: Do Nanoparticle Avoid Uptake by Red Blood Cells? *J. Drug Deliv. Sci. Technol.* **2020**, *56*, 101552. [CrossRef]
33. Sobhy, Y.; Mady, M.; Mina, S.; Abo-zeid, Y. Phytochemical and Pharmacological Values of Two Major Constituents of *Asparagus* Species and Their Nano Formulations: A Review. *J. Adv. Pharm. Res.* **2022**, *6*, 94–106. [CrossRef]
34. Burgess, K.; Li, H.; Abo-Zeid, Y.; Fatimah; Williams, G.R. The Effect of Molecular Properties on Active Ingredient Release from Electrospun Eudragit Fibers. *Pharmaceutics* **2018**, *10*, 103. [CrossRef] [PubMed]
35. Wang, Y.; Zhang, Z.; Abo-zeid, Y.; Bear, J.C.; Davies, G.; Lei, X.; Williams, G.R. SiO_2-Coated Layered Gadolinium Hydroxides for Simultaneous Drug Delivery and Magnetic Resonance Imaging. *J. Solid State Chem.* **2020**, *286*, 121291. [CrossRef]
36. Rangayasami, A.; Kannan, K.; Murugesan, S.; Radhika, D.; Sadasivuni, K.K.; Reddy, K.R.; Raghu, A.V. Influence of Nanotechnology to Combat against COVID-19 for Global Health Emergency: A Review. *Sensors Int.* **2021**, *2*, 100079. [CrossRef]
37. Abo-Zeid, Y.; Ismail, N.S.; McLean, G.R.; Hamdy, N.M. A Molecular Docking Study Repurposes FDA Approved Iron Oxide Nanoparticles to Treat and Control COVID-19 Infection. *Eur. J. Pharm. Sci.* **2020**, *153*, 105465. [CrossRef]
38. Abo-zeid, Y.; Urbanowicz, R.A.; Thomson, B.J.; William, L.; Tarr, A.W.; Garnett, M.C. Enhanced Nanoparticle Uptake into Virus Infected Cells: Could Nanoparticles Be Useful in Antiviral Therapy? *Int. J. Pharm.* **2018**, *547*, 572–581. [CrossRef]
39. Chintagunta, A.D.; Sai Krishna, M.; Nalluru, S.; Sampath, S.K. Nanotechnology: An Emerging Approach to Combat COVID-19. *Emergent Mater.* **2021**, *4*, 119–130. [CrossRef]
40. Abo-zeid, Y.; Amer, A.; El-Houssieny, B.; Bakkar, M.; Sakran, W. Overview on Bacterial Resistance and Nanoparticles to Overcome Bacterial Resistance. *J. Adv. Pharm. Res.* **2021**, *5*, 312–326. [CrossRef]
41. Lotfabad, T.B.; Shahcheraghi, F.; Shooraj, F. Assessment of Antibacterial Capability of Rhamnolipids Produced by Two Indigenous *Pseudomonas aeruginosa* Strains. *Jundishapur J. Microbiol.* **2013**, *6*, 29–35. [CrossRef]
42. Carrazco-Palafox, J.; Rivera-Chavira, B.E.; Adame-Gallegos, J.R.; Rodríguez-Valdez, L.M.; Orrantia-Borunda, E.; Nevárez-Moorillón, G.V. Rhamnolipids from *Pseudomonas aeruginosa* Rn19a Modifies the Biofilm Formation over a Borosilicate Surface by Clinical Isolates. *Coatings* **2021**, *11*, 136. [CrossRef]

3. Sana, S.; Datta, S.; Biswas, D.; Sengupta, D. Assessment of Synergistic Antibacterial Activity of Combined Biosurfactants Revealed by Bacterial Cell Envelop Damage. *Biochim. Biophys. Acta-Biomembr.* **2018**, *1860*, 579–585. [CrossRef]
4. Remichkova, M.; Galabova, D.; Roeva, I.; Karpenko, E.; Shulga, A.; Galabov, A.S. Anti-Herpesvirus Activities of *Pseudomonas* sp. S-17 Rhamnolipid and Its Complex with Alginate. *Z. Nat. C* **2008**, *63*, 75–81. [CrossRef]
5. Buonocore, C.; Tedesco, P.; Vitale, G.A.; Esposito, F.P.; Giugliano, R.; Monti, M.C.; D'Auria, M.V.; de Pascale, D. Characterization of a New Mixture of Mono-Rhamnolipids Produced by *Pseudomonas gessardii* Isolated from Edmonson Point (Antarctica). *Mar. Drugs* **2020**, *18*, 269. [CrossRef]
6. Akbari, B.; Tavandashti, M.P.; Zandrahimi, M. Particle Size Characterization of Nanoparticles—A Practicalapproach. *Iran. J. Mater. Sci. Eng.* **2011**, *8*, 48–56.
7. Putri, D.C.A.; Dwiastuti, R.; Marchaban, M.; Nugroho, A.K. Optimization of Mixing Temperature and Sonication Duration in Liposomes Preparation. *J. Pharm. Sci. Community* **2017**, *14*, 79–85. [CrossRef]
8. Piazzini, V.; D'Ambrosio, M.; Luceri, C.; Cinci, L.; Landucci, E.; Bilia, A.R.; Bergonzi, M.C. Formulation of Nanomicelles to Improve the Solubility and the Oral Absorption of Silymarin. *Molecules* **2019**, *24*, 1688. [CrossRef]
9. Das, P.; Yang, X.P.; Ma, L.Z. Analysis of Biosurfactants from Industrially Viable *Pseudomonas* Strain Isolated from Crude Oil Suggests How Rhamnolipids Congeners Affect Emulsification Property and Antimicrobial Activity. *Front. Microbiol.* **2014**, *5*, 696. [CrossRef]
0. Surini, S.; Amirtha, N.I.; Lestari, D.C. Formulation and Effectiveness of a Hand Sanitizer Gel Produced Using Salam Bark Extract. *Int. J. Appl. Pharm.* **2018**, *10*, 216–220. [CrossRef]
1. Ali, S.M.; Yosipovitch, G. Skin PH: From Basic Science to Basic Skin Care. *Acta Derm. Venereol.* **2013**, *93*, 261–267. [CrossRef]
2. Aodah, A.H.; Bakr, A.A.; Booq, R.Y.; Rahman, M.J.; Alzahrani, D.A.; Alsulami, K.A.; Alshaya, H.A.; Alsuabeyl, M.S.; Alyamani, E.J.; Tawfik, E.A. Preparation and Evaluation of Benzalkonium Chloride Hand Sanitizer as a Potential Alternative for Alcohol-Based Hand Gels. *Saudi Pharm. J.* **2021**, *29*, 807–814. [CrossRef]
3. Booq, R.Y.; Alshehri, A.A.; Almughem, F.A.; Zaidan, N.M.; Aburayan, W.S.; Bakr, A.A.; Kabli, S.H.; Alshaya, H.A.; Alsuabeyl, M.S.; Alyamani, E.J.; et al. Formulation and Evaluation of Alcohol-Free Hand Sanitizer Gels to Prevent the Spread of Infections during Pandemics. *Int. J. Environ. Res. Public Health* **2021**, *18*, 6252. [CrossRef]
4. Lukić, M.; Pantelić, I.; Savić, S.D. Towards Optimal Ph of the Skin and Topical Formulations: From the Current State of the Art to Tailored Products. *Cosmetics* **2021**, *8*, 69. [CrossRef]
5. Choudhary, S.; Malik, Y.S.; Tomar, S. Identification of SARS-CoV-2 Cell Entry Inhibitors by Drug Repurposing Using in Silico Structure-Based Virtual Screening Approach. *Front. Immunol.* **2020**, *11*, 1664. [CrossRef] [PubMed]
6. Kaczorek, E.; Chrzanowski; Pijanowska, A.; Olszanowski, A. Yeast and Bacteria Cell Hydrophobicity and Hydrocarbon Biodegradation in the Presence of Natural Surfactants: Rhamnolipides and Saponins. *Bioresour. Technol.* **2008**, *99*, 4285–4291. [CrossRef] [PubMed]
7. Zeng, G.M.; Shi, J.G.; Yuan, X.Z.; Liu, J.; Zhang, Z.B.; Huang, G.H.; Li, J.B.; Xi, B.D.; Liu, H.L. Effects of Tween 80 and Rhamnolipid on the Extracellular Enzymes of *Penicillium simplicissimum* Isolated from Compost. *Enzyme Microb. Technol.* **2006**, *39*, 1451–1456. [CrossRef]
8. Zeng, G.; Liu, Z.; Zhong, H.; Li, J.; Yuan, X.; Fu, H.; Ding, Y.; Wang, J.; Zhou, M. Effect of Monorhamnolipid on the Degradation of N-Hexadecane by Candida Tropicalis and the Association with Cell Surface Properties. *Appl. Microbiol. Biotechnol.* **2011**, *90*, 1155–1161. [CrossRef]
9. Shao, B.; Liu, Z.; Zhong, H.; Zeng, G.; Liu, G.; Yu, M.; Liu, Y.; Yang, X.; Li, Z.; Fang, Z.; et al. Effects of Rhamnolipids on Microorganism Characteristics and Applications in Composting: A Review. *Microbiol. Res.* **2017**, *200*, 33–44. [CrossRef]
0. Koynova, R.; Tenchov, B. Interactions of Surfactants and Fatty Acids with Lipids. *Curr. Opin. Colloid Interface Sci.* **2001**, *6*, 277–286. [CrossRef]
1. Chaudhary, N.K.; Chaudhary, N.; Dahal, M.; Guragain, B.; Rai, S.; Chaudhary, R.; Sachin, K.; Lamichhane-Khadka, R.; Bhattarai, A. Fighting the SARS CoV-2 (COVID-19) Pandemic with Soap. *Preprints* **2020**, 2020050060. [CrossRef]
2. Bharali, P.; Das, S.; Ray, A.; Pradeep Singh, S.; Bora, U.; Kumar Konwar, B.; Singh, C.B.; Sahoo, D. Biocompatibility Natural Effect of Rhamnolipids in Bioremediation Process on Different Biological Systems at the Site of Contamination. *Bioremediat. J.* **2018**, *22*, 91–102. [CrossRef]
3. Monteiro, S.A.; Sassaki, G.L.; de Souza, L.M.; Meira, J.A.; de Araújo, J.M.; Mitchell, D.A.; Ramos, L.P.; Krieger, N. Molecular and Structural Characterization of the Biosurfactant Produced by *Pseudomonas aeruginosa* DAUPE 614. *Chem. Phys. Lipids* **2007**, *147*, 1–13. [CrossRef]
4. Rahman, P.K.S.M.; Pasirayi, G.; Auger, V.; Ali, Z.; Rahman, K.S.M. Production of Rhamnolipid Biosurfactants by *Pseudomonas aeruginosa* DS10-129 in a Microfluidic Bioreactor. *Biotechnol. Appl. Biochem.* **2010**, *55*, 45–52. [CrossRef]
5. Abraham, M.H.; Hassanisadi, M.; Jalali-Heravi, M.; Ghafourian, T.; Cain, W.S.; Cometto-Muñiz, J.E. Draize Rabbit Eye Test Compatibility with Eye Irritation Thresholds in Humans: A Quantitative Structure-Activity Relationship Analysis. *Toxicol. Sci.* **2003**, *76*, 384–391. [CrossRef]
6. Ameeduzzafar; Syed Sarim, I.; Abbas Bukhari, S.N.; Ahmad, J.; Ali, A. Formulation and Optimization of Levofloxacin Loaded Chitosan Nanoparticle for Ocular Delivery: In-Vitro Characterization, Ocular Tolerance and Antibacterial Activity. *Int. J. Biol. Macromol.* **2018**, *108*, 650–659. [CrossRef]

67. Guideline, O.; The, F.O.R.; Of, T. *OECD 404 Guidelines for Testing of Chemicals—Acute Dermal Irritation/Corrotion*; Organization fo Economic Cooperation and Development: Paris, France, 2002; pp. 1–13.
68. Suvarna, S.K.; Layton, C.; Bancroft, J.D. *Bancroft's Theory and Practice of Histological Techniques*; Elsevier Health Sciences: Amster dam, The Netherlands, 2019.
69. Hirasawa, Y.; Ohtsu, S.; Matsui, Y.; Nagase, T.; Shimizu, M.; Masaaki, O.; Kyuki, K.; Takahashi, T.; Takahashi, K. Assessing Effects of a Product Containing Crude Drugs Including Scutellaria Root, Phellodendron Bark, Coptis Rhizome and Produc Containing Crude Drugs Including Lithospermum Root, Japanese Angelica Root, Sesame Oil in Atopic Dermatitis NC/Nga Mice *Pharmacometrics* **2000**, *59*, 123–134.

antibiotics

Article

Antibiofilm and Antivirulence Activities of Gold and Zinc Oxide Nanoparticles Synthesized from Kimchi-Isolated *Leuconostoc* sp. Strain C2

Min-Gyun Kang [1], Fazlurrahman Khan [2,3,*], Du-Min Jo [1,2,3], DoKyung Oh [1,2,3], Nazia Tabassum [2,3] and Young-Mog Kim [1,2,3,*]

[1] Department of Food Science and Technology, Pukyong National University, Busan 48513, Korea
[2] Marine Integrated Biomedical Technology Center, The National Key Research Institutes in Universities, Pukyong National University, Busan 48513, Korea
[3] Research Center for Marine Integrated Bionics Technology, Pukyong National University, Busan 48513, Korea
* Correspondence: fkhan055@pknu.ac.kr (F.K.); ymkim@pknu.ac.kr (Y.-M.K.); Tel.: +82-51-629-5832 (Y.-M.K.); Fax: +82-51-629-5824 (Y.-M.K.)

Abstract: The rapid emergence of antimicrobial resistance (AMR) among bacterial pathogens results in antimicrobial treatment failure and the high mortality rate associated with AMR. The application of nanoparticles synthesized from probiotics will be widely accepted due to their efficacy and biocompatibility in treating microbial infections in humans. The current work sought to isolate and identify lactic acid bacteria (LAB) from Kimchi. Based on 16S rRNA gene sequencing, the LAB isolate C2 was identified as a member of the genus *Leuconostoc*. The obtained supernatant from *Leuconostoc* sp. strain C2 was employed for the green synthesis of metal (AuNPs) and metal oxide (ZnONPs) nanoparticles. UV–vis absorption spectra, FTIR analysis, XRD, DLS, FE-TEM, and EDS mapping were used to fully characterize these C2-AuNPs and C2-ZnONPs. The C2-AuNPs were found to be spherical in shape, with a size of 47.77 ± 5.7 nm and zeta potential of −19.35 ± 0.67 mV. The C2-ZnONPs were observed to be rod-shaped and 173.77 ± 14.53 nm in size. The C2-ZnONPs zeta potential was determined to be 26.62 ± 0.35 mV. The C2-AuNPs and C2-ZnONPs were shown to have antimicrobial activity against different pathogens. Furthermore, these nanoparticles inhibited the growth of *Candida albicans*. The antibiofilm and antivirulence properties of these NPs against *Pseudomonas aeruginosa* and *Staphylococcus aureus* were thoroughly investigated. C2-AuNPs were reported to be antibiofilm and antivirulence against *P. aeruginosa*, whereas C2-ZnONPs were antibiofilm and antivirulence against both *P. aeruginosa* and *S. aureus*. Furthermore, these nanoparticles disrupted the preformed mature biofilm of *P. aeruginosa* and *S. aureus*. The inhibitory impact was discovered to be concentration-dependent. The current research demonstrated that C2-AuNPs and C2-ZnONPs exhibited potential inhibitory effects on the biofilm and virulence features of bacterial pathogens. Further studies are needed to unravel the molecular mechanism behind biofilm inhibition and virulence attenuation.

Keywords: *Leuconostoc* sp. strain C2; supernatant; nanoparticles; *Pseudomonas aeruginosa*; *Staphylococcus aureus*; antibiofilm; antivirulence

1. Introduction

The emergence of antimicrobial resistance (AMR) among bacterial pathogens has become a major concern globally [1]. The problem arises due to bacterial resistance, imposes a significant economic burden, and necessitates large sums of money to treat the patients [2]. According to a recent study, around 4.95 million individuals died as a result of bacterial AMR [3]. The ESKAPE (*Enterococcus faecium*, *Staphylococcus aureus*, *Klebsiella pneumoniae*, *Acinetobacter baumannii*, *Pseudomonas aeruginosa*, and *Enterobacter* spp.) are the top priority bacterial pathogens on the WHO's comprehensive list of which new antimicrobial agents must be identified urgently [1,4]. Pathogens have several antimicrobial resistance mechanisms, including drug target alteration, cell permeability changes, increased efflux pump, genetic mutations, and the acquisition of mobile genetic elements [5,6]. Furthermore, these pathogens exhibit an adaptive mechanism in the form of biofilm development, which is a collection of microbial populations covered by self-produced polymeric substances [7]. The biofilms serve as a physical barrier against antimicrobial agents [8]. Biofilms also cause the formation of a small population of persister cells, which are the primary source of persistent and recurring infection [9]. Biofilm formation has been seen in human tissues such as skin, teeth, trachea, and urinary system surfaces, as well as the surfaces of various medical devices [10]. With the growing concern over bacterial pathogen infection and resistance, there is always a high demand for new drugs to tackle these diseases. Much progress has been made in treating infectious and non-infectious disorders due to the advancement of nanotechnology [11]. The use of nanoparticles has grown due to various benefits, including a large surface area, small size, stability, and biocompatibility [12]. Furthermore, the green technique for nanoparticle synthesis offers significant benefits over the chemical-based technique [13]. Green synthesis NPs have several benefits over chemical and physical approaches, including being easier to synthesize, eco-friendly, cost-effective, minimizing the release of harmful compounds into the environment, and producing stable and biocompatible NPs [14,15]. There are more and more reports on the production of nanoparticles from biologically derived materials such as plants, algae, fungi, bacteria, and animals [13]. Since probiotics, particularly lactic acid bacteria, have health benefits, they provide a number of benefits to human health and well-being. Several studies have found that LAB in the gut microbiota utilize a competitive exclusion strategy to avoid the colonization of pathogenic bacteria [16]. As a result, using LAB in nanoparticle synthesis would be another viable technique with highly biocompatible applicability for treating infections caused by bacterial pathogens. The current study is divided into three sections: (1) the isolation and identification of LAB from Kimchi; (2) the synthesis of metal (e.g., gold) and metal oxide (e.g., zinc oxide) nanoparticles using the LAB's cell-free supernatant; and (3) the application of these nanoparticles to inhibit the biofilm formation and virulence properties of *S. aureus* and *P. aeruginosa* bacterial pathogens.

2. Results

2.1. Isolation and Identification of the Lactic Acid Bacteria

A total of 15 LAB strains were isolated from the fermented food product Kimchi from the Korean island of Jeju. The isolates were screened for antioxidant activity, and the best strain was chosen for the metal and metal–oxide nanoparticle synthesis. The isolate C2 was identified using 16S rRNA gene sequencing. The 16S rRNA gene sequence of strain C2 was determined to be most similar to the genus *Leuconostoc*. A blast search also reveals that it closely resembles several LAB strains from the genus *Weissella*. The phylogenetic tree analysis also reveals the presence of strain C2 between the branches of *Weissella paramesenteroides* and *Leuconostoc inhae* (Figure 1). We identified it as *Leuconostoc* sp. strain C2 based on the highest similarity with several LAB strains of the genus *Leconostoc*. The 16S rRNA gene sequence was submitted to GenBank under accession number OP482275.

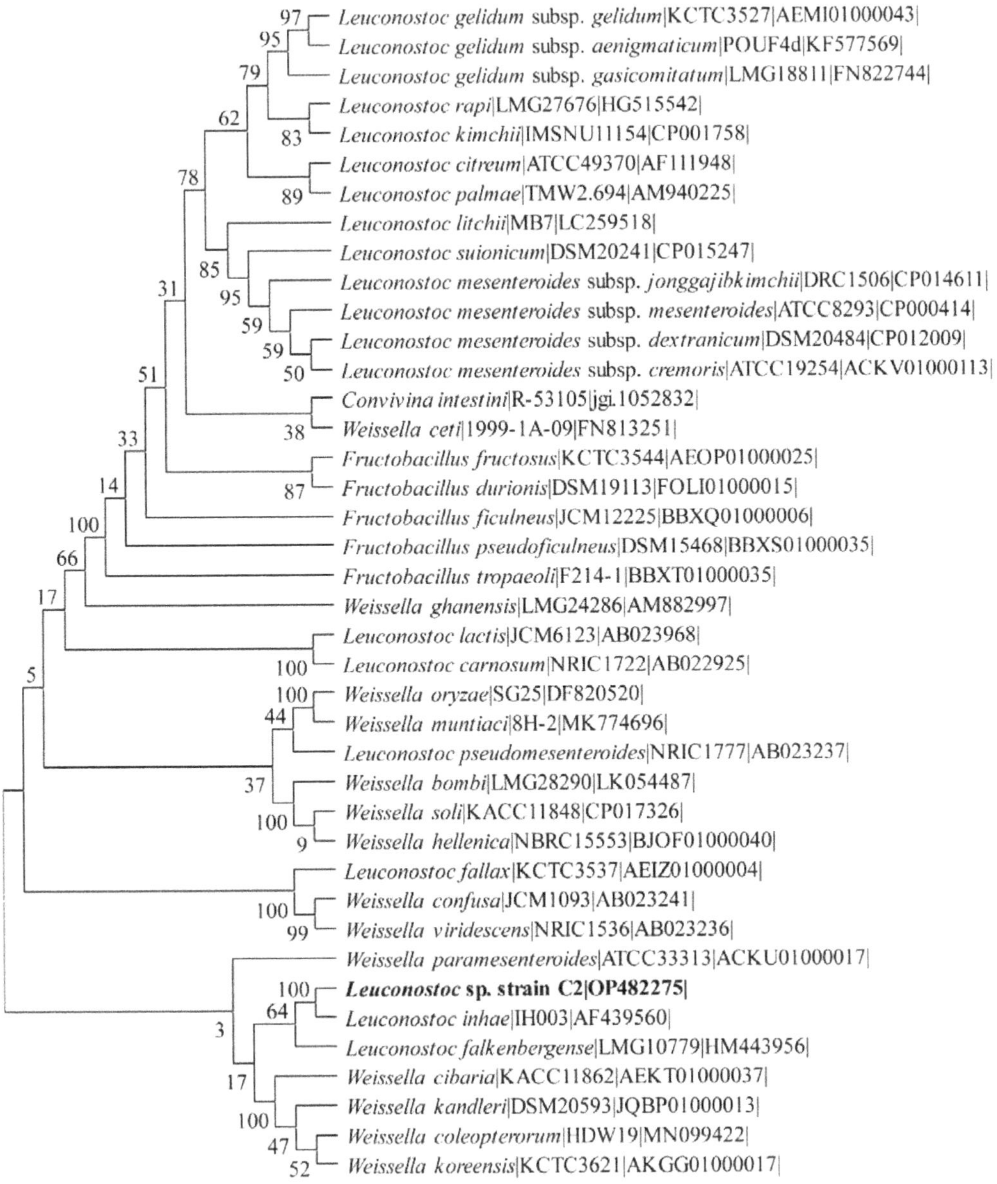

Figure 1. Phylogenetic tree of strain C2 based on the 16S rRNA gene sequence. The tree shows relations with the other LAB-type strain. The neighbor-joining tree was constructed using the bootstrap value of 1000 under the MEGA 11 package [17].

2.2. Biosynthesis and Characterization of the Gold and Zinc Oxide Nanoparticles

The supernatant of the LAB strain C2 was used as a reducing biological material for the synthesis of both metal and metal–oxide nanoparticles in a green synthesis strategy. The formation of AuNPs is indicated by the visible appearance of the final red-wine color from the initial yellow color of the reaction. In contrast, a white precipitated formation suggests the formation of ZnONPs. The manifestation of nanoparticles was further examined under the UV–vis absorption spectra, which show that particular specific absorption peaks increase with the reaction time. The absorption peak at 540 nm was found to constantly grow with the reaction in the case of AuNPs (Figure 2A), whereas the absorption peak at 330 nm was observed to increase over the course of the reaction in the case of ZnONPs (Figure 2B). Furthermore, the FTIR analysis of these nanoparticles was performed to record some of the distinctive vibration band that is particular to the nanoparticles (Figure 2C,D). The typical absorption peaks in the FTIR of C2-AuNPs are seen at 3253, 2935, 1584, 1407, and 1037 cm^{-1} (Figure 2C). Similar bands of absorption have previously been discovered in the FTIR, indicating their uniqueness to AuNPs [18]. Similarly, there are certain distinctive bands in the FTIR of C2-ZnONPs at 3396, 2948, 1578, 1406, and 1033 cm^{-1} (Figure 2D). Similar absorption bands have been observed in ZnONPs synthesized from *Marinobacter* sp. 2C8 and *Vibrio* sp. VLA [19]. The stretching of hydroxyl bonds is recognized as having an absorption peak at 3396 cm^{-1}, a peak at 2948 cm^{-1} for the C-H, a peak at 1578 cm^{-1} for the C-N, a peak at 1406 cm^{-1} for the C-H, and a peak at 1033 cm^{-1} for the C-O stretching of amide II.

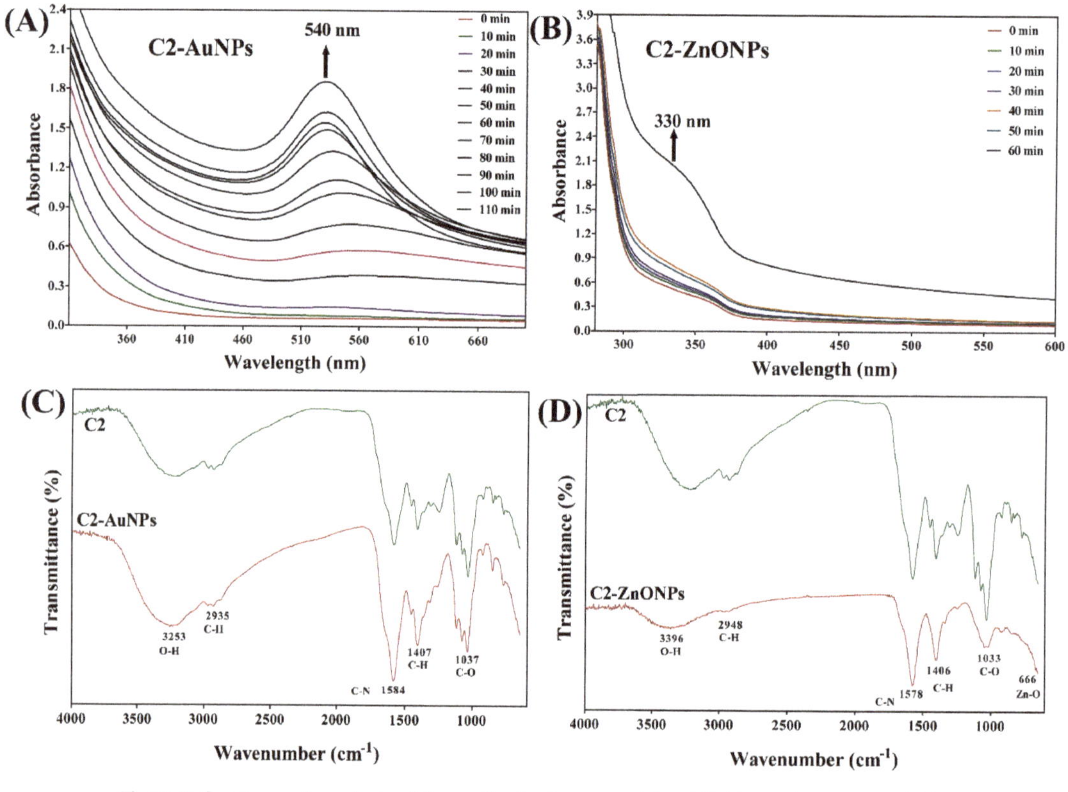

Figure 2. Spectroscopic analysis of the synthesized nanoparticles. (**A**) Absorption spectra of C2-AuNPs; (**B**) Absorption spectra of C2-ZnONPs; (**C**) FTIR spectra of the C2-supernatant and C2-AuNPs; and (**D**) FTIR spectra of C2-supernatant and C2-ZnONPs.

The morphology of C2-AuNPs and C2-ZnONPs was examined with the assistance of field-emission transmission electron microscopy (FE-TEM) (Figure 3). The C2-AuNPs have a spherical shape, as seen in images acquired by FE-TEM with resolutions of 100, 50, and 10 nm (Figure 3A–C). C2-ZnONPs were discovered to be rod-shaped, as seen in FE-TEM images with resolutions of 200, 100, and 50 nm (Figure 3E–G). The size of C2-AuNPs was calculated by dynamic light scattering (DLS) to be 47.77 ± 5.7 nm, with a polydispersity index (PDI) of 0.40 (Figure 4A).

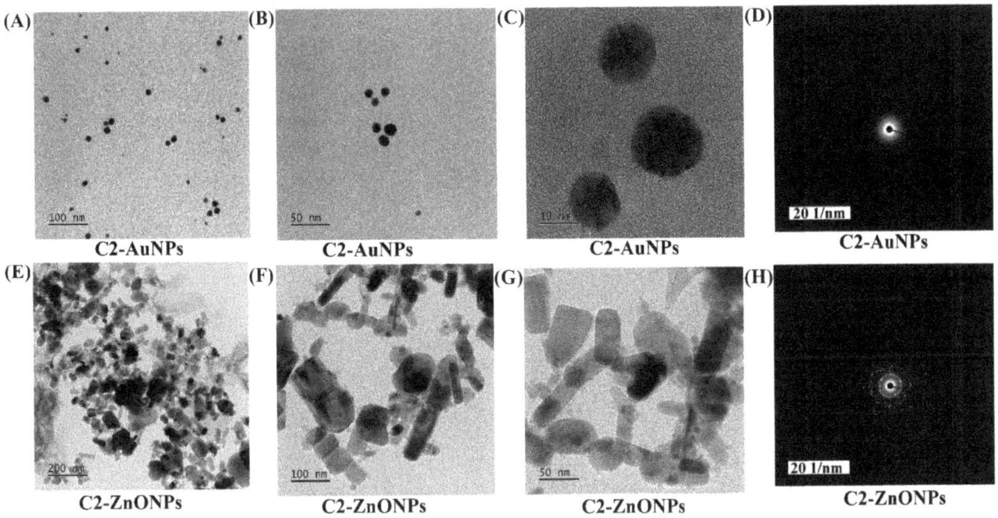

Figure 3. The morphology of nanoparticles was examined using FE-TEM. (**A**) FE-TEM image of C2-AuNPs at 100 nm resolution; (**B**) FE-TEM image of C2-AuNPs at 50 nm resolution; (**C**) FE-TEM image of C2-AuNPs at 10 nm resolution; (**D**) SAED of C2-AuNPs; (**E**) FE-TEM image of C2-ZnONPs at 200 nm resolution; (**F**) FE-TEM image of C2-ZnONPs at 100 nm resolution; (**G**) FE-TEM image of C2-ZnONPs at 50 nm resolution; and (**H**) SAED of C2-ZnONPs.

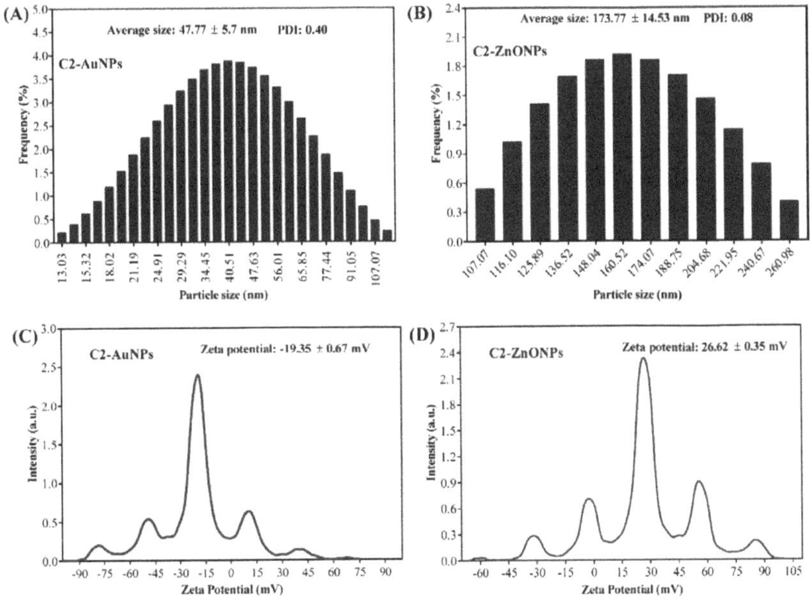

Figure 4. (**A**) DLS of C2-AuNPs; (**B**) DLS of C2-ZnONPs; (**C**) Zeta potential of C2-AuNPs; and (**D**) zeta potential of C2-ZnONPs.

Similarly, the size distribution of C2-ZnONPs determined by DLS was found to be with an average value of 173.77 ± 14.53 nm and PDI value of 0.08 (Figure 4B). The overall net charge on the surface of C2-AuNPs was found to be −19.35 ± 0.67 mV, while the net charge on the surface of C2-ZnONPs was 26.62 ± 0.35 mV (Figure 4C,D). EDS analysis was used to determine the elemental composition and mapping in C2-AuNPs and C2-ZnONPs (Figure 5). The EDS spectra and elemental mapping findings showed the presence of Au in C2-AuNPs (Figure 5B,C), as well as Zn and O in C2-ZnONPs (Figure 5E–G). XRD analysis was used to assess the crystallinity of these nanoparticles (Figure 6). Although no identifiable peaks were identified in the case of the C2-supernatant, a wide peak between 16 and 25° was observed (Figure 6A). The C2-AuNP diffraction peaks were seen at 2θ values of 28.40, 31.76, 38.22, 40.59, 44.58, 50.28, 56.53, 64.75, 66.35, 75.25, 77.61, and 84.68 (Figure 6B). Similarly, the C2-ZnONP diffraction peaks were detected at 2θ values of 34.28, 36.10, 47.36, 56.40, 58.97, 62.70, 67.69, 68.6, and 76.6 (Figure 6C).

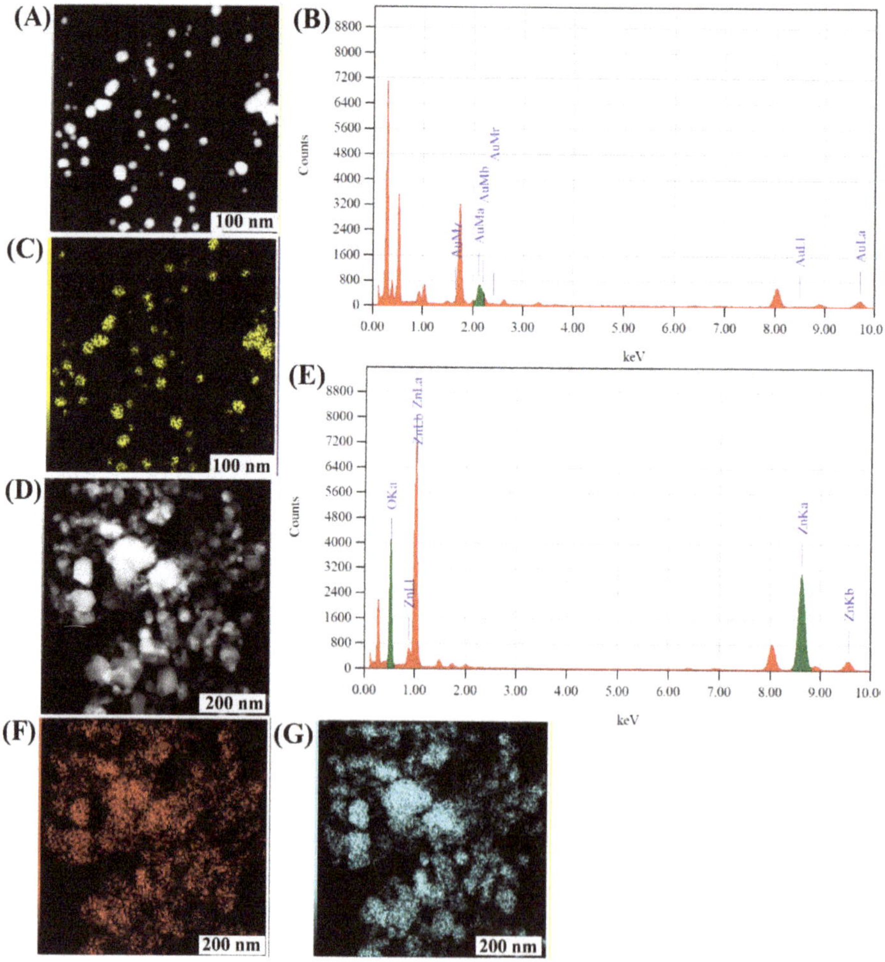

Figure 5. Energy-dispersive X-ray diffraction spectrum and EDS mapping of nanoparticles. (**A**) SEM elemental mapping of C2-AuNPs; (**B**) EDS spectrum of C2-AuNPs; (**C**) Au elemental map from C2-AuNPs; (**D**) SEM elemental mapping of C2-ZnONPs; (**E**) EDS spectrum of C2-ZnONPs; (**F**) O elemental map from C2-ZnONPs; and (**G**) Zn elemental map from C2-ZnONPs.

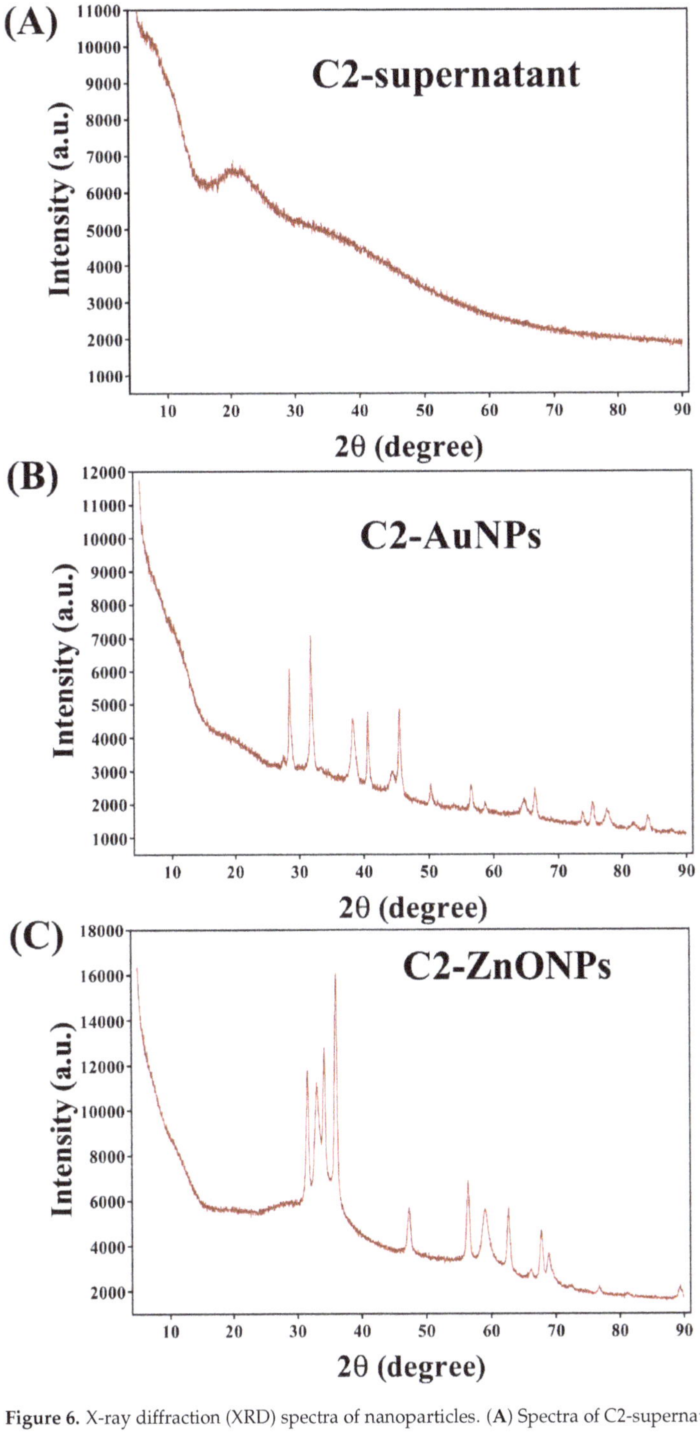

Figure 6. X-ray diffraction (XRD) spectra of nanoparticles. (**A**) Spectra of C2-supernatant; (**B**) Spectra of C2-AuNPs; and (**C**) Spectra of C2-ZnONPs.

2.3. Antimicrobial Effect of LAB Supernatant and Synthesized Nanoparticles

The growth inhibitory effects of strain C2 and the synthesized nanoparticles (C2-AuNPs and C2-ZnONPs) against different bacterial species, including the fungal pathogen *Candida albicans*, were investigated utilizing microbroth dilution methods. The antimicrobial results obtained in this study are reported in Table 1. The antimicrobial activity of the C2 supernatant against *S. aureus*, *K. pneumoniae*, *S. mutans*, *E. coli*, and *L. monocytogenes* was effective (>90% cell growth inhibition) at a 1/4-dilution of the prepared supernatant stock. Whereas *C. albicans* was found to be effective at a 1/2-dilution, *V. vulnificus* was effective at a 1/8-dilution of the supernatant.

Table 1. MIC values of C2-supernatant, C2-AuNPs, and C2-ZnONPs against different microbial pathogens.

Sample	PA	SA	SM	CA	EC	VV	KP	LM
C-2 supernatant (%)	ND	25	25	50	25	12.5	25	25
C2-AuNPs (μg/mL)	>2048	1024	512	2048	1024	ND	512	2048
C2-ZnONPs (μg/mL)	>2048	512	ND	256	>2048	>2048	>2048	>2048

ND, not determined. PA, *Pseudomonas aeruginosa*; SA, *Staphylococcus aureus*; SM, *Streptococcus mutans*; CA, *Candida albicans*; EC, *Escherichia coli*; VV, *Vibrio vulnificus*; KP, *Klebsiella pneumoniae*; LM, *Listeria monocytogenes*.

The minimum inhibitory concentration (MIC) values of C2-AuNPs and C2-ZnONPs were discovered to be variable against these microbial pathogens. The MIC values of C2-AuNPs against Gram-negative bacteria such as *P. aeruginosa*, *L. monocytogenes*, and *E. coli* were determined to be >2048, 2048, and 1024 μg/mL, respectively. In contrast, the MIC values of C2-AuNPs against Gram-positive bacteria such as *S. aureus*, *S. mutans*, and *K. pneumonaie* were determined to be 1024, 512, and 512 μg/mL, respectively. Similarly, the MIC values of C2-ZnONPs against *P. aeruginosa* were >2048 μg/mL, whereas the MIC values against *S. aureus* were (512 μg/mL), which is four times lower than *P. aeruginosa*. The current investigation found that the MIC values of C2-AuNPs and C2-ZnONPs against Gram-negative bacteria were 2- and 4-fold higher, respectively, than for Gram-positive bacteria.

2.4. Antibiofilm Properties of the Nanoparticles toward Bacterial Pathogens

Crystal violet assays were used to assess the inhibitory impact of C2-AuNPs and C2-ZnONPs on *P. aeruginosa* and *S. aureus* biofilm formation at the early stage. The biofilm inhibition effect was achieved using sub-MIC values of C2-AuNPs and C2-ZnONPs. C2-AuNPs inhibited the biofilm formation solely against *P. aeruginosa*, whereas C2-ZnONPs inhibited biofilm formation against *P. aeruginosa* and *S. aureus*, respectively (Figure 7). C2-AuNPs and C2-ZnONPs dose-dependently inhibited these pathogens (Figure 7A,C). The maximal biofilm inhibition by C2-AuNPs against *P. aeruginosa* was determined to be 84.62% at 256 μg/mL. At a concentration of 512 μg/mL, the inhibitory effect of C2-ZnONPs on *P. aeruginosa* and *S. aureus* biofilms was 84.69% and 82.42%, respectively.

The cell growth of planktonic and adherent cells was measured at $OD_{600\ nm}$ to confirm that 100% of the cells were involved in biofilm formation in the presence of sub-MIC levels of C2-AuNPs and C2-ZnONPs. The growth findings demonstrated that *P. aeruginosa* and *S. aureus* cells were unaffected by the various concentration ranges of C2-AuNPs and C2-ZnONPs (Figure 7B,D). The biofilm inhibitory effects of C2-AuNPs and C2-ZnONPs on the biofilms of *P. aeruginosa* and *S. aureus* were also validated by examining the NPs in treated and untreated cells using scanning electron microscopy (SEM).

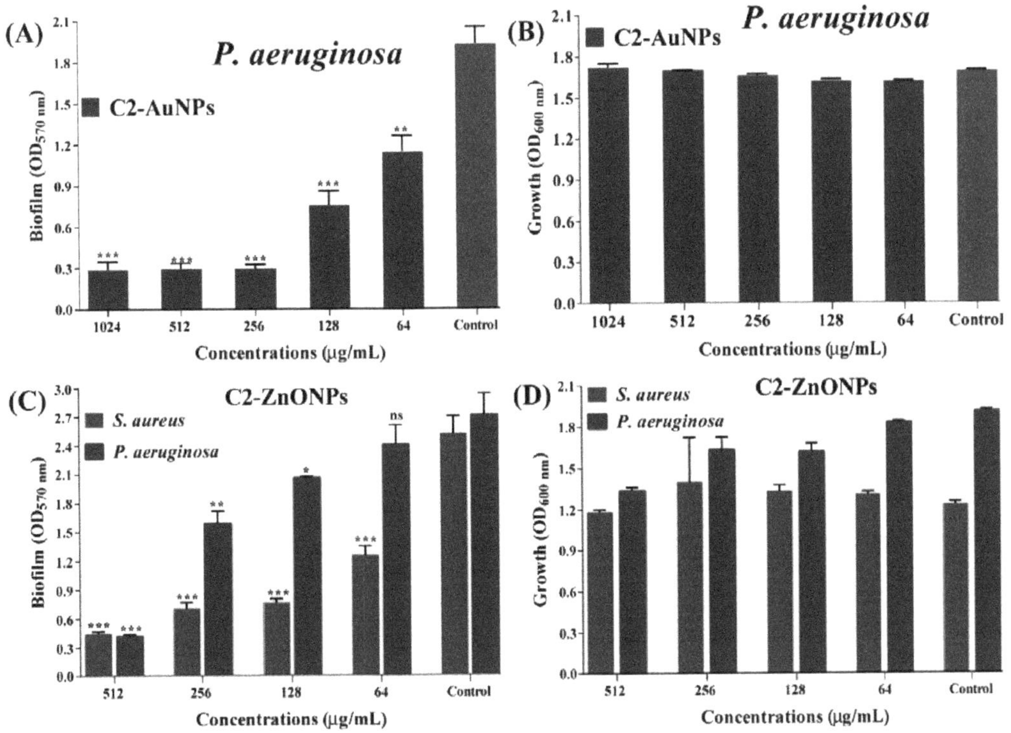

Figure 7. Biofilm inhibitory effect of nanoparticles towards *P. aeruginosa* and *S. aureus*. (**A**) Inhibition of *P. aeruginosa* biofilms by C2-AuNPs; (**B**) Cell growth of *P. aeruginosa* incubated with C2-AuNPs; (**C**) Inhibition of *P. aeruginosa* and *S. aureus* biofilms by C2-ZnONPs; and (**D**) Cell growth of *P. aeruginosa* and *S. aureus* incubated with C2-ZnONPs. The experiments were carried out using the crystal violet staining assays followed by measuring OD_{570}, whereas the cell growth of bacteria incubated with nanoparticles was estimated at OD_{600}. The statistical significance values ***, **, and * denote significance at $p < 0.0001$, $p < 0.01$, and $p < 0.05$—whereas ns denotes non-significance.

The SEM image showed that *P. aeruginosa* cells treated with C2-AuNPs inhibit cell adhesion on the nylon membrane (Figure 8A), but untreated (control) cells displayed a thick biofilm cell adhering to the membrane (Figure 8B). Similarly, C2-ZnONPs was discovered to have an effect on the biofilm of *P. aeruginosa* and *S. aureus*. *P. aeruginosa* cells treated with C2-ZnONPs stopped adhering and ceased the biofilm formation on the surface of the nylon membrane (Figure 8C). In contrast, a few cells of *S. aureus* were observed to be attached in the presence of C2-ZnONPs (Figure 8D). Untreated (control) *S. aureus* cells were shown to produce highly thick biofilm architecture on the nylon membrane, similar to *P. aeruginosa*. The SEM imaging of cells treated with NPs demonstrated that these NPs inhibit biofilm formation.

2.5. Eradication of the Established Mature Biofilm

The in vitro preformed-mature biofilms of *P. aeruginosa* and *S. aureus* were treated with C2-AuNPs and C2-ZnONPs at their above MIC, MIC, and sub-MIC concentrations. The eradication efficiency was assessed by quantifying the adherent cells using the crystal violet staining method. The results demonstrated that at the above MIC values, these NPs substantially eradicate the mature biofilm of *P. aeruginosa* and *S. aureus* (Figure 9A,B). The highest percentage of the eradication of mature *P. aeruginosa* biofilm by C2-AuNPs at a concentration of 2048 µg/mL was reported to be 87.75% (Figure 9A). Although C2-

ZnONPs inhibit the biofilm formation in both *P. aeruginosa* and *S. aureus*, it is most effective in *S. aureus* mature biofilm. The maximal eradication of the mature *S. aureus* biofilm by C2-ZnONPs at a concentration of 2048 μg/mL was reported to be 66.8% (Figure 9B). Furthermore, the *S. aureus* mature biofilm eradication by C2-ZnONPs was concentration-dependent.

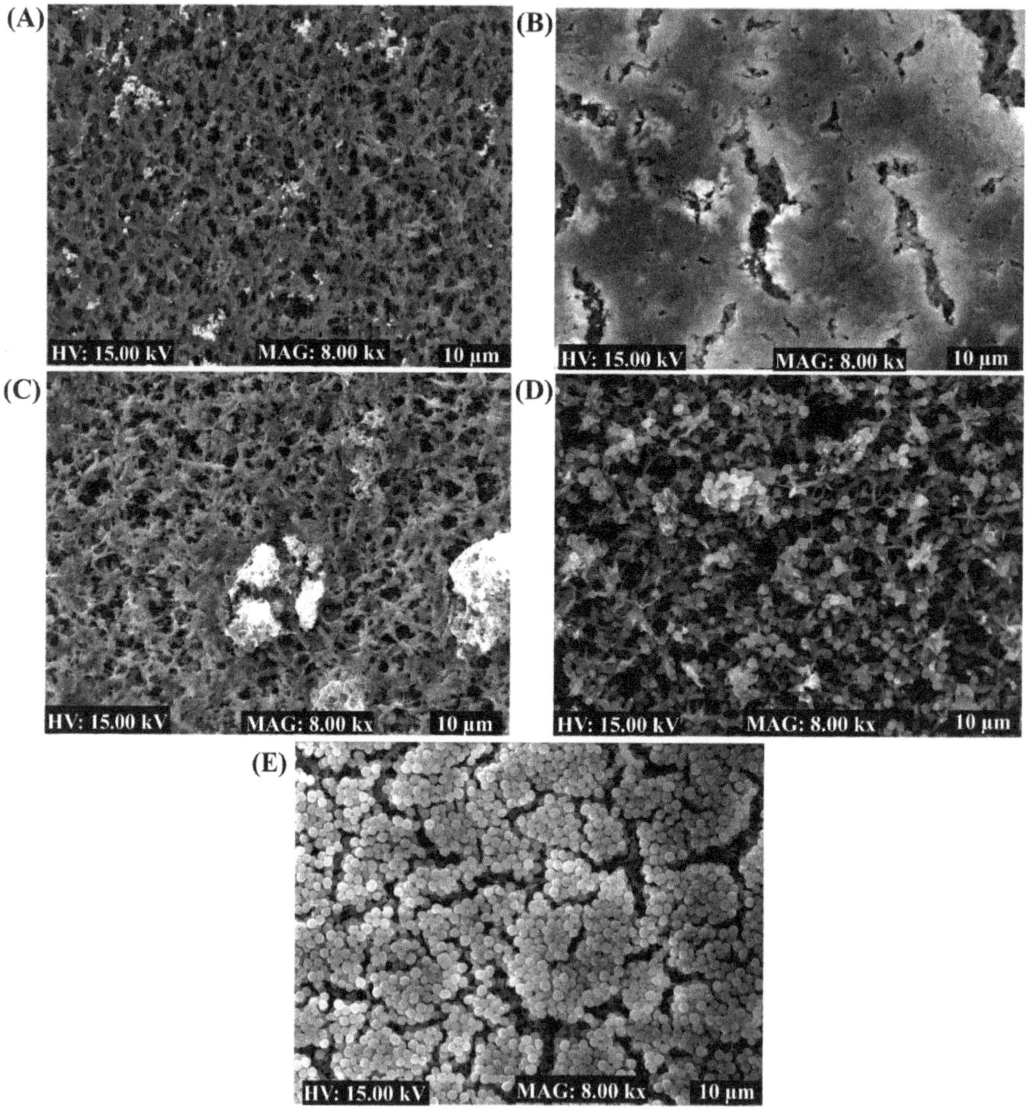

Figure 8. Imaging of biofilm cells of *P. aeruginosa* and *S. aureus* treated with C2-AuNPs and C2-ZnONPs using scanning electron microscopy. (**A**) Biofilm of *P. aeruginosa* treated with C2-AuNPs; (**B**) Biofilm of *P. aeruginosa* (untreated-control); (**C**) Biofilm of *P. aeruginosa* treated with C2-ZnONPs; (**D**) Biofilm of *S. aureus* treated with C2-ZnONPs; and (**E**) Biofilm of *S. aureus* (untreated-control).

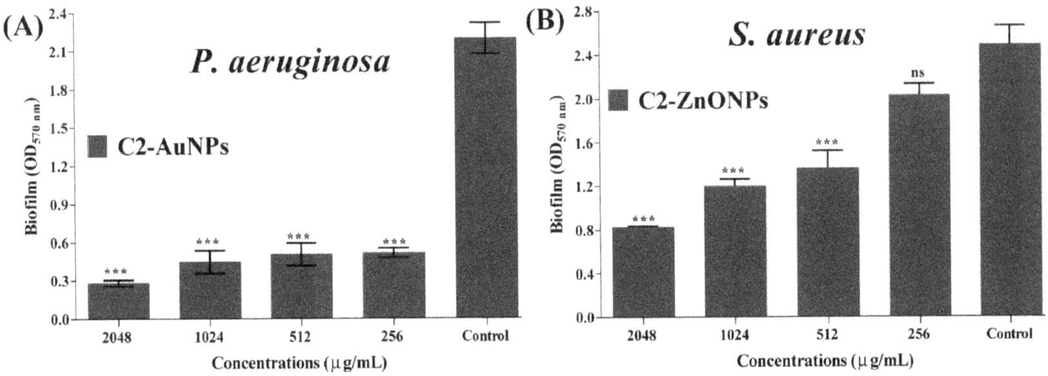

Figure 9. Analysis of the eradication of mature biofilm cells of *P. aeruginosa* and *S. aureus* by nanoparticles. (**A**) Eradication of the *P. aeruginosa* mature biofilm using C2-AuNPs and (**B**) eradication of *S. aureus* mature biofilm using C2-ZnONPs. Biofilm stained with crystal violet was used to determine eradication. The statistical significance *** denotes significance at $p < 0.0001$, and ns denotes non-significance.

2.6. Inhibitory Effect of NPs on *P. aeruginosa* Motilities

Different motility patterns in *P. aeruginosa*, such as swimming, swarming, and twitching, play a significant role in biofilm development and pathogenicity. Inhibiting various forms of motility will be an essential step in reducing *P. aeruginosa* biofilm formation and virulence activities. The sub-MIC value of C2-AuNPs and C2-ZnONPs was utilized to assess the impairment of these motilities (Figure 10). C2-AuNPs (Figure 10A) and C2-ZnONPs (Figure 10B) both significantly inhibited the swarming motility on the agar plate surface when compared to the control (Figure 10C). Swarming inhibition by C2-AuNPs and C2-ZnONPs was determined to be 44.31% and 43.18%, respectively (Figure 10D).

Both C2-AuNPs and C2-ZnONPs significantly inhibited the swimming motility of *P. aeruginosa* at sub-MIC concentrations (Figure 10E,F), whereas the control cells displayed considerable swimming motility (Figure 10G). Swimming motility inhibition by C2-AuNPs and C2-ZnONPs was shown to be 73.13% and 85.82% effective, respectively, when compared to the control (Figure 10H). C2-AuNPs and C2-ZnONPs were similarly observed to drastically reduce twitching motility when compared to the controls (Figure 10I–K). The twitching motility inhibition by C2-AuNPs and C2-ZnONPs was calculated to be 77.77% and 84.12%, respectively (Figure 10L).

2.7. Virulence-Inhibitory Properties of the Nanoparticles

A variety of virulence factors play a role in the pathogenesis of *P. aeruginosa* and *S. aureus* [20,21]. The effect of C2-AuNPs and C2-ZnONPs on *P. aeruginosa* pathogenicity characteristics such as hemolytic activity, pyocyanin, and pyoverdine production, and protease activity was studied at sub-MIC levels (Figure 11). In the case of *S. aureus*, we solely looked at the hemolytic effect of these NPs. In the case of *S. aureus*, however, we solely investigated the hemolytic effect of these NPs. However, there was no concentration-dependent inhibition of *P. aeruginosa* hemolytic activities by C2-AuNPs or C2-ZnONPs. However, at a concentration of 1024 μg/mL, C2-AuNPs and C2-ZnONPs inhibited the hemolytic activity by ~25% on *P. aeruginosa* (Figure 11A). The inhibition of *S. aureus* hemolytic activity by C2-AuNPs and C2-ZnONPs was highly significant and dose-dependent (Figure 11B). At a 512 μg/mL concentration, C2-AuNPs and C2-ZnONPs inhibited the hemolytic activity by 78.87% and 77.27%, respectively. C2-AuNPs and C2-ZnONPs were likewise discovered to dramatically inhibit pyocyanin virulence factor production from *P. aeruginosa* (Figure 11C). The inhibitory impact of C2-AuNPs and C2-ZnONPs on pyocyanin production was 71.07% and 82.15%, respectively, at 1024 μg/mL concentrations. Similarly, both C2-AuNPs and C2-ZnONPs inhibited the production of the *P. aeruginosa* pyoverdine virulence factor (Figure 11D). The inhibition was dose-dependent, with C2-AuNPs and C2-ZnONPs exhibiting a maximal inhibition of 55.1% and 85.24% at 1024 μg/mL, respectively.

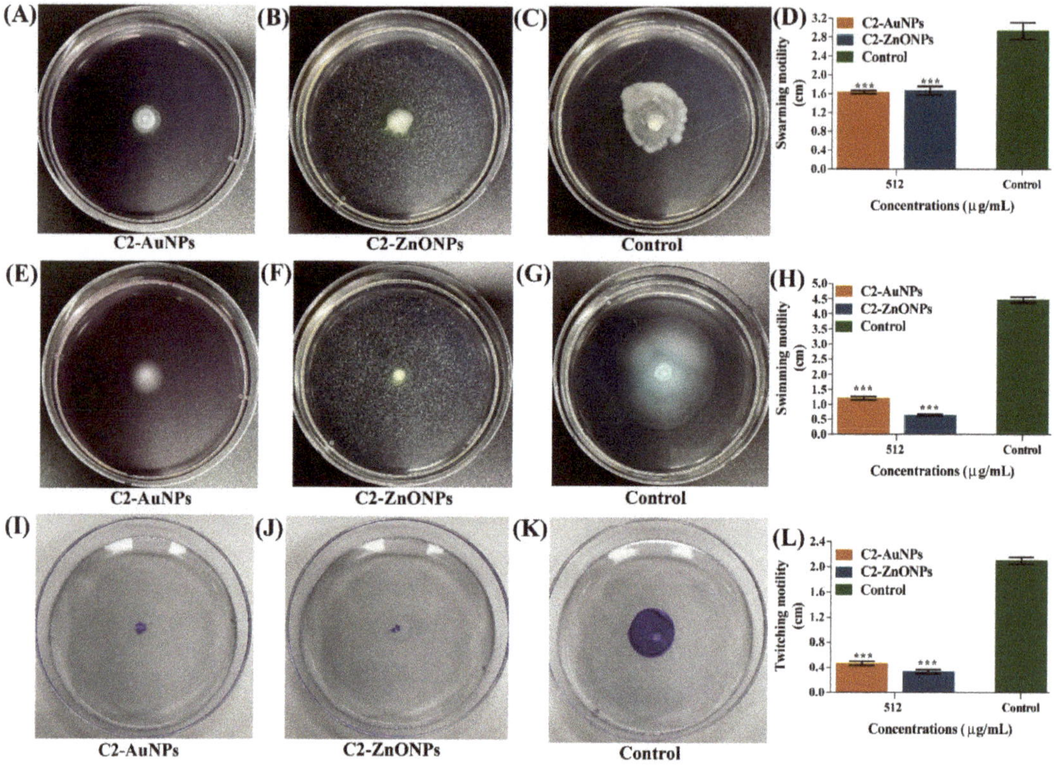

Figure 10. Inhibition of different types motilities of *P. aeruginosa*. (**A**) Swarming in the presence of C2-AuNPs; (**B**) Swarming in the presence of C2-ZnONPs; (**C**) Swarming of non-treated cells; (**D**) Bar graph showing swarming inhibitory values C2-AuNPs; (**E**) Swimming in the presence of C2-AuNPs; (**F**) Swimming in the presence of C2-ZnONPs; (**G**) Swimming of non-treated cells; (**H**) Bar graph showing swimming inhibitory values C2-AuNPs; (**I**) Twitching in the presence of C2-AuNPs; (**J**) Twitching in the presence of C2-ZnONPs; (**K**) Twitching of non-treated cells; and (**L**) Bar graph showing twitching inhibitory values C2-AuNPs. *** denotes significance at $p < 0.0001$.

In a casein agar plate assay, the *P. aeruginosa* protease activity was dramatically inhibited in the presence of C2-AuNPs and C2-ZnONPs at their sub-MIC value (Figure 11E,F). The supernatant produced from cells treated with C2-AuNPs and C2-ZnONPs at 1024 and 512 μg/mL concentrations had a negligible zone of casein protein digestion. The finding of these inhibitory effects by these NPs predicts that *P. aeruginosa* and *S. aureus* would be non-virulent in the presence of C2-AuNPs and C2-ZnONPs.

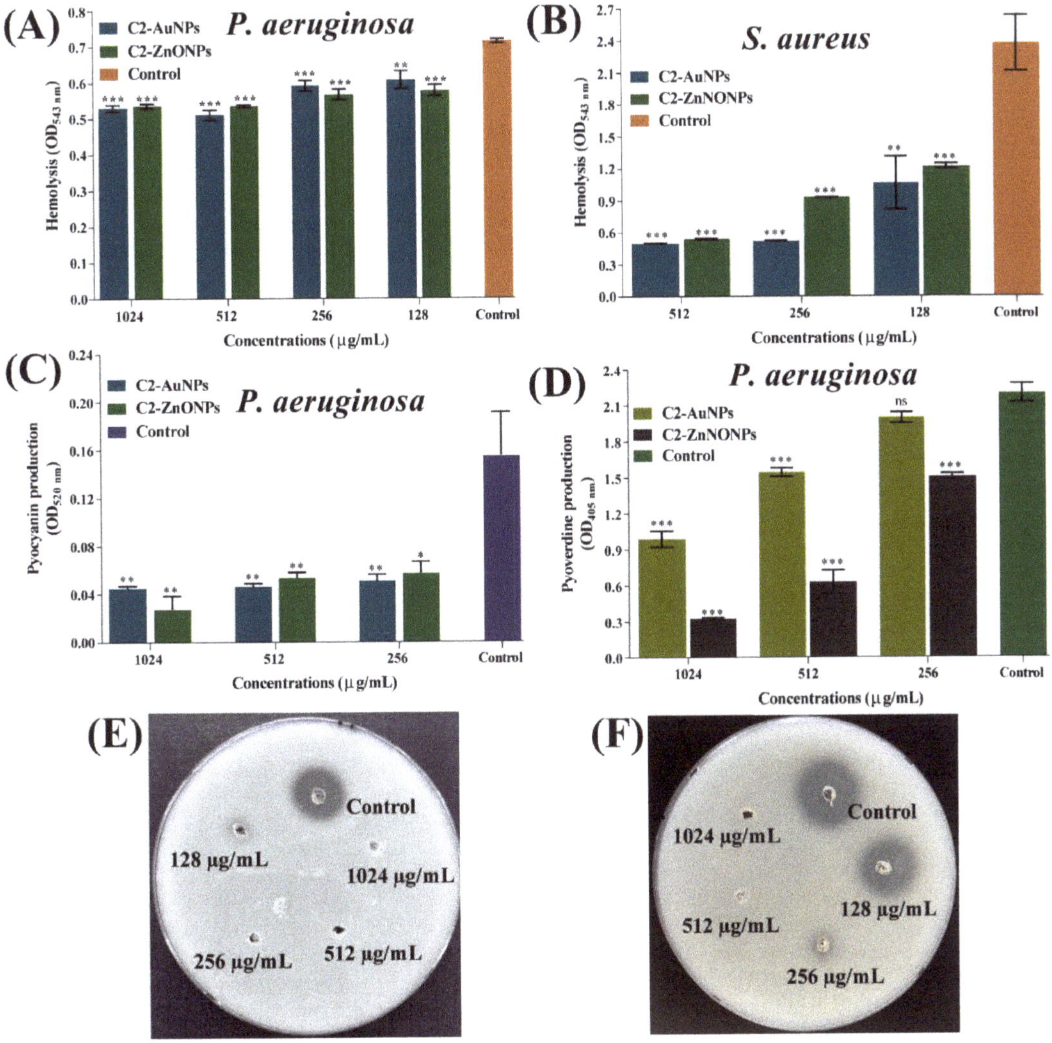

Figure 11. Inhibition of different virulence properties of *P. aeruginosa* and *S. aureus* by C2-AuNPs and C2-ZnONPs. (**A**) Inhibition of the hemolytic activity of *P. aeruginosa* by C2-AuNPs and C2-ZnONPs; (**B**) Inhibition of the hemolytic activity of *S. aureus* by C2-AuNPs and C2-ZnONPs, (**C**) Inhibition of the production of pyocyanin from *P. aeruginosa* by C2-AuNPs and C2-ZnONPs; (**D**) Inhibition of pyoverdine production from *P. aeruginosa* by C2-AuNPs and C2-ZnONPs; (**E**) Inhibition of protease activity of *P. aeruginosa* by C2-AuNPs; and (**F**) Inhibition of protease activity by C2-ZnONPs. ***, ** and * denote significance at $p < 0.0001$, $p < 0.01$, and $p < 0.05$.

3. Discussion

For instance, the biofilm formed by the microbial pathogens represents a key barrier in the failure of antimicrobial therapy [22]. Furthermore, these biofilm-forming bacterial pathogens produce hyper-mutative cells and persister cells, which are more persistent and have more resistance mechanisms than normal cells [9,23]. Hence, the infection issue associated with biofilm-forming bacterial pathogens always encourages the researcher to develop alternative strategies to combat the infection. With the several advantages of the application of nanoparticles in treating infectious and non-infectious diseases [11,24], we aimed to synthesize AuNPs and ZnONPs using biologically derived extracts, particularly from probiotic bacteria.

Since probiotic bacteria are widely accepted and have been used to colonize the intestines of both humans and animals for a variety of health benefits [25]. Several LAB were isolated from Korean Kimchi in this investigation (collected from Jeju). Among the 10 isolates, isolate C2 was chosen for further characterization due to its high antioxidant activity. C2 was identified as *Leuconostoc* sp. strain C2 and was selected for the synthesis and physiochemical characterization of metal and metal–oxide nanoparticles.

The supernatant from isolate C2 was prepared and employed in the green synthesis of AuNPs and ZnONPs. Several instrumental approaches were used to characterize the produced AuNPs and ZnONPs in detail. Previous work corroborates the preliminary indication of the formation of C2-AuNPs and C2-ZnONPs using the change in color from yellow to dark red wine [26]. Furthermore, increasing the UV–visible absorption peak at 540 nm throughout the reaction was used to confirm the formation of AuNPs and was found to be in close agreement with the prior studies [27,28]. Although no color changes were seen during the synthesis of C2-ZnONPs, the formation of a white precipitate implies the production of nanoparticles, as previously described [26]. The formation of C2-ZnONPs is also indicated by an increase in the UV–vis absorption peak at 330 nm, which is corroborated by prior work [26].

The shape, size, and zeta potential of C2-AuNPs were spherical, 47.77 ± 5.7 nm, and -19.35 ± 0.67 mV, respectively. Previous reports showed that the AuNPs prepared from different bacterial species, such as *Streptomyces* sp. strain NH21 and *Rhodopseudomonas capsulata* have also been reported to be spherical [28,29]. Although the size of AuNPs prepared from the different bacterial species was varied, the size of C2-AuNPs falls in those reported size ranges [28]. The zeta potential of C2-AuNPs demonstrates marginally stable NPs, with values close to previous reports [30]. Similarly, the shape, size, and net charge of C2-ZnONPs were determined to be a rod, 173.77 ± 14.53 nm and 26.62 ± 0.35 mV, respectively. Similar positive zeta potential values have been reported for ZnONPs synthesized employing bacteria such as *Pseudomonas hibiscicola* and *Alkalibacillus* sp.W7 [31,32]. The hexagonal, nanoflower, rod, cube, and spherical shapes of ZnONPs produced using bacterial extract have also been described [33,34]. C2-ZnONPs have a smaller size than ZnONPs-CFS (291.1 nm), which was synthesized using cell-free supernatant from *L. plantarum* TA [33]. The high average zeta potential shows that the synthesized C2-ZnONPs are stable and comparable to previously reported microbially synthesized ZnONPs [19].

According to reports, the conversion of metal ions into metallic nanoparticles requires reducing and capping the agents found in natural extracts [14]. Several metabolites were identified during GC–MS profiling, some of which were functionally described (Supplementary Materials Figure S1 and Table S1). These components have previously been shown to have antioxidant, antimicrobial, and other biological activities [35–37]. Based on the GC–MS profiling of the metabolites present in the supernatant, it is inferred that it includes numerous reducing molecules that may be responsible for the formation of nanoparticles. Future research should focus on synthesizing metal and metal–oxide nanoparticles utilizing pure active compounds. C2-AuNPs inhibit the biofilm of *P. aeruginosa*, whereas C2-ZnONPs have antibiofilm efficacy against both *P. aeruginosa* and *S. aureus*. C2-AuNPs and C2-ZnONPs were shown to have a concentration-dependent biofilm inhibitory effect. AuNPs and ZnONPs produced from various bacterial species, as well as pure natural compounds, have been shown to inhibit *P. aeruginosa* and *S. aureus* biofilms in a dose-dependent manner [19]. The maximal biofilm inhibition by C2-ZnONPs against *P. aeruginosa* at 512 μg/mL was equivalent to the previously reported biofilm inhibition by ZnONPs [19,38]. The anti-biofilm effect of these NPs was further validated by employing SEM imaging to examine the biofilm architecture, where cells treated with C2-AuNPs and C2-ZnONPs were either entirely detached or just a few cells adhered to the surface.

Similar studies on the effect of these types of NPs on biofilm architecture have previously been reported [26,39], indicating that the synthesized C2-AuNPs and C2-ZnONPs are potent antibiofilm drugs. The biofilms developed on the surfaces of medical devices and host tissues has become one of the major challenges in the battle against microbial infection [40,41]. The biofilm matrix is extremely thick and composed of extracellular polymeric substances (EPS), and it has been found to function as a barrier to antimicrobial agent entry [42]. The MIC and above MIC of C2-AuNPs and C2-ZnONPs successfully eradicated *P. aeruginosa*, and *S. aureus* developed mature biofilms. The results are consistent with the eradication efficiency of AuNPs and ZnONPs synthesized from various materials against the mature biofilms of *P. aeruginosa* and *S. aureus* [26,43]. Another alternate strategy for controlling pathogenicity and chronic infection caused by *P. aeruginosa* and *S. aureus* has been the diminution of their virulence features [44–46]. The sub-MIC of C2-AuNPs and C2-ZnONPs strongly inhibited *P. aeruginosa* motility properties, which is consistent with earlier findings of the nanoparticle-mediated inhibition of virulence features [26,39]. We can decrease the early stage of biofilm formation and the dissemination of bacterial cells throughout the dispersal of a mature biofilm by limiting the motility properties using these NPs [47]. Several additional virulence features of *P. aeruginosa*, such as pyocyanin production, pyoverdine formation, protease activity, and hemolytic activity, were significant at the sub-MIC levels of C2-AuNPs and C2-ZnONPs. The suppression of *P. aeruginosa* virulence properties by C2-AuNPs and C2-ZnONPs was comparable with prior virulence attenuation findings by similar types of nanoparticles [26,39]. The hemolytic activity of *S. aureus* was dose-dependently inhibited by both C2-AuNPs and C2-ZnONPs, which is consistent with prior findings on the inhibitory impact of such NPs. Although the specific mechanism involved in suppressing the biofilm and eradicating the mature biofilm by these nanoparticles is unknown in this study, very little information on the action mechanism of the nanoparticles is available in the literature. The cell membrane breakdown, inhibition of genes associated with biofilm and virulence features, and the formation of cellular reactive oxygen species, which can interfere with cellular metabolism, are the principal effects of NPs on bacterial pathogens [48–52]. To offer clear proof of the mechanism of action of these metal and metal–oxide NPs, more research into the gene expression of genes involved in biofilm development and virulence is necessary.

4. Materials and Methods

4.1. Microbial Pathogens, Culture Media, and Growth Conditions

Escherichia coli (KCTC1682), *Pseudomonas aeruginosa* PAO1 (KCTC1637), *Listeria monocytogenes* (KCTC3569), and *Staphylococcus aureus* (KCTC 1916) were purchased from the Korean Collection for Type Cultures (KCTC, Daejeon, Korea). *Klebsiella pneumoniae* (ATCC4352) was purchased from American Type Culture Collection (ATCC). *Streptococcus mutans* (KCCM 40105) and *Candida albicans* (KCCM11282) were purchased from the Korean Culture Center of Microorganisms (KCCM; Seodaemun-gu, Seoul, Korea). Tryptic soy broth (TSB; Difco Laboratory Inc., Detroit, MI, USA) was used to grow *P. aeruginosa*, *S. aureus*, *E. coli*, *L. monocytogenes*, and *Vibrio vulnificus*. For culturing *C. albicans*, the potato dextrose broth (PDB) containing glucose (5 %) was used, whereas for culturing the *S. mutans*, the brain heart infusion (BHI) broth (Difco Laboratory Inc., Detroit, MI, USA) was used. A deMan Rogosa Sharpe Medium (MRS; Difco Laboratory Inc., Detroit, MI, USA) was used to cultivate LAB. Gold (III) chloride trihydrate (CAS# 16961-25-4; purity 99%) and zinc acetate ($Zn(CH_3\text{-}COO)_2{\cdot}2H_2O$) (CAS # 5970-45-6) were obtained from Sigma-Aldrich Co. (St. Louis, MO, USA). The temperature for the growth of all microbes was 37 °C.

4.2. Isolation and Identification of LAB Strains

The LAB strain was isolated from the Kimchi sample using the previously described method with a small modification [53]. In brief, the 25 g Kimchi sample was diluted 10 times with 0.1 M phosphate-buffered saline (PBS; pH 7.2), and then became homogenous before being employed as an experiment sample. The homogenous sample was inoculated to the surface of MRS agar containing 0.002% (*w*/*v*) Bromocresol purple (BCP; Duksan Pure Chemical Co., Ltd., Ansan-si, Korea). It was cultured at 37 °C for 24 h–48 h. A single colony was isolated from the agar plate based on the change in yellow color caused by the formation of organic acid by LAB. As described earlier, the isolated LAB strain was taxonomically identified using 16S rRNA gene sequencing [53,54]. The genomic DNA of the LAB isolates was extracted using the extraction Kit (Bioneer, Daejeon, Korea). The universal primer used for sequencing the 16S rRNA gene was 27F (5′-AGAGTTTGATCCTGGCTCAG-3′) and 1492R (5′-GTTACCTGTTACGACTT-3′). The sequencing of the 16S rRNA gene was carried out by the company Bionics (Seoul, Korea). The 16S rRNA gene sequence of strain C2 and a closely related type strain (retrieved from the EzBiOCloud 16S database) were aligned using the multiple sequence alignment program ClustalX version 2.0.12. The phylogenetic tree was built with the neighbor-joining algorithm and the distance estimation method in MEGA 11 [17].

4.3. Preparation of LAB Cell-Free Culture Supernatant (CFCS)

LAB strains were cultivated in an MRS medium for 24 h at 37 °C to reach a cell growth value of ~9 log CFU/mL. The supernatant from the growing cell culture was separated by centrifugation (10,000× *g* for 20 min) at 4 °C. The supernatant was filtered via a filter membrane with a pore size of 0.20 μm (DISMIC 25AS020AN, Advantec Toyo, Tokyo, Japan). The resulting supernatant was utilized to test the antibacterial activity and nanoparticle synthesis. In addition, the LAB supernatant was lyophilized using a freeze-dryer (FD8518, ilShinBiobase Co. Ltd., Yangju-si, Korea), and the dried powder was employed for metabolomics analysis using GC–MS.

4.4. Synthesis of LAB CFCS-AuNPs and ZnONPs

The previously described technique was used to synthesize C2-AuNPs and C2-ZnONPs [26]. Prior to synthesis, the pH of the CFCS was adjusted to 8.5 using 0.1 M NaOH. To synthesize C2-AuNPs, 1 mM $HAuCl_4 \cdot 3H_2O$ was dissolved in 200 mL of deionized water and constantly agitated at 70 °C. The prepared CFCS (3 mL) was added dropwise into the heated AuNPs solution and stirred for 2 h at 60 °C. The synthesis of the C2-AuNPs was confirmed by the visual observation of the color changing from yellow to resembling dark red wine. To synthesize C2-ZnONPs, 20 mm $Zn(CH_3\text{-}COO)_2 \cdot 2H_2O$ was dissolved in deionized water (200 mL) and stirred at 70 °C. The CFCS (3 mL) was added dropwise into the heated solution. After 1 h of continuous stirring at 70 °C, a solution of NaOH (0.1 N) was added dropwise to the solution and left to stir for 2 h. The appearance of a white precipitate indicates the production of C2-ZnONPs. Aside from the visual inspection of the color change, another method used was to scan the UV–vis absorption spectra of the produced NPs from 200 to 800 nm using a microplate reader (BioTek, Winooski, VT, USA). The entire C2-AuNPs solution was first frozen for 30 min at −70 °C before being freeze-dried. On the other hand, the C2-ZnONPs solution was centrifuged (13,000 rpm; 30 min) at 4 °C. Before freeze-drying, the white pellet of C2-ZnONPs was washed and frozen at −70 °C.

4.5. Metabolomic Analysis of LAB-Supernatant by GC–MS

The GC–MS analysis of LAB CFCS was carried out with changes to previous reports [55]. LAB CFCS freeze-dried powder was dissolved in methanol. The presence of metabolites in the samples was determined using gas chromatography–mass spectrometry (GC–MS, Agilent Technologies 7890A-5975C) using helium as the carrier gas at a flow rate of 1.0 mL/min. One microliter of the sample was injected with a 1 min split flow delay and resolved on a 30 m × 0.25 mm × 0.25 μm DB-5MS column (Agilent Technologies, Palo Alto, CA, USA). Temperatures at the inlet, interface, and ion source were 250, 250, and 230 °C, respectively. The oven's beginning and end temperatures were set at 50 and 230 °C, respectively, at a rate of 5 °C/min for 36 min and then for 2 min at a constant temperature. At 70 eV, electron impact mass spectra from m/z 50 to 550 were observed. Metabolite identification was accomplished by comparing the mass spectra to analytical standards in the NIST14 database (National Institute of Standards and Technology, Gaithersburg, MD, USA).

4.6. Physiochemical Characterization of C2-AuNPs and C2-ZnONPs

In addition to developing dark red wine-colored (C2-AuNPs) and white precipitates (C2-ZnONPs) as preliminary signs, the synthesized NPs were validated by measuring typical UV–vis absorption spectra from 200 to 800 nm. In addition, as previously stated, various other instrumental analyses of the NPs were conducted [26]. Fourier transform infrared spectroscopy (FTIR, JASCO (FT-4100), Tokyo, Japan) was utilized to characterize C2-AuNPs and C2-ZnONPs at frequencies ranging from 4000 to 400 cm^{-1}. The morphological shape of the C2-AuNPs and C2-ZnONPs was evaluated using field emission transmission electron microscopy (FE-TEM; JEM-F200, JEOL, Tokyo, Japan). Similarly, the sizes of these NPs were assessed using dynamic light scattering (DLS) and a particle analyzer Litesizer 500 (Anton Paar, GmbH, Graz, Austria). The zeta potentials of C2-AuNPs and C2-ZnONPs were also measured using the particle analyzer Litesizer 500. The energy dispersive X-ray spectroscopy (EDS) and mapping of the metal present in the synthesized C2-AuNPs and C2-ZnONPs were also determined using the FE-TEM instrument, which was also employed to evaluate the morphology. The crystalline nature of the nanoparticles was examined using an X-ray diffractometer (XRD; X-ray diffractometer, Rigaku (Tokyo, Japan), Ultima IV). These completely characterized NPs were employed in antibacterial, antibiofilm, and antivirulence studies against microbial pathogens.

4.7. Antibacterial Effects of LAB-Supernatant and Nanoparticles

The supernatant obtained from the LAB isolate was tested for antimicrobial activity against different microbial pathogens (given in Table 1), as previously described [56]. Furthermore, *Candida albicans* was utilized as a fungal pathogen to test the antifungal efficacy of the LAB supernatant. These microbial pathogens' seed cultures were cultivated in their respective growth conditions at 37 °C overnight. The culture with an OD_{600} value of 0.05 was put on a 96-well microtiter plate containing CFCS at dilutions ranging from 10^{-1} to 10^{-6}. Similarly, the antimicrobial activity of C2-AuNPs and C2-ZnONPs was tested by incubating cell cultures ($OD_{600\ nm}$ = 0.05) on a 96-well microtiter plate containing varied nanoparticle concentrations (ranging from 64 to 2048 μg/mL) [26]. The microbial cell culture in the microtiter plate was incubated for 24 h at 37 °C. The OD_{600} value of microbe cell growth was checked using a microtiter plate reader. The OD_{600} value from the test sample (treated with NPs) was subtracted from the OD_{600} value obtained from the control group (containing NPs in growth media without bacterial cells).

4.8. Biofilm Inhibition and Eradication of Established Mature Biofilm

The inhibitory effects of C2-AuNPs and C2-ZnONPs on biofilms of *S. aureus* and *P. aeruginosa* were checked the same way as the previously reported method [26]. As a surface for the biofilm formation in the presence of NPs, 96-well microplates were used. The cell culture ($OD_{600\ nm}$ = 0.05) of *P. aeruginosa* and *S. aureus* were put on microplates with NPs (ranging from 64 to 1024 μg/mL). Microplates containing cell culture and various doses of NP were incubated at 37 °C for 24 h. The planktonic cells were discarded entirely, and the remaining biofilm cells on the surface were washed thrice using distilled water. Biofilm cells were stained for 20 min with 0.1% crystal violet. These stained cells were again washed three times with distilled water and air-dried. After dissolving the stained cells with 95% alcohol, quantification was carried out at 570 nm. The eradication of the established mature biofilm was accomplished as described earlier [26]. We first established a biofilm in a microplate by adding cell culture ($OD_{600\ nm}$ = 0.05) and incubating for 24 h at 37 °C. The planktonic cells were removed aseptically, and the adherent cells were rinsed using sterile TSB. The adhered cells were treated with different concentrations of NPs (ranging from 256 to 2048 μg/mL) and further incubated overnight at 37 °C. The growth media containing free-floating cells from each well was removed, and the stained biofilm cells were quantified at OD at 570 nm. Experiments on initial-stage biofilm inhibition and mature biofilm eradication were carried out in triplicate.

4.9. Examination of Biofilms Cells Using SEM

The nanoparticle-affected biofilm cells were examined using advanced microscopy, namely scanning electron microscopy (SEM). The sample that was incubated with nanoparticles was prepared using the methodology described before [26]. In brief, following overnight growth, *P. aeruginosa* and *S. aureus* cell cultures were diluted in fresh TSB to an OD_{600} value of 0.05. These cell cultures were put in a 24-well microplate that had already been fitted with a 0.5 cm × 0.5 cm nylon membrane. C2-AuNPs and C2-ZnONPs at a concentration of 512 μg/mL were used to treat these cell cultures. As a control group, a well-containing nylon membrane was left untreated. After 24 h, the cells were immediately fixed with formaldehyde and glutaraldehyde, then incubated overnight at 4 °C. Unfixed and floating bacterial cells were thrown out, while the cells adhered to the surface were rinsed three times with PBS (pH 7.4). These surface-attached cells were then subjected to varying amounts of ethyl alcohol (50–100%). Following dehydration, these biofilm cells were freeze-dried. Finally, the SEM TESCAN (Vega II LSU), Brno, Czech Republic) was used to investigate the profile of biofilm cells.

4.10. Assays for the Inhibition of Multiple Virulence Properties

The quorum sensing (QS)-mediated virulence features of *P. aeruginosa*, such as pyocyanin, pyoverdine, hemolytic, and as well as the protease and hemolytic activity of *S. aureus* in the presence and absence of NPs were determined [57]. The pyocyanin production experiment was started by incubating *P. aeruginosa* cell culture ($OD_{600\ nm}$ = 0.05) with varying concentrations (512–64 μg/mL) of C2-AuNPs and C2-ZnONPs. After 24 h incubation at 37 °C, the pink-colored pyocyanin was extracted with chloroform and 0.1 N HCl, and the $OD_{520\ nm}$ was measured [58]. Pyoverdine production was measured in the presence and absence of C2-AuNPs and C2-ZnONPs by growing the cell culture in minimal salt media containing 2% sodium succinate. The cell culture was centrifuged (13,000 rpm; 10 min), and pyoverdine was determined by measuring the $OD_{405\ nm}$. The attenuation of *P. aeruginosa* and *S. aureus* hemolytic activity by C2-AuNPs and C2-ZnONPs was accomplished by incubating the cell culture ($OD_{600\ nm}$ = 0.05) for 24 h at 37 °C in the same manner as previously described [57]. The NPs treated and untreated cell cultures (50 μL) were inoculated into a tube containing 950 μL sheep red blood cells (RBCs; MBcell Ltd., Seoul, Korea) for 2 h at 37 °C. After incubation, the cell culture was centrifuged, and the obtained supernatant was measured at 534 nm. As previously described, the activity of the protein-digesting enzyme protease was investigated using Bacto agar containing casein (5%) [57]. To check the protease activity, the cell culture ($OD_{600\ nm}$ = 0.05) was incubated with varying concentrations (128–1024 μg/mL) of C2-AuNPs and C2-ZnONPs. After centrifugation, the collected supernatant was filter sterilized using a 0.2 μ filter. The cell-free supernatant (30 μL) was placed into the hole formed in the casein agar plate and incubated at 37 °C for 24 h. The cleared zone surrounding the hole was measured and compared to the measured zone obtained from the control group. As previously described, *P. aeruginosa* motilities were assessed in the presence or absence of C2-AuNPs and C2-ZnONPs [57]. Swarming was studied by making an agar plate with 0.5% Bacto agar, 0.4% casamino acid, and 0.4% glucose in Luria Bertani (LB) broth. Similarly, for swimming, 0.3% Bacto agar, 1% tryptone, and 0.25% NaCl agar media were made in LB broth and autoclaved. After autoclaving, the swarming and swimming media, C2-AuNPs, and C2-ZnONPs (512 μg/mL) were added to the agar medium and poured into Petri dishes. The overnight grown seed culture (5 μL) of *P. aeruginosa* was placed in the center of the agar plate and incubated for 24 h at 37 °C. The twitching agar medium was made by autoclaving 1.5% Bacto agar, 30 mM glucose, and 0.2% casamino acid (0.2%) in LB. NPs were added to the agar media after autoclaving. Before pouring the agar media, the cell culture was placed in the center of the Petri dishes using a sterile toothpick and aseptically air dried. These Petri plates were filled with twitching agar media. Each of the agar plates was incubated at 37 °C for 24 h. In the case of the swarming and swimming agar plates containing NPs, the surface area traversed by the cells was assessed and compared to the control plates. The twitching motility of the cells on the surface of the Petri dishes was examined by staining with crystal violet after removing the solid agar media, and the dye-stained surface area of the cells was measured.

4.11. Statistical Analysis

GraphPad Prism 7.0 (GraphPad Software Inc., San Diego, CA, USA) was used to plot all graphs. The statistical analysis was carried out using one-way ANOVA. *** $p < 0.0001$, ** $p < 0.01$, and * $p < 0.05$ were considered significant. Experiments were repeated three times with triplicate samples.

5. Conclusions

In this work, we isolated LAB strain C2 from Kimchi samples, and the isolate was identified as *Leuconostoc* sp. strain C2 based on 16S rRNA gene sequencing. The green synthesis method was used to synthesize AuNPs and ZnONPs with antibiofilm and antivirulence activities against *P. aeruginosa* and *S. aureus* utilizing LAB extract. The biofilm inhibitory activity of C2-AuNPs was shown to be concentration-dependent against *P. aeruginosa*, but not against *S. aureus*. Interestingly, C2-ZnONPs inhibit biofilm development in both *P. aeruginosa* and *S. aureus*. Aside from inhibiting biofilm, these NPs significantly eradicated pre-existing mature biofilms of *P. aeruginosa* and *S. aureus*, and the inhibition was dose-dependent. C2-AuNPs and C2-ZnONPs significantly reduced *P. aeruginosa* virulence features such as hemolytic activity, protease activity, and pyocyanin and pyoverdine production. In contrast, both C2-AuNPs and C2-ZnONPs have been shown to inhibit *S. aureus* hemolytic activity in a dose-dependent manner. Furthermore, these nanoparticles effectively suppress multiple forms of motility in *P. aeruginosa*, including swarming, swimming, and twitching. C2-ZnONPs more effectively inhibit motility (e.g., swimming and twitching) and pyocyanin and pyoverdine production than C2-AuNPs at the same dose. As a result, C2-AuNPs would be a better choice as an antibiofilm and antivirulence agent against human pathogenic bacteria. The current study concludes that these synthesized LAB-based nanoparticles have potential inhibitory effects on the biofilm and virulence features of bacterial pathogens. As a result, we may employ these nanoparticles as possible therapeutic drugs to control infections caused by biofilm-forming bacterial pathogens. A deeper analysis of the molecular action mechanism by undertaking the gene expression studies of the genes associated with the biofilm and virulence features of these bacterial pathogens is necessary in the future.

Supplementary Materials: The following are available online at https://www.mdpi.com/article/10.3390/antibiotics11111524/s1, Figure S1: GC–MS chromatogram of the supernatant obtained from the cell culture of the *Leuconostoc* sp. strain C2. Table S1: Identification of secondary metabolites from *Leuconostoc* sp. strain C2 supernatant using gas chromatograph–mass spectroscopy (GC–MS). References [36,37,59–102] are cited in the supplementary materials.

Author Contributions: Methodology, investigation, data curation, M.-G.K.; Methodology, D.-M.J., D.O. and N.T. Conceptualization, supervision, funding, writing—review and editing, F.K. and Y.-M.K. All authors have read and agreed to the published version of the manuscript.

Funding: This research was supported by Basic Science Research Program through the National Research Foundation of Korea (NRF) grant funded by the Ministry of Education (2021R1A6A1A03039211 and 2022R1A2B5B01001998). This research was also supported by the Korea Institute of Marine Science & Technology Promotion (KIMST) and funded by the Ministry of Oceans and Fisheries (20220319).

Institutional Review Board Statement: Not applicable.

Informed Consent Statement: Not applicable.

Data Availability Statement: All data are included in the manuscript or supplementary material.

Conflicts of Interest: The authors declare no conflict of interest.

References

1. De Oliveira, D.M.P.; Forde, B.M.; Kidd, T.J.; Harris, P.N.A.; Schembri, M.A.; Beatson, S.A.; Paterson, D.L.; Walker, M.J. Antimicrobial Resistance in ESKAPE Pathogens. *Clin. Microbiol. Rev.* **2020**, *33*, e00181-19. [CrossRef] [PubMed]
2. Laxminarayan, R.; Duse, A.; Wattal, C.; Zaidi, A.K.; Wertheim, H.F.; Sumpradit, N.; Vlieghe, E.; Hara, G.L.; Gould, I.M.; Goossens, H.; et al. Antibiotic resistance-the need for global solutions. *Lancet Infect. Dis.* **2013**, *13*, 1057–1098. [CrossRef]
3. Global burden of bacterial antimicrobial resistance in 2019: A systematic analysis. *Lancet* **2022**, *399*, 629–655. [CrossRef]
4. Rice, L.B. Federal funding for the study of antimicrobial resistance in nosocomial pathogens: No ESKAPE. *J. Infect. Dis.* **2008**, *197*, 1079–1081. [CrossRef]
5. Beatson, S.A.; Walker, M.J. Microbiology. Tracking antibiotic resistance. *Science* **2014**, *345*, 1454–1455. [CrossRef]
6. Santajit, S.; Indrawattana, N. Mechanisms of Antimicrobial Resistance in ESKAPE Pathogens. *Biomed Res. Int.* **2016**, *2016*, 2475067. [CrossRef]
7. de la Fuente-Núñez, C.; Reffuveille, F.; Fernández, L.; Hancock, R.E. Bacterial biofilm development as a multicellular adaptation: Antibiotic resistance and new therapeutic strategies. *Curr. Opin. Microbiol.* **2013**, *16*, 580–589. [CrossRef]
8. Mah, T.-F.C.; O'Toole, G.A. Mechanisms of biofilm resistance to antimicrobial agents. *Trends Microbiol.* **2001**, *9*, 34–39. [CrossRef]
9. Khan, F.; Pham, D.T.N.; Tabassum, N.; Oloketuyi, S.F.; Kim, Y.M. Treatment strategies targeting persister cell formation in bacterial pathogens. *Crit. Rev. Microbiol.* **2020**, *46*, 665–688. [CrossRef]
10. Lin, Q.; Deslouches, B.; Montelaro, R.C.; Di, Y.P. Prevention of ESKAPE pathogen biofilm formation by antimicrobial peptides WLBU2 and LL37. *Int. J. Antimicrob. Agents* **2018**, *52*, 667–672. [CrossRef]
11. Jeong, G.J.; Khan, S.; Tabassum, N.; Khan, F.; Kim, Y.M. Marine-Bioinspired Nanoparticles as Potential Drugs for Multiple Biological Roles. *Mar. Drugs* **2022**, *20*, 527. [CrossRef]
12. Cheng, Z.; Li, M.; Dey, R.; Chen, Y. Nanomaterials for cancer therapy: Current progress and perspectives. *J. Hematol. Oncol.* **2021**, *14*, 85. [CrossRef]
13. Pandit, C.; Roy, A.; Ghotekar, S.; Khusro, A.; Islam, M.N.; Emran, T.B.; Lam, S.E.; Khandaker, M.U.; Bradley, D.A. Biological agents for synthesis of nanoparticles and their applications. *J. King Saud Univ.-Sci.* **2022**, *34*, 101869. [CrossRef]
14. Singh, J.; Dutta, T.; Kim, K.-H.; Rawat, M.; Samddar, P.; Kumar, P. 'Green' synthesis of metals and their oxide nanoparticles: Applications for environmental remediation. *J. Nanobiotechnol.* **2018**, *16*, 84. [CrossRef] [PubMed]
15. Khan, F.; Jeong, G.-J.; Singh, P.; Tabassum, N.; Mijakovic, I.; Kim, Y.-M. Retrospective analysis of the key molecules involved in the green synthesis of nanoparticles. *Nanoscale* **2022**, *14*, 14824–14857. [CrossRef] [PubMed]
16. Khan, I.; Bai, Y.; Zha, L.; Ullah, N.; Ullah, H.; Shah, S.R.H.; Sun, H.; Zhang, C. Mechanism of the Gut Microbiota Colonization Resistance and Enteric Pathogen Infection. *Front. Cell. Infect. Microbiol.* **2021**, *11*, 716299. [CrossRef]
17. Tamura, K.; Stecher, G.; Kumar, S. MEGA11: Molecular Evolutionary Genetics Analysis Version 11. *Mol. Biol. Evol.* **2021**, *38*, 3022–3027. [CrossRef] [PubMed]
18. Liu, Y.; Kim, S.; Kim, Y.J.; Perumalsamy, H.; Lee, S.; Hwang, E.; Yi, T.H. Green synthesis of gold nanoparticles using *Euphrasia officinalis* leaf extract to inhibit lipopolysaccharide-induced inflammation through NF-κB and JAK/STAT pathways in RAW 264.7 macrophages. *Int. J. Nanomed.* **2019**, *14*, 2945–2959. [CrossRef]
19. Barani, M.; Masoudi, M.; Mashreghi, M.; Makhdoumi, A.; Eshghi, H. Cell-free extract assisted synthesis of ZnO nanoparticles using aquatic bacterial strains: Biological activities and toxicological evaluation. *Int. J. Pharm.* **2021**, *606*, 120878. [CrossRef]
20. Qin, S.; Xiao, W.; Zhou, C.; Pu, Q.; Deng, X.; Lan, L.; Liang, H.; Song, X.; Wu, M. *Pseudomonas aeruginosa*: Pathogenesis, virulence factors, antibiotic resistance, interaction with host, technology advances and emerging therapeutics. *Signal Transduct. Target. Ther.* **2022**, *7*, 199. [CrossRef]
21. Cheung, G.Y.C.; Bae, J.S.; Otto, M. Pathogenicity and virulence of *Staphylococcus aureus*. *Virulence* **2021**, *12*, 547–569. [CrossRef] [PubMed]
22. Stewart, P.S. Mechanisms of antibiotic resistance in bacterial biofilms. *Int. J. Med. Microbiol.* **2002**, *292*, 107–113. [CrossRef] [PubMed]
23. Driffield, K.; Miller, K.; Bostock, J.M.; O'Neill, A.J.; Chopra, I. Increased mutability of *Pseudomonas aeruginosa* in biofilms. *J. Antimicrob. Chemother.* **2008**, *61*, 1053–1056. [CrossRef] [PubMed]
24. Kirtane, A.R.; Verma, M.; Karandikar, P.; Furin, J.; Langer, R.; Traverso, G. Nanotechnology approaches for global infectious diseases. *Nat. Nanotechnol.* **2021**, *16*, 369–384. [CrossRef]
25. Kechagia, M.; Basoulis, D.; Konstantopoulou, S.; Dimitriadi, D.; Gyftopoulou, K.; Skarmoutsou, N.; Fakiri, E.M. Health benefits of probiotics: A review. *ISRN Nutr.* **2013**, *2013*, 481651. [CrossRef]
26. Khan, F.; Kang, M.G.; Jo, D.M.; Chandika, P.; Jung, W.K.; Kang, H.W.; Kim, Y.M. Phloroglucinol-Gold and -Zinc Oxide Nanoparticles: Antibiofilm and Antivirulence Activities towards *Pseudomonas aeruginosa* PAO1. *Mar. Drugs* **2021**, *19*, 601. [CrossRef]
27. Singh, M.; Kalaivani, R.; Manikandan, S.; Sangeetha, N.; Kumaraguru, A.K. Facile green synthesis of variable metallic gold nanoparticle using *Padina gymnospora*, a brown marine macroalga. *Appl. Nanosci.* **2013**, *3*, 145–151. [CrossRef]

3. He, S.; Guo, Z.; Zhang, Y.; Zhang, S.; Wang, J.; Gu, N. Biosynthesis of gold nanoparticles using the bacteria *Rhodopseudomonas capsulata*. *Mater. Lett.* **2007**, *61*, 3984–3987. [CrossRef]
). Składanowski, M.; Wypij, M.; Laskowski, D.; Golińska, P.; Dahm, H.; Rai, M. Silver and gold nanoparticles synthesized from *Streptomyces* sp. isolated from acid forest soil with special reference to its antibacterial activity against pathogens. *J. Clust. Sci.* **2017**, *28*, 59–79. [CrossRef]
). Li, J.; Li, Q.; Ma, X.; Tian, B.; Li, T.; Yu, J.; Dai, S.; Weng, Y.; Hua, Y. Biosynthesis of gold nanoparticles by the extreme bacterium *Deinococcus radiodurans* and an evaluation of their antibacterial properties. *Int. J. Nanomed.* **2016**, *11*, 5931–5944. [CrossRef]
1. Al-Kordy, H.M.H.; Sabry, S.A.; Mabrouk, M.E.M. Statistical optimization of experimental parameters for extracellular synthesis of zinc oxide nanoparticles by a novel haloalaliphilic *Alkalibacillus* sp.W7. *Sci. Rep.* **2021**, *11*, 10924. [CrossRef] [PubMed]
2. Punjabi, K.; Mehta, S.; Chavan, R.; Chitalia, V.; Deogharkar, D.; Deshpande, S. Efficiency of Biosynthesized Silver and Zinc Nanoparticles Against Multi-Drug Resistant Pathogens. *Front. Microbiol.* **2018**, *9*, 2207. [CrossRef] [PubMed]
3. Mohd Yusof, H.; Abdul Rahman, N.A.; Mohamad, R.; Zaidan, U.H.; Samsudin, A.A. Biosynthesis of zinc oxide nanoparticles by cell-biomass and supernatant of *Lactobacillus plantarum* TA4 and its antibacterial and biocompatibility properties. *Sci. Rep.* **2020**, *10*, 19996. [CrossRef] [PubMed]
4. Jain, D.; Shivani; Bhojiya, A.A.; Singh, H.; Daima, H.K.; Singh, M.; Mohanty, S.R.; Stephen, B.J.; Singh, A. Microbial Fabrication of Zinc Oxide Nanoparticles and Evaluation of Their Antimicrobial and Photocatalytic Properties. *Front. Chem.* **2020**, *8*, 778. [CrossRef]
5. Ziklo, N.; Bibi, M.; Salama, P. The Antimicrobial Mode of Action of Maltol and Its Synergistic Efficacy with Selected Cationic Surfactants. *Cosmetics* **2021**, *8*, 86. [CrossRef]
6. Guo, L.; Wang, A.; Sun, Y.; Xu, C. Evaluation of antioxidant and immunity function of tetramethylpyrazine phosphate tablets in vivo. *Molecules* **2012**, *17*, 5412–5421. [CrossRef]
7. Edeeva, S.E.; Kopylova, G.N.; Bakaeva, Z.V.; Samonina, G.E.; Umarova, B.A.; Guseva, A.A. Protective and therapeutic effects of glyprolines in psychoemotional stress induced by cholecystokinin-4 injection. *Bull. Exp. Biol. Med.* **2008**, *145*, 302–306. [CrossRef]
8. Abinaya, M.; Vaseeharan, B.; Divya, M.; Sharmili, A.; Govindarajan, M.; Alharbi, N.S.; Kadaikunnan, S.; Khaled, J.M.; Benelli, G. Bacterial exopolysaccharide (EPS)-coated ZnO nanoparticles showed high antibiofilm activity and larvicidal toxicity against malaria and Zika virus vectors. *J. Trace Elem. Med. Biol.* **2018**, *45*, 93–103. [CrossRef]
9. Khan, F.; Manivasagan, P.; Lee, J.W.; Pham, D.T.N.; Oh, J.; Kim, Y.M. Fucoidan-Stabilized Gold Nanoparticle-Mediated Biofilm Inhibition, Attenuation of Virulence and Motility Properties in *Pseudomonas aeruginosa* PAO1. *Mar. Drugs* **2019**, *17*, 208. [CrossRef]
0. Khatoon, Z.; McTiernan, C.D.; Suuronen, E.J.; Mah, T.F.; Alarcon, E.I. Bacterial biofilm formation on implantable devices and approaches to its treatment and prevention. *Heliyon* **2018**, *4*, e01067. [CrossRef]
1. Dongari-Bagtzoglou, A. Pathogenesis of mucosal biofilm infections: Challenges and progress. *Expert Rev. Anti Infect.* **2008**, *6*, 201–208. [CrossRef] [PubMed]
2. Flemming, H.-C.; Wingender, J. The biofilm matrix. *Nat. Rev. Microbiol.* **2010**, *8*, 623–633. [CrossRef]
3. Husain, F.M.; Qais, F.A.; Ahmad, I.; Hakeem, M.J.; Baig, M.H.; Masood Khan, J.; Al-Shabib, N.A. Biosynthesized Zinc Oxide Nanoparticles Disrupt Established Biofilms of Pathogenic Bacteria. *Appl. Sci.* **2022**, *12*, 710. [CrossRef]
4. Hentzer, M.; Wu, H.; Andersen, J.B.; Riedel, K.; Rasmussen, T.B.; Bagge, N.; Kumar, N.; Schembri, M.A.; Song, Z.; Kristoffersen, P.; et al. Attenuation of *Pseudomonas aeruginosa* virulence by quorum sensing inhibitors. *EMBO J.* **2003**, *22*, 3803–3815. [CrossRef]
5. Lee, J.-H.; Cho, H.S.; Kim, Y.; Kim, J.-A.; Banskota, S.; Cho, M.H.; Lee, J. Indole and 7-benzyloxyindole attenuate the virulence of *Staphylococcus aureus*. *Appl. Microbiol. Biotechnol.* **2013**, *97*, 4543–4552. [CrossRef]
6. Salam, A.M.; Quave, C.L. Targeting Virulence in *Staphylococcus aureus* by Chemical Inhibition of the Accessory Gene Regulator System In Vivo. *mSphere* **2018**, *3*, e00500-17. [CrossRef]
7. Khan, F.; Pham, D.T.N.; Oloketuyi, S.F.; Kim, Y.M. Regulation and controlling the motility properties of *Pseudomonas aeruginosa*. *Appl. Microbiol. Biotechnol.* **2020**, *104*, 33–49. [CrossRef]
8. Lee, J.-H.; Kim, Y.-G.; Cho, M.H.; Lee, J. ZnO nanoparticles inhibit *Pseudomonas aeruginosa* biofilm formation and virulence factor production. *Microbiol. Res.* **2014**, *169*, 888–896. [CrossRef]
9. Dwivedi, S.; Wahab, R.; Khan, F.; Mishra, Y.K.; Musarrat, J.; Al-Khedhairy, A.A. Reactive oxygen species mediated bacterial biofilm inhibition via zinc oxide nanoparticles and their statistical determination. *PLoS ONE* **2014**, *9*, e111289. [CrossRef]
0. Kaur, T.; Putatunda, C.; Vyas, A.; Kumar, G. Zinc oxide nanoparticles inhibit bacterial biofilm formation via altering cell membrane permeability. *Prep. Biochem. Biotechnol.* **2021**, *51*, 309–319. [CrossRef]
1. Perveen, K.; Husain, F.M.; Qais, F.A.; Khan, A.; Razak, S.; Afsar, T.; Alam, P.; Almajwal, A.M.; Abulmeaty, M.M.A. Microwave-Assisted Rapid Green Synthesis of Gold Nanoparticles Using Seed Extract of *Trachyspermum ammi*: ROS Mediated Biofilm Inhibition and Anticancer Activity. *Biomolecules* **2021**, *11*, 197. [CrossRef]
2. Lin, Y.K.; Yang, S.C.; Hsu, C.Y.; Sung, J.T.; Fang, J.Y. The Antibiofilm Nanosystems for Improved Infection Inhibition of Microbes in Skin. *Molecules* **2021**, *26*, 6392. [CrossRef] [PubMed]
3. Won, S.-M.; Chen, S.; Park, K.W.; Yoon, J.-H. Isolation of lactic acid bacteria from kimchi and screening of *Lactobacillus sakei* ADM14 with anti-adipogenic effect and potential probiotic properties. *LWT* **2020**, *126*, 109296. [CrossRef]
4. Fazlurrahman; Batra, M.; Pandey, J.; Suri, C.R.; Jain, R.K. Isolation and characterization of an atrazine-degrading *Rhodococcus* sp. strain MB-P1 from contaminated soil. *Lett. Appl. Microbiol.* **2009**, *49*, 721–729. [CrossRef] [PubMed]

55. Mao, Y.; Zhang, X.; Xu, Z. Identification of antibacterial substances of Lactobacillus plantarum DY-6 for bacteriostatic action. *Food Sci. Nutr.* **2020**, *8*, 2854–2863. [CrossRef] [PubMed]
56. Jo, D.-M.; Park, S.-K.; Khan, F.; Kang, M.-G.; Lee, J.-H.; Kim, Y.-M. An approach to extend the shelf life of ribbonfish fillet using lactic acid bacteria cell-free culture supernatant. *Food Control* **2021**, *123*, 107731. [CrossRef]
57. Khan, F.; Lee, J.W.; Pham, D.T.N.; Lee, J.H.; Kim, H.W.; Kim, Y.K.; Kim, Y.M. Streptomycin mediated biofilm inhibition and suppression of virulence properties in *Pseudomonas aeruginosa* PAO1. *Appl. Microbiol. Biotechnol.* **2020**, *104*, 799–816. [CrossRef]
58. Essar, D.W.; Eberly, L.; Hadero, A.; Crawford, I.P. Identification and characterization of genes for a second anthranilate synthase in *Pseudomonas aeruginosa*: Interchangeability of the two anthranilate synthases and evolutionary implications. *J. Bacteriol.* **1990**, *172*, 884–900. [CrossRef] [PubMed]
59. Gilardoni, G.; Clericuzio, M.; Tosi, S.; Zanoni, G.; Vidari, G. Antifungal acylcyclopentenediones from fruiting bodies of *Hygrophorus chrysodon*. *J. Nat. Prod.* **2007**, *70*, 137–139. [CrossRef]
60. Mokhtari, M.; Jackson, M.D.; Brown, A.S.; Ackerley, D.F.; Ritson, N.J.; Keyzers, R.A.; Munkacsi, A.B. Bioactivity-Guided Metabolite Profiling of Feijoa (*Acca sellowiana*) Cultivars Identifies 4-Cyclopentene-1,3-dione as a Potent Antifungal Inhibitor of Chitin Synthesis. *J. Agric. Food Chem.* **2018**, *66*, 5531–5539. [CrossRef]
61. Maier, N.K.; Leppla, S.H.; Moayeri, M. The cyclopentenone prostaglandin 15d-PGJ2 inhibits the NLRP1 and NLRP3 inflammasomes. *J. Immunol.* **2015**, *194*, 2776–2785. [CrossRef] [PubMed]
62. Martinez, A.; Alonso, M.; Castro, A.; Dorronsoro, I.; Gelpí, J.L.; Luque, F.J.; Pérez, C.; Moreno, F.J. SAR and 3D-QSAR studies on thiadiazolidinone derivatives: Exploration of structural requirements for glycogen synthase kinase 3 inhibitors. *J. Med. Chem.* **2005**, *48*, 7103–7112. [CrossRef] [PubMed]
63. Lizardi, L.; Garcia, M.C.; Sanchez, J.A.; Zuazaga, C. Sulfhydryl alkylating agents induce calcium current in skeletal muscle fibers of a crustacean (*Atya lanipes*). *J. Membr. Biol.* **1992**, *129*, 167–178. [CrossRef] [PubMed]
64. Bui, T.; Straus, D.S. Effects of cyclopentenone prostaglandins and related compounds on insulin-like growth factor-I and Waf1 gene expression. *Biochim. Biophys. Acta* **1998**, *1397*, 31–42. [CrossRef]
65. Uto, Y.; Nagasawa, H.; Jin, C.Z.; Nakayama, S.; Tanaka, A.; Kiyoi, S.; Nakashima, H.; Shimamura, M.; Inayama, S.; Fujiwara, T.; et al. Design of antiangiogenic hypoxic cell radiosensitizers: 2-nitroimidazoles containing a 2-aminomethylene-4-cyclopentene-1,3-dione moiety. *Bioorg. Med. Chem.* **2008**, *16*, 6042–6053. [CrossRef] [PubMed]
66. Du, L.; King, J.B.; Morrow, B.H.; Shen, J.K.; Miller, A.N.; Cichewicz, R.H. Diarylcyclopentendione metabolite obtained from a *Preussia typharum* isolate procured using an unconventional cultivation approach. *J. Nat. Prod.* **2012**, *75*, 1819–1823. [CrossRef]
67. Neumann, P.; Brodhun, F.; Sauer, K.; Herrfurth, C.; Hamberg, M.; Brinkmann, J.; Scholz, J.; Dickmanns, A.; Feussner, I.; Ficner, R. Crystal structures of *Physcomitrella patens* AOC1 and AOC2: Insights into the enzyme mechanism and differences in substrate specificity. *Plant Physiol.* **2012**, *160*, 1251–1266. [CrossRef]
68. Braga, S.F.; Alves, É.V.; Ferreira, R.S.; Fradico, J.R.; Lage, P.S.; Duarte, M.C.; Ribeiro, T.G.; Júnior, P.A.; Romanha, A.J.; Tonini, M.L.; et al. Synthesis and evaluation of the antiparasitic activity of bis-(arylmethylidene) cycloalkanones. *Eur. J. Med. Chem.* **2014**, *71*, 282–289. [CrossRef]
69. Beranič, N.; Stefane, B.; Brus, B.; Gobec, S.; Rižner, T.L. New enzymatic assay for the AKR1C enzymes. *Chem. Biol. Interact.* **2013**, *202*, 204–209. [CrossRef]
70. Li, J.; Zhang, D.; Wu, X. Synthesis and biological evaluation of novel exo-methylene cyclopentanone tetracyclic diterpenoids as antitumor agents. *Bioorg. Med. Chem. Lett.* **2011**, *21*, 130–132. [CrossRef]
71. Yamada, K.; Shimizu, A.; Komatsu, H.; Sakata, R.; Ohta, A. Effects of 2,5-dimethylpyrazine on plasma testosterone and polyamines- and acid phosphatase-levels in the rat prostate. *Biol. Pharm. Bull.* **1994**, *17*, 730–731. [CrossRef] [PubMed]
72. Yamada, K.; Watanabe, Y.; Aoyagi, Y.; Ohta, A. Effect of alkylpyrazine derivatives on the duration of pentobarbital-induced sleep, picrotoxicin-induced convulsion and gamma-aminobutyric acid (GABA) levels in the mouse brain. *Biol. Pharm. Bull.* **2001**, *24*, 1068–1071. [CrossRef] [PubMed]
73. Yamada, K.; Kobayashi, Y.; Fujihara, H.; Ohta, A. Inhibitory effect of 2,5-dimethylpyrazine on oxytocic agent-induced uterine hypercontraction of normal or pregnant female rats. *Biol. Pharm. Bull.* **1998**, *21*, 538–540. [CrossRef] [PubMed]
74. Chen, Z.; Liu, Q.; Zhao, Z.; Bai, B.; Sun, Z.; Cai, L.; Fu, Y.; Ma, Y.; Wang, Q.; Xi, G. Effect of hydroxyl on antioxidant properties of 2,3-dihydro-3,5-dihydroxy-6-methyl-4H-pyran-4-one to scavenge free radicals. *RSC Adv.* **2021**, *11*, 34456–34461. [CrossRef]
75. Yu, X.; Zhao, M.; Liu, F.; Zeng, S.; Hu, J. Identification of 2,3-dihydro-3,5-dihydroxy-6-methyl-4H-pyran-4-one as a strong antioxidant in glucose–histidine Maillard reaction products. *Food Res. Int.* **2013**, *51*, 397–403. [CrossRef]
76. Čechovská, L.; Cejpek, K.; Konečný, M.; Velíšek, J. On the role of 2,3-dihydro-3,5-dihydroxy-6-methyl-(4H)-pyran-4-one in antioxidant capacity of prunes. *Eur. Food Res. Technol.* **2011**, *233*, 367–376. [CrossRef]
77. Suganuma, H.; Inakuma, T.; Kikuchi, Y. Amelioratory Effect of Barley Tea Drinking on Blood Fluidity. *J. Nutr. Sci. Vitaminol.* **2002**, *48*, 165–168. [CrossRef]
78. Yanagimoto, K.; Lee, K.G.; Ochi, H.; Shibamoto, T. Antioxidative activity of heterocyclic compounds found in coffee volatiles produced by Maillard reaction. *J. Agric. Food Chem.* **2002**, *50*, 5480–5484. [CrossRef]
79. Hidalgo, F.J.; Nogales, F.; Zamora, R. Effect of the Pyrrole Polymerization Mechanism on the Antioxidative Activity of Nonenzymatic Browning Reactions. *J. Agric. Food Chem.* **2003**, *51*, 5703–5708. [CrossRef]
80. Devadas, S.M.; Nayak, U.Y.; Narayan, R.; Hande, M.H.; Ballal, M. 2,5-Dimethyl-4-hydroxy-3(2H)-furanone as an Anti-biofilm Agent Against Non-*Candida albicans Candida* Species. *Mycopathologia* **2019**, *184*, 403–411. [CrossRef]

1. Sung, W.S.; Jung, H.J.; Park, K.; Kim, H.S.; Lee, I.-S.; Lee, D.G. 2,5-dimethyl-4-hydroxy-3(2H)-furanone (DMHF); antimicrobial compound with cell cycle arrest in nosocomial pathogens. *Life Sci.* **2007**, *80*, 586–591. [CrossRef] [PubMed]

2. Chuang, P.-H.; Lee, C.-W.; Chou, J.-Y.; Murugan, M.; Shieh, B.-J.; Chen, H.-M. Anti-fungal activity of crude extracts and essential oil of *Moringa oleifera* Lam. *Bioresour. Technol.* **2007**, *98*, 232–236. [CrossRef] [PubMed]

3. Wang, J.-q.; Zhang, L.; Tao, X.-g.; Wei, L.; Liu, B.; Huang, L.-l.; Chen, Y.-g. Tetramethylpyrazine upregulates the aquaporin 8 expression of hepatocellular mitochondria in septic rats. *J. Surg. Res.* **2013**, *185*, 286–293. [CrossRef] [PubMed]

4. Sun, Y.; Jiang, J.; Zhang, Z.; Yu, P.; Wang, L.; Xu, C.; Liu, W.; Wang, Y. Antioxidative and thrombolytic TMP nitrone for treatment of ischemic stroke. *Bioorganic Med. Chem.* **2008**, *16*, 8868–8874. [CrossRef]

5. Sun, Y.; Song, M.; Niu, L.; Bai, X.; Sun, N.; Zhao, X.; Jiang, J.; He, J.; Li, H. Antiviral effects of the constituents derived from Chinese herb medicines on infectious bursal disease virus. *Pharm. Biol.* **2013**, *51*, 1137–1143. [CrossRef]

6. Ren, Z.; Ma, J.; Zhang, P.; Luo, A.; Zhang, S.; Kong, L.; Qian, C. The effect of ligustrazine on L-type calcium current, calcium transient and contractility in rabbit ventricular myocytes. *J. Ethnopharmacol.* **2012**, *144*, 555–561. [CrossRef]

7. Wang, H.; Jenner, A.M.; Lee, C.-Y.J.; Shui, G.; Tang, S.Y.; Whiteman, M.; Wenk, M.R.; Halliwell, B. The identification of antioxidants in dark soy sauce. *Free Radic. Res.* **2007**, *41*, 479–488. [CrossRef]

8. Chaves, S.; Canário, S.; Carrasco, M.P.; Mira, L.; Santos, M.A. Hydroxy(thio)pyrone and hydroxy(thio)pyridinone iron chelators: Physico-chemical properties and anti-oxidant activity. *J. Inorg. Biochem.* **2012**, *114*, 38–46. [CrossRef]

9. Yang, M.-L.; Kuo, P.-C.; Hwang, T.-L.; Wu, T.-S. Anti-inflammatory Principles from *Cordyceps sinensis*. *J. Nat. Prod.* **2011**, *74*, 1996–2000. [CrossRef]

0. Kandioller, W.; Hartinger, C.G.; Nazarov, A.A.; Bartel, C.; Skocic, M.; Jakupec, M.A.; Arion, V.B.; Keppler, B.K. Maltol-Derived Ruthenium–Cymene Complexes with Tumor Inhibiting Properties: The Impact of Ligand–Metal Bond Stability on Anticancer Activity In Vitro. *Chem.-Eur. J.* **2009**, *15*, 12283–12291. [CrossRef]

1. Amatori, S.; Ambrosi, G.; Fanelli, M.; Formica, M.; Fusi, V.; Giorgi, L.; Macedi, E.; Micheloni, M.; Paoli, P.; Pontellini, R.; et al. Synthesis, Basicity, Structural Characterization, and Biochemical Properties of Two [(3-Hydroxy-4-pyron-2-yl)methyl]amine Derivatives Showing Antineoplastic Features. *J. Org. Chem.* **2012**, *77*, 2207–2218. [CrossRef] [PubMed]

2. Watanabe-Akanuma, M.; Inaba, Y.; Ohta, T. Mutagenicity of UV-irradiated maltol in *Salmonella typhimurium*. *Mutagenesis* **2007**, *22*, 43–47. [CrossRef] [PubMed]

3. Kang, K.S.; Tanaka, T.; Cho, E.J.; Yokozawa, T. Evaluation of the Peroxynitrite Scavenging Activity of Heat-Processed Ginseng. *J. Med. Food* **2009**, *12*, 124–130. [CrossRef]

4. Lalitha, N.; Sadashivaiah, B.; Ramaprasad, T.R.; Singh, S.A. Anti-hyperglycemic activity of myricetin, through inhibition of DPP-4 and enhanced GLP-1 levels, is attenuated by co-ingestion with lectin-rich protein. *PLoS ONE* **2020**, *15*, e0231543. [CrossRef] [PubMed]

5. Ichimura, T.; Yamanaka, A.; Otsuka, T.; Yamashita, E.; Maruyama, S. Antihypertensive Effect of Enzymatic Hydrolysate of Collagen and Gly-Pro in Spontaneously Hypertensive Rats. *Biosci. Biotechnol. Biochem.* **2009**, *73*, 2317–2319. [CrossRef]

6. Abdurraafi, M.; Dermawan, A.; Julianti, E.; Putra, M.; Karim, F. Identification and Evaluation of Antibacterial Compounds from the Vibrio sp. associated with the Ascidian Pycnoclavella diminuta. *Pharm. Sci. Res.* **2019**, *6*, 2.

7. Lee, S.; Eom, S.; Nguyen, K.V.A.; Lee, J.; Park, Y.; Yeom, H.D.; Lee, J.H. The Application of the Neuroprotective and Potential Antioxidant Effect of Ergotamine Mediated by Targeting N-Methyl-D-Aspartate Receptors. *Antioxidants* **2022**, *11*, 1471. [CrossRef]

8. Vesely, D.L. Ergotamine and dihydroergotamine enhance guanylate cyclase activity. *Res. Commun. Chem. Pathol. Pharm.* **1983**, *40*, 245–254.

9. Kruk-Slomka, M.; Michalak, A.; Biala, G. Antidepressant-like effects of the cannabinoid receptor ligands in the forced swimming test in mice: Mechanism of action and possible interactions with cholinergic system. *Behav. Brain Res.* **2015**, *284*, 24–36. [CrossRef]

00. Nojima, H.; Ohba, Y.; Kita, Y. Oleamide derivatives are prototypical anti-metastasis drugs that act by inhibiting Connexin 26. *Curr. Drug Saf.* **2007**, *2*, 204–211. [CrossRef]

01. Lechin, F.; Van Der Dijs, B.; Bentolila, A.; Peña, F. Antidiarrheal effects of dihydroergotamine. *J. Clin. Pharm.* **1977**, *17*, 339–349. [CrossRef] [PubMed]

02. Silberstein, S.D. The pharmacology of ergotamine and dihydroergotamine. *Headache* **1997**, *37* (Suppl. 1), S15–S25. [PubMed]

antibiotics MDPI

Review

A Mini-Review of Anti-Listerial Compounds from Marine Actinobacteria (1990–2023)

Siyanda S. Ngema and Evelyn Madoroba *

Department of Biochemistry and Microbiology, University of Zululand, KwaDlangezwa 3886, South Africa; ngemasiyanda05@gmail.com
* Correspondence: evelyn.madoroba@gmail.com or madorobae@unizulu.ac.za

Abstract: Among the foodborne illnesses, listeriosis has the third highest case mortality rate (20–30% or higher). Emerging drug-resistant strains of *Listeria monocytogenes*, a causative bacterium of listeriosis, exacerbate the seriousness of this public health concern. Novel anti-Listerial compounds are therefore needed to combat this challenge. In recent years, marine actinobacteria have come to be regarded as a promising source of novel antimicrobials. Hence, our aim was to provide a narrative of the available literature and discuss trends regarding bioprospecting marine actinobacteria for new anti-Listerial compounds. Four databases were searched for the review: Academic Search Ultimate, Google Scholar, ScienceDirect, and South African Thesis and Dissertations. The search was restricted to peer-reviewed full-text manuscripts that discussed marine actinobacteria as a source of antimicrobials and were written in English from 1990 to December 2023. In total, for the past three decades (1990–December 2023), only 23 compounds from marine actinobacteria have been tested for their anti-Listerial potential. Out of the 23 reported compounds, only 2-allyoxyphenol, adipostatins E–G, 4-bromophenol, and ansamycins (seco-geldanamycin B, 4.5-dihydro-17-O-demethylgeldanamycin, and seco-geldanamycin) have been found to possess anti-Listerial activity. Thus, our literature survey reveals the scarcity of published assays testing the anti-Listerial capacity of bioactive compounds sourced from marine actinobacteria during this period.

Keywords: anti-*Listeria* compounds; marine actinobacteria; natural antimicrobials; antimicrobial resistance

1. Introduction

Listeria monocytogenes is a major foodborne pathogen that can cause severe listeriosis in humans [1]. Severe listeriosis may be characterized by meningitis, septicaemia, meningoencephalitis in immunocompromised people, invasive infections in newborns and the elderly, fetal malformations, and serious complications in pregnant women (abortion and stillbirth), with a case mortality rate that can range from 20% to 30% [2–5]. The hospitalization rate of the infection is more than 95% [1]. Although potentially deadly, the disease is relatively rare, with 0.1 to 10 cases per 1 million people annually, depending on the country and area of the world [3]. In contrast to high socio-economic regions of the world, Africa has a comparatively lower prevalence of severe listeriosis, despite the continent generally having a higher burden of foodborne infections [2,5]. Annually, 23,150 illnesses and 5463 deaths result from listeriosis worldwide [2,5]. However, the African continent contributes to 16% of the prevalence [2,5]. Despite this relatively low prevalence, the largest *L. monocytogenes* outbreak in the world happened in South Africa in 2017–2018, with 1060 reported cases [4,6,7]. The occurrence of an outbreak of this magnitude against the backdrop of a historically low prevalence of the disease in Africa is probably due to specific risk factors, and less awareness of *Listeria* transmission and risk factors along the food value chain [5].

Drug-resistant strains of *Listeria* have been documented [8,9], contributing to an escalation in morbidity and mortality rates associated with listeriosis [10]. Consequently, the pursuit of novel anti-Listerial compounds becomes a crucial strategy in addressing this crisis. Nature serves as the primary source of biotechnologically important molecules applicable across diverse fields [11]. Microorganisms, particularly within the *Actinobacteria* phylum, exhibit a notable proficiency in producing a variety of bioactive compounds due to their unique genetic composition [12,13]. While the extensive bioprospecting of terrestrial actinomycetes has diminished the probability of discovering novel bioactive compounds from this source, the largely unexplored marine environment is promising for identifying new actinomycetes with distinctive bioactive secondary metabolites. This potential is attributed to the harsh physicochemical conditions in the sea, such as high salinity, pressure, and low temperature, which create a conducive environment for the microbial synthesis of structurally and functionally unique molecules [14].

This review aimed to discuss the anti-Listerial compounds obtained from marine actinobacteria from the years 1990 to 2023 in the context of the sources of bioactive compounds, their structures, and antimicrobial activities. Furthermore, the review highlights the challenges and opportunities in the search for new anti-Listerial compounds from marine actinobacteria. Some reviews have summarized the sources, chemical structures, and anti-Listerial activities of bioactive compounds derived from marine actinobacteria [15,16]. In contrast, our review is unique as it focusses on consolidating information specifically and exclusively related to anti-Listerial compounds sourced from marine actinobacteria. By concentrating solely on the reporting of these compounds, our review identifies and emphasizes the gaps and trends observed from 1990 to 2023 in anti-Listerial compounds research. This focused analysis serves to guide future research directions, aiding researchers in pinpointing areas that need deeper exploration.

This literature review was conducted using four databases: Academic Search Ultimate, Google Scholar, ScienceDirect, and South African Thesis and Dissertations. The literature was accessed through the EBSCOhost research platform. The search terms used were "Marine actinomycetes", "OR", "actinobacteria", "AND", "antimicrobials", "OR", "antibiotics", "OR", "anti-listerial", "AND", "Listeria"." Individual genera also replaced "actinomycetes" and "actinobacteria" as search terms. Only English-published full-text and peer-reviewed manuscripts were selected for the review. Those manuscripts discussing anti-Listerial compounds obtained from terrestrial actinobacteria or from plant extracts were excluded (Figure 1).

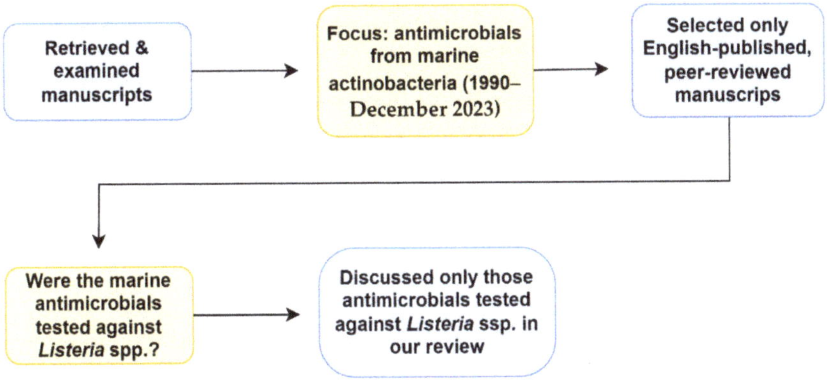

Figure 1. Relevant data retrieval process.

A limitation of this review is that the literature regarding alternative treatments for listeriosis is outside the scope of our discussion.

2. Background

2.1. Foodborne Listeriosis

Among the foodborne illnesses, listeriosis has the third highest case mortality rate (20–30% or higher) [17]. Thus, listeriosis is a serious public health concern [3,18].

2.1.1. Clinical Features

The clinical signs and symptoms of listeriosis can be differentiated into two categories, namely perinatal (i.e., feto-maternal and neonatal) listeriosis and listeriosis in adults. Perinatal listeriosis involves the infection of the fetus via the placenta. This may lead to abortion, stillbirth, or the baby being born with granulomatosis infantiseptica (a generalized infection) presenting as pyogranulomatous micro-abscesses [18–20]. For the mother, the infection may be asymptomatic or may present as a flu-like illness characterized by fatigue, headache, chills, and painful joints and muscles around 2 to 14 days prior to the miscarriage [21,22]. For reasons that are not clear, the mother's central nervous system (CNS) is rarely infected [20–23]. In 10–15% of perinatal cases, the aspiration of maternal fluids during delivery may cause late-onset neonatal listeriosis [24,25]. The clinical features may include flu-like symptoms accompanied by meningitis, 1–8 weeks postpartum [24,25].

2.1.2. Listeriosis Outbreaks

Listeriosis was first recognized as a foodborne disease in 1981 [26]. Thereafter, deadly outbreaks of the disease have been recorded worldwide [4,6,27–32]. The foods commonly implicated in the outbreaks include ready-to-eat (RTE) foods, unpasteurized milk, dairy products (yogurt, cheese, etc.), raw and unprocessed meats, salads, and fresh produce [3,7,33,34]. However, meat and its products have been responsible for all the major foodborne listeriosis outbreaks worldwide [7]. Table 1 summarizes some of the disease outbreaks associated with various meat types. Between 2017 and 2018, South Africa experienced the largest listeriosis outbreak in the world (Table 1) [4,6,7]. Two hundred and sixteen people (out of 1060 patients that were traced) died, resulting in a case mortality rate of 20.4% [4,31]. Whole-genome sequencing (WGS) was employed to track the source of the outbreak. Ready-to-eat-meat products (mainly polony) were linked to the outbreak [4]. The implicated bacterial strain was confirmed to be *L. monocytogenes* sequence type (ST) 6 [4]. Consequently, products were recalled from retailers across the country and from 15 importing African countries [35], and litigation was initiated against the RTE meat manufacturing company, leading to financial losses [31]. Therefore, beyond just the impacted customers, the aftermaths of these incidents have wider socioeconomic effects [31].

Table 1. Selected foodborne listeriosis outbreaks from meat and associated products from 1987 to 2019.

Year	Location	No. of Cases (Death)	Type of Meat	Serotype	References
1987–1989	United Kingdom	366 (ND)	Paté	4b	[7]
1900	Australia	9 (6)	Processed meats, paté	—	[7]
1992	France	279 (85)	Pork tongue in jelly	4b	[7]
1993	France	38 (10)	Rillettes	4b	[7]
1996	Australia	5 (1)	Diced, cooked chicken	—	[7]
1998–1999	United States	108 (14)	Hot dogs	4b	[7,31,36]
1999	United States	11 (ND)	Paté	—	[7,29]
1999–2000	France	10 (3)	Rillettes	4b	[7,29]
1999–2000	France	32 (10)	Pork tongue in aspic	4b	[7,29]
2000	United States	30 (7)	RTE deli turkey meat	1/2a	[7,29]
2000	New Zealand	30 (ND)	RET deli meats	1/2a	[7,29]
2001	United States	16 (ND)	Deli meats	1/2a	[7,29]
2002	United States	54 (8)	RET deli turkey meat	4b	[7,29]
2006–2007	Germany	16 (0)	RET scalded sausage	4b	[7,31]

Table 1. *Cont.*

Year	Location	No. of Cases (Death)	Type of Meat	Serotype	References
2008	Australia	13 (0)	Jellied pork	4b	[7,37]
2009	Denmark	8 (2)	Beef meat	—	[31,38]
2011	Switzerland	6 (ND)	Cooked ham	$\frac{1}{2}$a	[7]
2012	England	14 (0)	Pork pie	—	[31,39]
2012	New Zealand	4 (2)	RTE meat	—	[31,40]
2013	Scotland	3 (0)	Steak pie	$\frac{1}{2}$a	[31,41]
2013–2014	Denmark	41 (7)	Meat products	—	[7]
2014	Sweden	51 (0)	Cold cut ham	—	[31,42]
2014	Germany	39 (18)	RTE sausage (pork)	—	[31,43]
2015	Italy	35 (4)	Hog head cheese	$\frac{1}{2}$a *	[31,44]
2016	Italy	162 (0)	Cooked beef ham	$\frac{1}{2}$a **	[31,45]
2016	Switzerland	7 (0)	Meat pâté (beef)	4b (ST6)	[31,46]
2017–2018	South Africa	1060 (216)	Polony	4b (ST6)	[4,6,7]
2019	The Netherlands	21 (3)	RTE meat products	—	[31,47]

—: no record. * Only 24 of the cases were caused by *L. monocytogenes* serotype $\frac{1}{2}$a. ** All isolates except one were serotype $\frac{1}{2}$a.

2.1.3. Chemotherapeutic Treatment for Human Listeriosis

The feasible solution for the treatment of listeriosis is antimicrobial chemotherapy [48,49]. Usually, *Listeria* species are refractory to the lethal effects of many antibiotics. This is because the species are able to live and multiply within the host cells, where they remain hidden from the antibiotics in the extracellular fluid [50]. There are a limited number of antibiotics able to penetrate the host cells and reach the cytosol where the *Listeria* species normally reside in host cells [19,50]. Therefore, the antibiotic of choice for the treatment of listeriosis must be able to penetrate the host cell in sufficiently high concentrations for it to be efficacious [51–53]. Furthermore, the antibiotic must not undergo significant changes in its pH upon penetration into the host cell, as significant changes would hinder its efficacy. In addition, the antibiotic must bind to and block the penicillin-binding protein 3 (PBP 3) of *Listeria*, which result in a bactericidal effect (PBP 3 is an enzyme that catalyzes the last step of peptidoglycan synthesis) [48,52]. In the case of a pregnant woman, adequate concentrations of the antibiotic must be able to cross the placenta for fetal treatment [22,54]. Thus, the first choice of chemotherapy for treating listeriosis mainly includes the following antibiotics: penicillin G, amoxicillin, and ampicillin [54,55]. These antibiotics penetrate the host cells and exert their bactericidal effects by blocking a number of PBPs [50,52,56]. For synergy, the penicillins are commonly combined with an aminoglycoside, traditionally gentamicin [57]. In the second choice of therapy, trimethoprim is combined with a sulfonamide (e.g., sulfamethoxazole) in co-trimoxazole [56–59]. Other second-tier therapies include fluoroquinolones, vancomycin, and erythromycin (used in pregnancy) [57]. These second-choice therapies are normally reserved for people who are allergic to penicillin [54,57].

Prior to initiating treatment, it is imperative to conduct an in vitro assessment of the antimicrobial drugs used against the clinical *Listeria* isolates from a patient [60]. This is because there has been an increase in reports of antimicrobial-resistant *L. monocytogenes* strains from diverse sources such as meat and meat products [6,7], humans [9,61], animals [9,18,62], and food processing establishments [18,63]. The first strain of *L. monocytogenes* found to possess acquired antimicrobial resistance was a clinical isolate from France in 1988 [9,64,65]. The strain was multidrug resistant [9,64], resistant to aminoglycosides (erythromycin, and streptomycin), chloramphenicol and tetracyclines [64]. The antimicrobial resistance-encoding genes were found to be conferred by pIP811, a self-transferable 37-Kb plasmid [64]. Antimicrobial resistance has since spread, and is a serious public health challenge [8,9].

2.1.4. Resistance to Quinolones and Fluoroquinolones

Quinolones are a class of antimicrobial compounds that possess a 4-quinolone ring. Nearly all quinolones have a fluorine atom in their structure and are therefore called fluoroqinolones. Members of the quinolones and/or fluoroquinoles family include nalidixic acid, nafloxacin and ciprofloxacin. They exert a bactericidal effect by blocking bacterial DNA gyrase and topoisomerase IV, thereby inhibiting DNA and RNA syntheses [48,66,67]. However, *Listeria* spp. that are resistant to this class of antibiotics have been detected in food-producing animals and humans [68–70]. The mechanisms by which *Listeria* spp. resist fluoroquinolones include target gene alterations that include, for example, topoisomerase and gyrase gene mutations in some *Listeria* species [7,48,71]. These species also resist antibiotics via the overexpression of efflux pumps, namely Lde, MdrL, and FepA [7,72–75]. The pumps export an antimicrobial agent out of the cytosol and thus minimize its bactericidal or bacteriostatic effects [76,77]. Some *Listeria* spp. may harbor plasmid-mediated quinolone resistance (PMQR) genes, making them resistant to quinoline [48].

2.1.5. Resistance to Penicillins and Cephalosporins

Members of the penicillin group include penicillin G, penicillin V, ampicillin, methicillin, and oxacillin, and cephalosporins include the following broad-spectrum antibiotics: cefetamet, cefixime, ceftibuten, ceftazidime, cefdinir, cefpodoxime, cefotaxime, ceftriaxone, and cefuroxime [8,66,78,79]. Both penicillins and cephalosporins contain a β-lactam ring that is essential for their antimicrobial activity, which is the inhibition of bacterial cell wall synthesis [8,48,79]. These antimicrobials generally accomplish this inhibition by binding, through their β-lactam rings, to PBP 3, which inactivates the enzyme [80].

L. monocytogenes strains are naturally resistant to broad-spectrum cephalosporins due to the antimicrobials' low affinity for the PBB 3 of the bacteria [48,52,80,81]. In contrast, the bacterial strains are generally susceptible to penicillins, except for oxacillin [48]. However, some of the strains eventually become antimicrobial resistant, mainly through horizontal gene transfer, gene mutation, and biofilm formation [52,82–85]. The acquired antimicrobial resistance may involve various mechanisms [52]. For example, the efflux pump (AnrAB) in *L. monocytogenes* has been determined to confer resistance to β-lactam antibiotics (oxacillin, ampicillin, cephalosporins, and others) [52]. Another efflux pump (MdrL) makes *L. monocytogenes* resistant to cefotaxime [86]. According to Luque-Sastre et al. [48] and Srinivasan et al. [87], only the penicillin-binding protein gene (*penA*) has been linked with *L. monocytogenes* resistance to penicillin G.

2.1.6. Resistance to Aminoglycosides

Structurally, all aminoglycosides possess a cyclohexane ring and amino sugars [66]. Members of this class of antibiotics include gentamicin, kanamycin, streptomycin, and neomycin [66]. These antibiotics bind to the bacterial 30S ribosomal subunit, thereby blocking protein synthesis [66]. Aminoglycoside-resistant strains have been reported [9,52,53,57,61,87,88]. These strains normally emerge as a result of obtaining plasmid-borne genes and transposons encoding aminoglycoside-modifying enzymes [89]. The enzymes can be classified as follows: acetyltransferase, adenyltransferase, and phosphotransferase [89]. The gene (*aad6*) coding for 6-*N*-streptomycin adenylyltransferase (a streptomycin-modifying class of enzymes) has been detected in *L. monocytogenes* and *L. innocua* isolates [48,69]. Apart from *aad6*, no other aminoglycoside resistance genes have been identified to date in *Listeria* spp. [48,69].

2.1.7. Resistance to Tetracyclines

The tetracycline family includes naturally occurring antibiotics (tetracycline, chlortetracycline, and others) and semi-synthetic antibiotics (minocycline, doxycycline, etc.) [66]. Similar to aminoglycosides, tetracyclines bind to the bacterial 30S ribosomal subunit. This binding prevents the combination of aminoacyl-tRNA molecules with the A site of the ribosome. This family of antibiotics is a broad-spectrum bacteriostatic class [66]. However, the emergence of tetracycline resistance is the most common phenomenon among *Listeria* species [48,90]. The species obtain most of their resistance-encoding genes from *Enterococcus* and *Streptococcus* species through two types of mobile genetic elements: conjugative plasmids and transposons [48]. Usually, the genes [*tet* (S) and *tet* (L)] are transported by plasmids, whereas the *tet* (M) gene is carried by Tn916 (a conjugative transposon) [48,91]. The other tetracycline resistance-encoding genes found in *Listeria* spp. are *tet* (A) and *tet* (K) [48,90]. The genes *tet* (A), *tet* (K) and *tet* (L) code for proton antiporters that facilitate the efflux mechanism, which confers tetracycline resistance only. Meanwhile, the genes [*tet* (M) and *tet* (S)] code for ribosome protection proteins that confer resistance to both tetracycline and minocycline [48,90,91].

2.1.8. Resistance to Trimethoprim

Trimethoprim is a broad-spectrum synthetic antibiotic. It inhibits folic acid production by binding to dihydrofolate reductase (DHFR), an enzyme that catalyzes the conversion of dihydrofolic acid (DFA) to tetrahydrofolic acid during folic acid production. Due to its structural similarity to DFA (a DHFR substrate), trimethoprim competes with DFA for binding on the active site of the enzyme; consequently, the folic acid synthesis pathway is stopped or hampered [66].

Even though *L. monocytogenes* is commonly susceptible to trimethoprim [48], cases of resistance against the antibiotic have been published [57,69,92,93]. Usually, the mechanism of resistance against trimethoprim arises through the synthesis of trimethoprim-resistant DHFR in addition to the susceptible enzyme [90,94]. There are two resistance genes in *Listeria* that facilitate this synthesis: *dfr*D and *dfr*G; these are carried by plasmid pIP823 and transposon Tn6198, respectively [48].

2.1.9. Resistance to Chloramphenicol

Initially, *Streptomycess venezuelae* produced chloramphenicol. However, the chloramphenicol antibiotic is now chemically synthesized. Chloramphenical inhibits peptide elongation by binding to 23S rRNA of the 50S ribosomal subunit. However, this broad-spectrum antibiotic tends to have severe side-effects (e.g., lowering of bone marrow function, which may lead to complications such as aplastic anemia and a low white blood cell count) [66]. Therefore, this antibiotic is only employed when all other antibiotics fail in potentially fatal cases. In *Listeria* spp., resistance to chloramphenicol is encoded by *cat* (type A-8) and *floR* genes. The cat genes code for type A chloramphenicol acetyltransferases (Cat), which are responsible for enzymatic inactivation via the acetylation of the antibiotic [95]. In *L. monocytogenes*, the *floR* gene is linked to the export of florfenicol—a fluorinated derivative of chloramphenicol [87].

2.1.10. Resistance to Macrolides

Macrolide antibiotics include members such as erythromycin, clindamycin, and azithromycin. They are broad-spectrum and normally bacteriostatic antibiotics whose mode of action involves inhibiting protein synthesis by binding to the 23S rRNA of the 50S ribosomal subunit [48,66]. Resistance to macrolide is commonly facilitated by rRNA methylases. These enzymes are encoded by *erm* genes, specifically, *erm* (A), *erm* (B), and *erm* (C) in *Listeria* spp. The enzymes modify the target site (23S rRNA) by methylating its adenine base, consequently blocking the antibiotic from binding to the site [48,96,97].

2.2. Drivers of Antimicrobial Resistance

Some of the drivers of antimicrobial resistance include the indiscriminate use of antimicrobials in veterinary medicine for prophylaxis, growth promotion, and disease treatment [1,98]. Furthermore, the treatment of listeriosis using antibiotics in humans results in the selection of resistant *L. monocytogenes* strains [99]. Therefore, the emergence of multi-drug-resistant strains of *L. monocytogenes* poses a challenge due to potential treatment failure, limited treatment options, extended stays in medical care facilities and even fatalities [10,60,100]. For these reasons, novel anti-Listerial compounds are required.

2.3. *Actinobacteria as Potential Sources of Anti-Listerial Compounds*

Actinobacteria are known to be prolific producers of novel secondary metabolites with applications in diverse industries including the pharmaceutical industry [101,102]. The phylum *Actinobacteria* contains gram-positive bacteria with a high G + C content in general (>50 mol%) except *Tropheryma whipplei*. These bacteria can be filamentous or non-filamentous, and are found both in terrestrial and aquatic environments [103]. They grow via a combination of tip extension and hyphae branching. Their optimum growth conditions include a mesophilic temperature (ranging between 25 and 30 °C) and a pH between 6 and 9. However, some thermophilic strains have been isolated in high temperatures ranging from 50 to 60 °C, and others can grow in acidic conditions (pH 3.5) [104]. Many actinobacteria are of the order *Actinomycetales*. Members of this order are called actinomycetes. The suborders of *Actinomycetales* that show significance in bio-prospecting include *Streptomycineae, Micromonospineae, Corynebacterineae*, and *Streptosporangineae* [105].

2.3.1. *Streptomycineae*

The suborder *Streptomycineae* has a single family, *Streptomycetaceae*. Family members are often called streptomycetes and play a pivotal role in medicine by producing medicinally important compounds [105]. Among the streptomycetes, the genus *Streptomyces* produces the majority of the compounds [106]. It has produced 7600 compounds [107]. However, there may still be more undiscovered bioactive compounds. This is because when *Streptomyces coelicolor*'s genome was analyzed in 2002, genes coding to produce 20 bioactive compounds were discovered, even though the bacterium is known to only produce four antibiotics. Therefore, the challenge of finding other ways of stimulating the expression of the other genes remains [105]. Nonetheless, about 75% of all the bioactive compounds applied in the medical treatment of humans and animals are from *Streptomyces* spp. [108].

2.3.2. *Micromonosporineae*

The suborder *Micromonosporineae* has a single family, *Micromonosporaceae*, containing four genera: *Actinoplanes, Dactylosporangium, Micromonospora*, and *Pilimela*. Collectively, the genera are often called actinoplanetes. They form an elaborate, highly colored, substrate mycelium that lacks aerial hypha. However, commonly raised above the surface of the substrate mycelium are spore-containing-sporangia. The arrangement and morphology of the spores vary among the genera. For instance, in *Actinoplanes* and *Pilimela*, several spores are contained in cylindrical, spherical, or irregular sporangia. Meanwhile, for *Dactylosporangium*, one to six spores are contained in a finger-like sporangia. *Micromonospora* usually produce single spores, usually in branched sporophores [105].

Actinoplanetes are found in diverse habitats, including the soil, freshwater and sea. These bacteria are a source of bioactive metabolites. The genus *Micromonospora* is a prolific producer of antimicrobials, second only to the *Streptomyces* genus [109]. About 8% of the antibiotics used in medicine are from *Micromonospora* species [110].

2.3.3. *Corynebacterineae*

The suborder *Corynebacterineae* has six families: *Nocardiaceae, Gordoniaceae, Mycobacteriaceae, Dietziaceae, Tsukamurellaceae* and *Corynebacteriaceae* [105]. In bioprospection, the family *Nocardiaceae* has shown significance. For example, the two genera (*Nocardia* and *Rhodococcus*) belonging to this family contain species capable of producing bioactive compounds [111,112].

The genera *Nocardia* and *Rhodococcus* are largely strict aerobes. They are distributed widely in terrestrial and aquatic ecosystems, wherein they form substrate hypha that fragment into coccoid fragments and rods. The bacteria also develop aerial mycelium that may form conidia. Most of the bacterial strains are free-living saprophytes. However, some strains of *Nocardia* (e.g., *N. asteroids*) can cause nocardiosis, an opportunistic disease in humans and animals [105].

2.3.4. *Streptosporangineae*

One of the families of the *Streptosporangineae* suborder is *Nocardiopsaceae* [113]. The family contains a type genus, *Nocardiopsis* [114]. The type species of the genus is *Nocardiopsis dassonvillei,* and IMRU 509 (DSM 43111, ATCC 23218, JCM 7437) is the type strain [114]. Members of the genus are able to prevail under harsh environmental conditions but are generally found in locales with high salt concentrations [115]. Their survival in such conditions is mediated by secreting different and novel bioactive compounds and extracellular enzymes [115]. Thus, the genus is among the biotechnologically important genera of actinobacteria.

2.4. *Marine Environment as a Source of Microbes Harbouring Novel Bioactive Metabolites*

In the last two decades, the number of newly approved antibiotics has dwindled by 75% [116]. For example, since the year 2000, only 30 novel antibiotics have been issued globally. Of these 30, 16 were synthetic, 2 were natural products, and 12 were natural product derivatives. Forty percent of the 30 antibiotics were from actinomycetes, either as natural products or their derivatives [117]. One of the factors contributing to the decline is the rediscovery of known bioactive compounds from conventional microbial habitats, such as the terrestrial habitat. Terrestrial actinomycetes have been bio-prospected extensively; as a result, 70% of commercial antibiotics are extracted from them. Therefore, the likelihood of finding novel bioactive compounds from terrestrial actinomycetes has decreased [116,118]. Thus, the search for novel bioactive compounds has switched to actinomycetes occupying unexplored or underexploited habitats, such as the marine environment [103].

The marine microbial habitats include sea water and sand, coastal and deep sea sediments, mangrove sediments, and hydrothermal vents [119,120]. These habitats tend to have complex conditions such as low temperatures, salinity, and high pressures [120,121]. Therefore, marine microorganisms may develop novel metabolites (absent in terrestrial microbes) in order to survive in such conditions. As a result, a variety of unique bioactive compounds have been recovered from marine actinomycetes in recent times [119,122].

3. Anti-Listerial Compounds from Marine Actinobacteria

In 2006, Saha et al. [123] isolated a putatively novel *Streptomyces* MS1/7 from the marine sediments of the Bay of Bengal, India. The actinobacterium produced 2-Allyoxyphenol (**1**) (Figure 2) (molecular formula: $C_9H_{10}O_2$), a synthetic compound obtained as a natural product for the first time by Arumugam et al. [124]. The compound was found to have antimicrobial activity against a panel of 21 bacteria and 3 fungi, with MIC values ranging from 0.2 to 1.75 mg/mL [124]. Specifically, 2-Allyoxyphenol had an MIC value of 0.45 mg/mL against *L. monocytogenes* MTCC 657, which was determined by the agar dilution method [124].

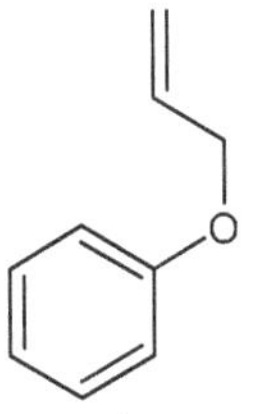

1

Figure 2. Structure of 2-Allyoxyphenol (**1**) [124].

Another study in India reported on the isolation of the rare actinobacterium *Nocardiopsis* sp. SCA21 from the marine sediment of Havelock Island, the Andaman Islands, and the Nicobar Islands [125]. A bromophenol derivative (4-bromophenol; Figures 2 and 3), and a phthalate ester (Bis (2-ethylhexyl) phthalate; Figure 3, **3**) were purified for the first time from the fermentation broth of the genus *Nocardiopsis* [125]. When tested using the disc diffusion method and the micro-dilution method, respectively, only 4-bromophenol demonstrated activity against *L. monocytogenes* 13932 (zone of inhibition of 24 ± 0.11 mm and MIC value of 62.5 µg/mL) [125].

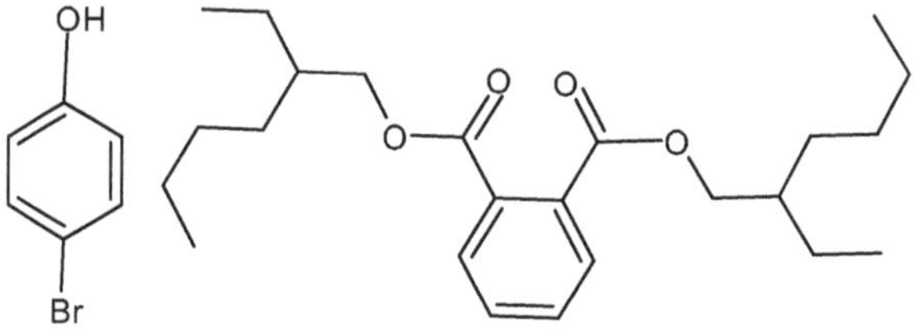

2 **3**

Figure 3. Structures of compounds **2–3** [125].

Streptomyces blancoensis 20733 was purified from the marine sediments of San Miguel, Costa Rica, 2020 [126]. Analysis of the bacterium's extracts led to the isolation of six novel phenolic lipids, adipostatins E–J (**4–9**) (Figure 4). Adipostatins E–G exhibited antagonistic activity against *L. monocytogenes* ATCC 19115, with IC50 values of 5.9 µM, 34.2 µM, and 20.3 µM, respectively. Adipostatins H–J showed no activity against the ATCC 19115 strain at the tested concentration [126].

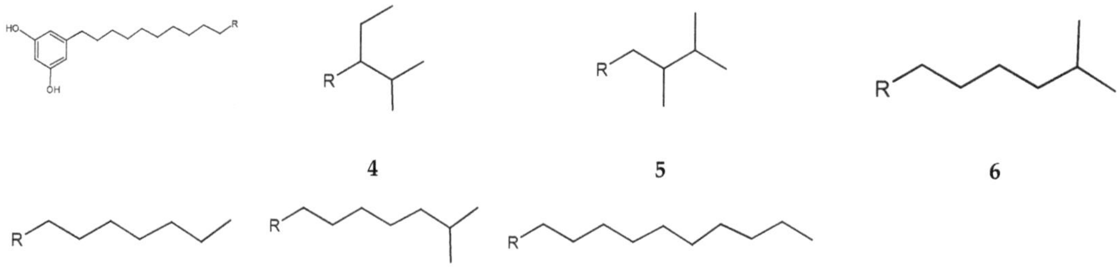

4 **5** **6** **7** **8** **9**

Figure 4. Structures of adipostatins E–J (**4–9**) [16,126].

Veratraldehyde (**10**) (Figure 5) was purified for the first time from the marine actinomycete *Streptomyces diastaticus* LC360811 (isolated from sediment samples of the Red Sea coast, Egypt) [127]. The compound was active against the tested gram-positive and gram-negative bacteria. However, no activity was detected against *L. monocytogenes* ATCC 35152 [127].

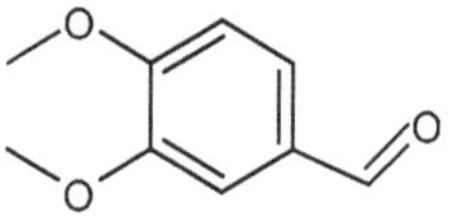

10

Figure 5. Structure of veratraldehyde (**10**) [127].

The marine sediment samples collected in July 2021 from Yongzing Island, China, yielded *Streptomyces* sp. ZYX-F-97 [128]. Four new ansamycin derivatives were then purified from the species' fermentation (Figure 6) [128]. These are 1,19-epithio-geldanamycin

A (**11**), 17-demethoxylherbimycin H (**12**), herbimycin M (**13**), and seco-geldanamycin B (**14**). Additionally, eight known ansamycin analogues were isolated (Figure 6) [128]: hervimycin H (**15**), herbimycin I (**16**), tetracyclic thiazinogeldanamycin (**17**), 4.5-dihydro-17-O-demethylgeldanamycin (**18**), 4.5-dihydro-17-S-demethylgeldanamycin (**19**), 19-S-methylgeldanamycin (**20**), 18-methylreblastatin (**21**), and seco-geldanamycin (**22**). But only compounds 14, 18, and 22 showed remarkable inhibition against *L. monocytogenes* ATCC 1911, with MIC values of 64 μg/mL, 8 μg/mL, and 16 μg/mL, respectively [128].

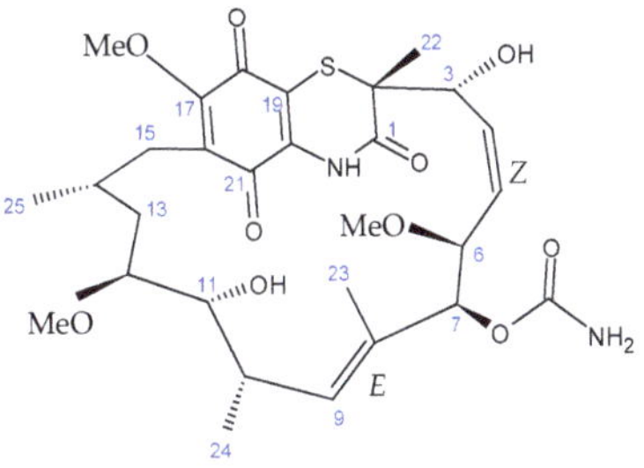
11

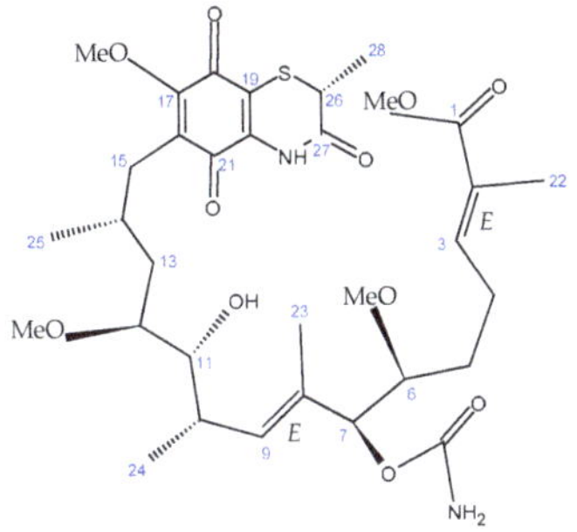
12: R=H
13: R=CHO
15: R=MeO

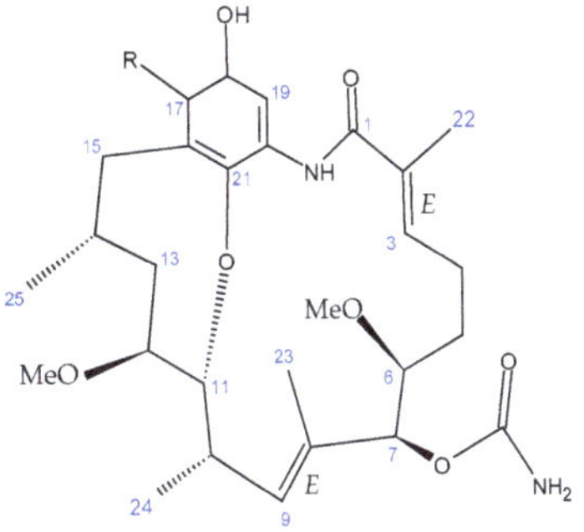
14

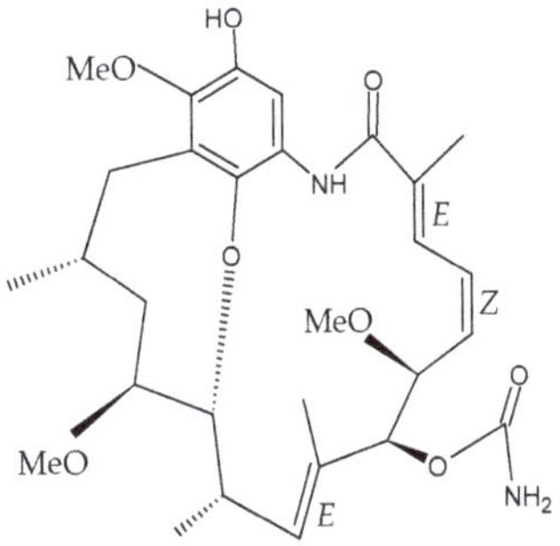
16

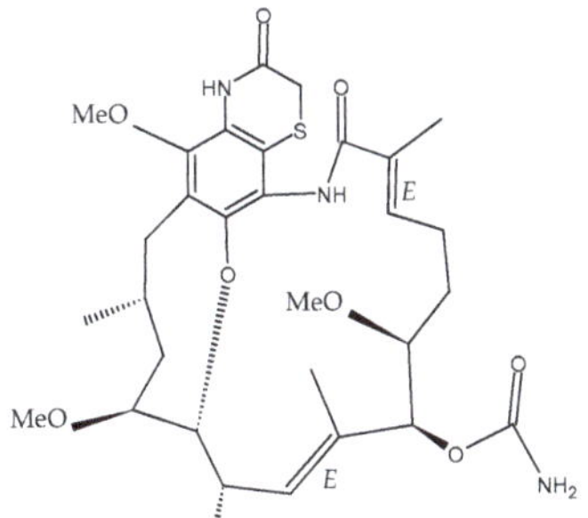
17

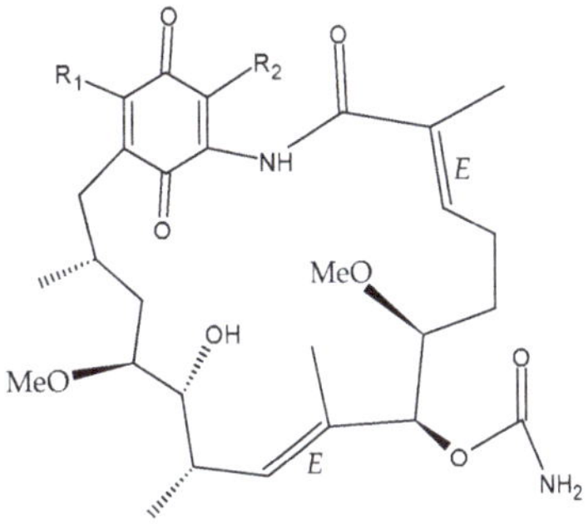
18: R_1=OH; R_2=H
19: R_1=OMe; R_2=SCH_3

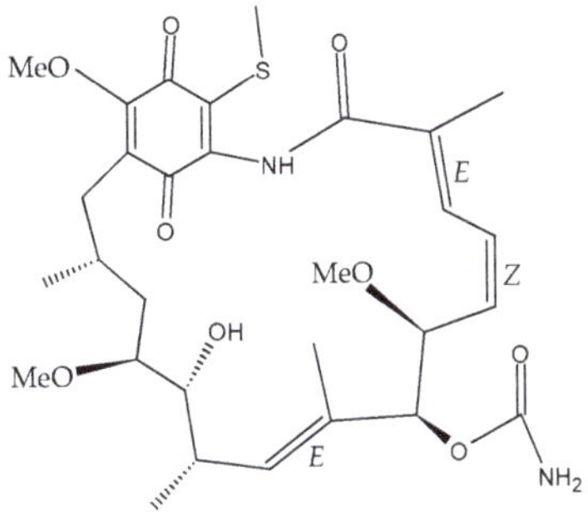
20

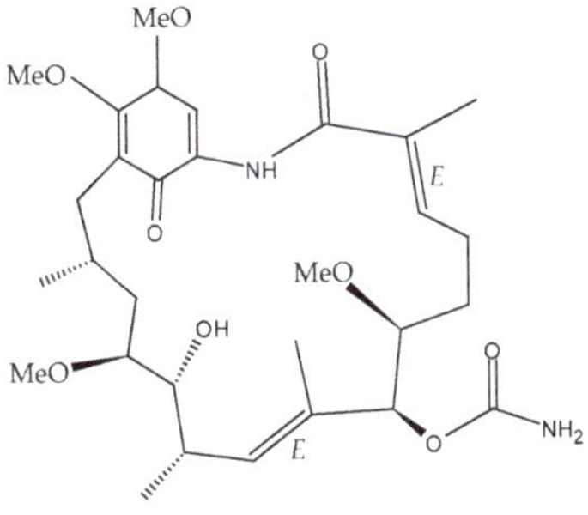
21

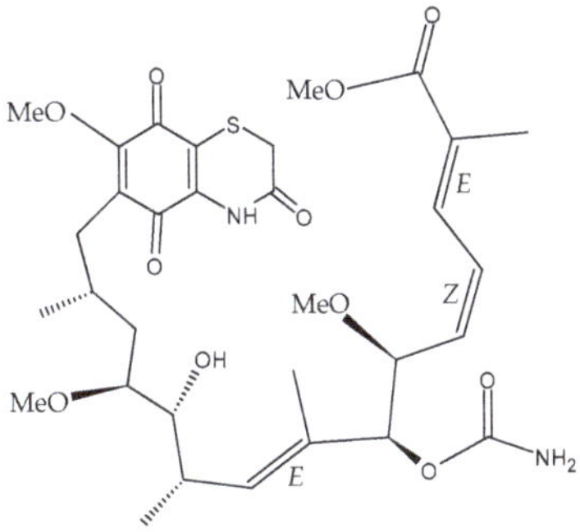
22

Figure 6. Structures of ansamycins (**11–22**) [128].

Another novel antibiotic, a 22-membered macrolide lactam named haneummycin (**23**) (Figure 7), was purified from *Streptomyces* sp. KM77-8; this was isolated from marine sediment collected from Tokyo Bay, Japan [129]. The antibiotic was then assayed against *Listeria innocua* ATCC 33090^{T} and displayed no anti-Listerial activity at the tested concentration [129].

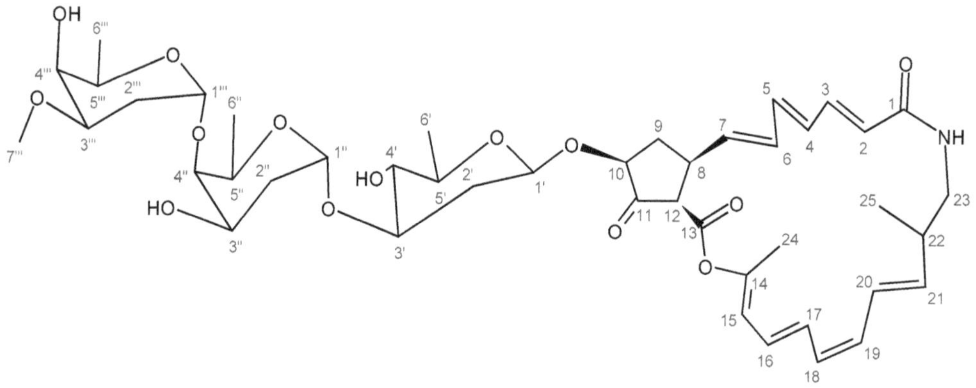

Figure 7. Structure of haneummycin (**23**) [129].

In summary, out of the 23 reported compounds, only eight of them possessed anti-Listerial activity. The eight compounds and their wide range of other biological activities are consolidated in Table 2. According to Arumugam et al. [124], the hydroxyl and allyloxy groups of compound **1** were responsible for its antimicrobial and antioxidant activities. Due to the absence of potential carcinogenicity and hemolytic toxicity, the cytotoxicity literature suggests the possible application of compound **1** as a food preservative and oral disinfectant [124].

Table 2. Anti-Listerial compounds from marine actinobacteria (1990–December 2023).

Compound	Producing Strain	Biological Activity	References
2-Allyoxyphenol (**1**)	*Streptomyces* MS1/7	Antibacterial; antifungal; antioxidative; food preservative; oral disinfectant	[123,124]
4-Bromophenol (**2**)	*Nocardiopsis* sp. SCA21	Antibacterial; antioxidant; metal chelating; enzyme inhibitory activity	[125]
Adipostatins E–G (**4–6**)	*Streptomyces blancoensis* 20733	Antibacterial	[126]
Seco-geldanamycin B (**14**) 4.5-Dihydro-17-O-demethylgeldanamycin (**18**) Seco-geldanamycin (**22**)	*Streptomyces* sp. ZYX-F-97	Antibacterial	[128]

Compound **2** had potent antibacterial activity against a panel of test clinical pathogens, with an MIC value ranging from 7.81 to 125 μg/mL [125]. The compound also showed remarkable free radical scavenging potential, with an IC_{50} value of 187.31 μg/mL against 2,2-diphenyl-1-picrylhydrazyl radicals and an IC_{50} value of 102.22 μg/mL against 2,2′-azinobis (3-ethylbenzothiazoline-6-sulfonic acid) radicals [125]. The iron metal chelating activity of compound **2** was less significant, with an IC_{50} value of >250 μg/mL. Furthermore, compound **2** exhibited remarkable enzyme inhibitory activities against α-glucosidase and α-amylase [125].

Adipostatins exhibited significant antibacterial activity against gram-positive bacteria through the inhibition of co-enzyme A biosynthesis [126]. None of the adipostatins possessed noticeable inhibition against gram-negative microbes (viz., *E. coli*, *Salmonella enterica*, and *Shigella flexneri*). It was suspected that by inhibiting CoA biosynthesis, the adipostatins might also be affecting the synthesis of peptidoglycans and fatty acids. Consequently, the compounds were more potent against gram-positive bacteria [126].

Similarly, the ansamycins (**14, 18, 22**) were reported to exhibit significant activity against a panel of gram-positive bacteria [128], but the compounds showed no activity against the gram-negative bacteria (*E. coli* and *Pseudomonas aruginosa*) [128].

4. Challenges and Opportunities Associated with Antimicrobials from Marine Actinobacteria

Our literature survey from 1990 to December 2023 reveals the scarcity of published assays testing the anti-Listerial capacity of bioactive compounds purified from marine actinobacteria during this period. For example, only one compound (2-Allyoxyphenol (**1**)) was tested for antibacterial activity against *L. monocytogenes* out of the 33 reported antibacterial compounds obtained from marine actinomycetes between 2005 and 2010 in a review by Subramani and Aalbersberg [15]. Furthermore, a review article by Schinke et al. [130] shows that the bioactive compounds derived from marine actinobacteria between 2010 and 2015 were not evaluated for anti-Listerial activity. However, it should be noted that the review only examined bioactive substances with MICs of less than or equal to 20 μg/mL [130]. According to Wang et al. [131], out of the 308 compounds derived from marine actinomycetes from 1990 to 2019, none of them were tested for anti-Listerial activity. Nevertheless, it is not apparent whether this review included studies from the entire year, because in the same year (2019), Siddharth and Rai [125] tested two compounds (4-bromophenol, 2, and bis (2-ethylhexyl) phthalate, **3**) against *L. monocytogenes* 13932 that were isolated from *Nocardiopsis* sp. SCA21. Likewise, from 1992 to 2020, none of the 127 novel halogenated compounds obtained from marine actinomycetes were examined for their anti-Listerial activities [132]. Only *Streptomyces blancoensis* strain 20733 was examined for the biosynthesis of new anti-Listerial compounds (**4–9**) out of more than nine marine streptomycete isolates assessed for the production of novel bioactive compounds in 2020 [16]. A literature review (from 2012 to 2022) by Liang et al. [133] shows that around 62 bioactive secondary metabolites were isolated from marine sponge-associated actinomycetes. Interestingly, according to the observed literature, none of the metabolites were examined for their anti-Listerial potential [133]. During these time periods, it appears that the antimicrobial activity assays used were mostly against microbes such as methicillin-resistant *Staphylococcus aureus* (MRSA), vancomycin-resistant *Enterococcus faecium* (VREF), *Bacillus subtilis*, *Candida* species, and others [15,130,134]. This is probably because the World Health Organization considers these bacteria to be the most dangerous group on its list of priority pathogens due to their link to multi-drug resistant nosocomial infections [135].

The research focus for novel anti-Listerial substances appears to be natural antimicrobial agents such as lactoperoxidase, lactoferrin, and lysozyme from animal sources, essential oils and herbal extracts from plant sources, and bacteriocin from microbial sources [136–139]. These chemicals are primarily being researched for their potential use as natural preservatives to prevent food spoilage and contamination, as *L. monocytogenes* is a major foodborne pathogen [139,140]. These innovative preservatives would aid in the prevention of *L. monocytogenes* growth in foods, while new anti-Listerial drug leads derived from marine actinobacteria could be developed and used to treat infected individuals. Thus, both research approaches are valuable.

The majority of the published literature is on preliminary screening assays for anti-Listerial compound production by marine actinobacteria that require further analysis to obtain pure compounds (Tables 3 and 4). For example, researchers [141–144] only used the cross-streak method to evaluate the anti-Listerial potential of various marine actinobacteria isolated from marine samples collected between 2013 and 2015 (Table 3). Eythorsdottir et al. [145] employed both the cross-streak and agar well diffusion methods to assess the anti-Listerial activity of marine actinobacterial symbionts recovered from shallow-water hydrothermal vents in Northern Iceland (Table 3). During the years 2016 to 2017, different fish were caught from the offshore waters of Ireland, Iceland, and international waters at depths of 1000 m and 850 m [146]. *Arthrobactor* sp. APC 3897 was then isolated from the skin of the deep-sea fish and tested for antibacterial activity by colony overlay assays [147]. The isolate showed activity against foodborne *L. innocua* and *L. monocytogenes* strains [147].

The in vitro anti-Listerial activities of ethyl acetate crude extracts of *Streptomyces cyaneofuscatus* isolates were determined by the disc diffusion method (Table 4) [148]. The agar well diffusion method was used to investigate the anti-Listerial capacity of ethyl acetate crude extracts of *Nocardiopsis alba* isolates (Table 4) [149].

The n-butanol crude extract of *Streptomyces* sp. Sp1 was evaluated for its anti-Listerial activity by the agar well diffusion method (Table 4) [150]. Ngema et al. [85] employed the 96-well microdilution and in silico molecular docking methods to examine the anti-Listerial potential of *Nocardiopsis dassonvillei* SOD(B)ST2SA2′s crude extract (Table 4). Such preliminary screening assays are a yet-to-be-tapped potential source of new anti-Listerial compounds, provided they are processed further. However, researchers may stop processing these crude extracts if the dereplication process fails to reveal novel compounds from the extracts [151].

Table 3. Preliminary assays of marine actinobacterial activity against *L. monocytogenes* (1990–December 2023).

Marine Actinomycete Strain	Year of Isolation	Country	Method Used for Antibacterial Activity Assay	Test Strain	Reference
Pseudonocardia carboxydivorans VO36-3 *Salinoactinospora qingdaoensis* VN6-2 *Microbacterium profundi* VP2-3 *Arthrobacter phenanthrenivorans* VO30-3 *Aeromicrobium alkaliterrae* V040-3 *Gordonia bronchialis* VO29-3 *Isoptericola halotolerans* VP3-3 *Streptomyces janthinus* VS4-2 *Streptomyces albogriseolus* VH47-3	2013	Chile	Cross-streak method	*L. monocytogenes* 07PF0776	[141]
Streptomyces sp. H-KF8	2013	Chile	Cross-streak method	*L. monocytogenes* 07PF0776	[144]
Streptomyces sp. H-KF8 *Rhodococcus* H-CA8F *Micrococcus* H-CD9b *Kocuria* H-KB6 *Curtobacterium* H-ED12 *Curtobacterium* H-BE10 *Corynebacterium* H-EH3 *Brachybacterium* H-CG1 *Brachybacterium* H-CD1 *Arthrobacter* H-CA8b	2013	Chile	Cross-streak method	*L. monocytogenes* 07PF0776	[143]
Actinomycete 111 Actinomycete 112 Actinomycete 115 Actinomycete 117 Actinomycete 127 Actinomycete 131 Actinomycete 135 Actinomycete 141 Actinomycete 275	2013	Iran	Cross-streak method	*L. monocytogenes* ATCC 1298	[142]
Tsukamurella strandjordii 101-1518	2015	Northern Iceland	Agar diffusion assay and cross-streak method	*L. monocytogenes*	[145]

Table 4. Anti-Listerial crude extracts from various marine actinobacteria (1990–December 2023).

Marine Actinomycete Strain	Year of Isolation	Country of Isolation	Crude Extract	Pathogen Target	Reference
Streptomyces cyaneofuscatus M-157 *Streptomyces cyaneofuscatus* M-169 *Streptomyces cyaneofuscatus* M-192 *Streptomyces cyaneofuscatus* M-207 *Streptomyces cyaneofuscatus* M-220 *Streptomyces cyaneofuscatus* M-231	2013	Spain	Ethyl acetate extract	*L. monocytogenes* 72964 *	[148]
Streptomyces sp. Sp1	2018	Egypt	n-butanol extract	*L. monocytogenes* 19115	[150]
Nocardiopsis alba PB-1 *Nocardiopsis alba* PB-3	2020	India	Ethyl acetate extract	*L. monocytogenes* ATCC 19112	[149]
N. dassonvillei SOD(B)ST2SA2	2021	South Africa	Chloroform extract	*L. monocytogenes* KGEO161 *L. monocytogenes* ILemanAP345 *L. monocytogenes* ILemanEO299 *L. monocytogenes* ILemanER317 *L. monocytogenes* ILestanBR361 *L. monocytogenes* ILestanBR363 *L. monocytogenes* ILestanGP395 *L. monocytogenes* ILestanGP400 *L. monocytogenes* ATCC 15313	[85]

* Cephalosporins-resistant *L. monocytogenes*.

5. Conclusions

Generally, new compounds from marine actinobacteria have rarely been tested for their anti-Listerial potential during the past three decades (1990–December 2023). Many countries have not conducted or published any research on the new anti-Listerial compounds purified from marine actinobacteria during this period. However, Costa Rica, Egypt, Japan, China, and India published data on 23 compounds from marine actinobacteria that were tested for their anti-Listerial potential. Eleven of the compounds (adipostatins E–J, ansamycins [11–14], and haneummycin) were novel. The other compounds (4-bromophenol and bis (2-ethylhexyl) phthalate) were purified for the first time from the genus *Nocardiopsis*. Similarily, veratraldehyde was extracted for the first time from a marine actinomycete. Moreover, a synthetic compound (2-Allyoxyphenol) was obtained as a natural product for the first time. Out of the 23 reported compounds, only 2-allyoxyphenol, adipostatins E–G, 4-bromophenol, and ansamycins (**14**, **18**, and **22**) were found to possess anti-Listerial activity. Considering the zoonotic nature of *L. monocytogenes*, the relatively high case fatality rate of listeriosis, the development and spread of antimicrobial resistance, and the unique nature of secondary metabolites from marine actinobacteria compared to their terrestrial counterparts, it is paramount to expand bioprospecting studies of bioactive compounds from marine actinobacteria, particularly from underexplored ocean environments on the African continent.

Author Contributions: Conceptualization, E.M. and S.S.N.; validation, E.M. and S.S.N.; writing—original draft preparation, S.S.N.; writing—review and editing, E.M.; supervision, E.M.; project administration, E.M.; funding acquisition, E.M. All authors have read and agreed to the published version of the manuscript.

Funding: This research was funded by Red Meat Research and Development Trust South Africa, and Department of Trade, Industry, and Competition grant number THRIP/22/30/11/2017 and E.M. is the recipient and Principal Investigator. S.S.N. acknowledges The Moses Kotane Institute (MKI) Research Fund (2022) and The FoodBev SETA Bursary (2023).

Institutional Review Board Statement: This study was approved by the University of Zululand's Research Ethics Committee (UZREC 171110-030 PGD 2021/26).

Informed Consent Statement: Not applicable.

Data Availability Statement: Not applicable.

Acknowledgments: The authors are grateful to the University of Zululand for their provision of library resources that were instrumental in the literature research.

Conflicts of Interest: The authors declare no conflicts of interest.

References

Matle, I.; Mbatha, K.R.; Lentsoane, O.; Magwedere, K.; Morey, L.; Madoroba, E. Occurrence, serotypes, and characteristics of *Listeria monocytogenes* in meat and meat products in South Africa between 2014 and 2016. *J. Food Saf.* **2019**, *39*, e12629. [CrossRef]

De Noordhout, C.M.; Devleesschauwer, B.; Angulo, F.J.; Verbeke, G.; Haagsma, J.; Kirk, M.; Havelaar, A.; Speybroeck, N. The global burden of listeriosis: A systematic review and meta-analysis. *Lancet Infect. Dis.* **2014**, *14*, 1073–1082. [CrossRef] [PubMed]

World Health Organization. Available online: https://www.who.int/news-room/fact-sheets/detail/listeriosis (accessed on 13 February 2024).

Smith, A.M.; Tau, N.P.; Smouse, S.L.; Allam, M.; Ismail, A.; Ramalwa, N.R.; Disenyeng, B.; Ngomane, M.; Thomas, J. Outbreak of *Listeria monocytogenes* in South Africa, 2017–2018: Laboratory activities and experiences associated with whole-genome sequencing analysis of isolates. *Foodborne Pathog. Dis.* **2019**, *16*, 524–530. [CrossRef] [PubMed]

Sibanda, T.; Ntuli, V.; Neetoo, S.H.; Habib, I.; Njage, P.M.K.; Parry-Hanson Kunadu, A.; Andoh, A.H.; Coorey, R.; Buys, E.M. *Listeria monocytogenes* at the food–human interface: A review of risk factors influencing transmission and consumer exposure in Africa. *Int. J. Food Sci. Technol.* **2023**, *58*, 4114–4126. [CrossRef]

Kaptchouang Tchatchouang, C.-D.; Fri, J.; De Santi, M.; Brandi, G.; Schiavano, G.F.; Amagliani, G.; Ateba, C.N. Listeriosis outbreak in South Africa: A comparative analysis with previously reported cases worldwide. *Microorganisms* **2020**, *8*, 135. [CrossRef] [PubMed]

Matle, I.; Mbatha, K.R.; Madoroba, E. A review of *Listeria monocytogenes* from meat and meat products: Epidemiology, virulence factors, antimicrobial resistance and diagnosis. *Onderstepoort J. Vet. Res.* **2020**, *87*, 1–20. [CrossRef]

8. Olaimat, A.N.; Al-Holy, M.A.; Shahbaz, H.M.; Al-Nabulsi, A.A.; Abu Ghoush, M.H.; Osaili, T.M.; Ayyash, M.M.; Holley, R.A Emergence of antibiotic resistance in *Listeria monocytogenes* isolated from food products: A comprehensive review. *Compr. Rev Food Sci. Food Saf.* **2018**, *17*, 1277–1292. [CrossRef] [PubMed]
9. Hanes, R.M.; Huang, Z. Investigation of antimicrobial resistance genes in *Listeria monocytogenes* from 2010 through to 2021. *Int. J. Environ. Res. Public Health* **2022**, *19*, 5506. [CrossRef] [PubMed]
10. World Health Organization. Antimicrobial Resistance. Available online: https://www.who.int/news-room/fact-sheets/detail/antimicrobial-resistance#:~:text=As%20a%20result%20of%20drug,severe%20illness,%20disability%20and%20death (accessed on 11 March 2024).
11. Paul, S.I.; Majumdar, B.C.; Ehsan, R.; Hasan, M.; Baidya, A.; Bakky, M.A.H. Bioprospecting potential of marine microbial natural bioactive compounds. *Appl. Biotechnol. Rep.* **2021**, *8*, 96–108.
12. Van Bergeijk, D.A.; Terlouw, B.R.; Medema, M.H.; van Wezel, G.P. Ecology and genomics of actinobacteria: New concepts for natural product discovery. *Nat. Rev. Microbiol.* **2020**, *18*, 546–558. [CrossRef]
13. Girão, M.; Ribeiro, I.; Carvalho, M.d.F. Actinobacteria from marine environments: A unique source of natural products. In *Natural Products from Actinomycetes: Diversity, Ecology and Drug Discovery*; Rai, R.V., Bai, J.A., Eds.; Springer: Singapore, 2022; pp. 1–45.
14. Sarkar, G.; Suthindhiran, K. Diversity and biotechnological potential of marine actinomycetes from India. *Indian J. Microbiol.* **2022** *62*, 475–493. [CrossRef] [PubMed]
15. Subramani, R.; Aalbersberg, W. Marine actinomycetes: An ongoing source of novel bioactive metabolites. *Microbiol. Res.* **2012**, *167*, 571–580. [CrossRef] [PubMed]
16. Lacey, H.J.; Rutledge, P.J. Recently discovered secondary metabolites from *Streptomyces* species. *Molecules* **2022**, *27*, 887. [CrossRef] [PubMed]
17. Centers for Disease Control and Prevention. Listeria (Listeriosis). Available online: https://www.cdc.gov/listeria/faq.html#:~text=Yes.,we%20call%20the%20infection%20invasive (accessed on 13 February 2023).
18. Vázquez-Boland, J.A.; Kuhn, M.; Berche, P.; Chakraborty, T.; Domínguez-Bernal, G.; Goebel, W.; González-Zorn, B.; Wehland, J.; Kreft, J. *Listeria* pathogenesis and molecular virulence determinants. *Clin. Microbiol. Rev.* **2001**, *14*, 584–640. [CrossRef] [PubMed]
19. Osek, J.; Wieczorek, K. *Listeria monocytogenes*—How this pathogen uses its virulence mechanisms to infect the hosts. *Pathogens* **2022**, *11*, 1491. [CrossRef] [PubMed]
20. Ramos Cela, M.; Medina Martínez, A.J.; Vera Martín, I. *Infection of Listeria monocytogenes Associated to the Central Nervous System: Cell Pathogenesis, Diagnosis and Risk Factors*; Archivos de Medicina Universitaria: Granada, Spain, 2022.
21. Donovan, S. Listeriosis: A rare but deadly disease. *Clin. Microbiol. Newsl.* **2015**, *37*, 135–140. [CrossRef]
22. Wang, Z.; Tao, X.; Liu, S.; Zhao, Y.; Yang, X. An update review on *Listeria* infection in pregnancy. *Infect. Drug Resist.* **2021**, *14*, 1967–1978. [CrossRef]
23. Olson-Chen, C. Neurologic infections in pregnancy. In *Neurological Diseases and Pregnancy: A Coordinated CARE model for Best Management*; Ciafaloni, E., Bushnell, C., Thornburg, L.L., Ciafaloni, E., Thornburg, L.L., Bushnell, C.D., Eds.; Oxford University Press: New York, NY, USA, 2018; pp. 95–104.
24. Farber, J.M.; Peterkin, P.I.; Carter, A.O.; Varughese, P.V.; Ashton, F.E.; Ewan, E.P. Neonatal listeriosis due to cross-infection confirmed by isoenzyme typing and DNA fingerprinting. *J. Infect. Dis.* **1991**, *163*, 927–928. [CrossRef] [PubMed]
25. Tesini, B.L. Neonatal Listeriosis. Available online: https://www.msdmanuals.com/professional/pediatrics/infections-in-neonates/neonatal-listeriosis (accessed on 13 February 2024).
26. Schlech III, W.F.; Lavigne, P.M.; Bortolussi, R.A.; Allen, A.C.; Haldane, E.V.; Wort, A.J.; Hightower, A.W.; Johnson, S.E.; King, S.H.; Nicholls, E.S. Epidemic listeriosis–evidence for transmission by food. *N. Engl. J. Med.* **1983**, *308*, 203–206. [CrossRef]
27. Fleming, D.W.; Cochi, S.L.; MacDonald, K.L.; Brondum, J.; Hayes, P.S.; Plikaytis, B.D.; Holmes, M.B.; Audurier, A.; Broome, C.V.; Reingold, A.L. Pasteurized milk as a vehicle of infection in an outbreak of listeriosis. *N. Engl. J. Med.* **1985**, *312*, 404–407. [CrossRef]
28. Büla, C.J.; Bille, J.; Glauser, M.P. An epidemic of food-borne listeriosis in western Switzerland: Description of 57 cases involving adults. *Clin. Infect. Dis.* **1995**, *20*, 66–72. [CrossRef] [PubMed]
29. Jadhav, S. *Detection, Subtyping and Control of Listeria monocytogenes in Food Processing Environments*; Swinburne University of Technology: Hawthorn, VIC, Australia, 2015.
30. Buchanan, R.L.; Gorris, L.G.; Hayman, M.M.; Jackson, T.C.; Whiting, R.C. A review of *Listeria monocytogenes*: An update on outbreaks, virulence, dose-response, ecology, and risk assessments. *Food Control* **2017**, *75*, 1–13. [CrossRef]
31. Warmate, D.; Onarinde, B.A. Food safety incidents in the red meat industry: A review of foodborne disease outbreaks linked to the consumption of red meat and its products, 1991 to 2021. *Int. J. Food Microbiol.* **2023**, *398*, 110240. [CrossRef] [PubMed]
32. Centers for Disease Control and Prevention. Listeria (Listeriosis)–Listeria Outbreaks. Available online: https://www.cdc.gov/listeria/outbreaks/index.html (accessed on 14 February 2024).
33. Zhang, Y. *Antimicrobial Resistance of Listeria monocytogenes and Enterococcus faecium from Food and Animal Sources*; University of Maryland: College Park, MD, USA, 2005.
34. Kananub, S.; Lertsakkongkul, P.; Aryatawong, P.; Horhirunkhajohn, W.; Pinniam, N.; Krajanglikit, P.; Sonthong, K.; Kasemsuwan, S. *Listeria* contamination in milk-processing chain and proficiency in *Listeria monocytogenes* decontamination of small-scale milk retailers. *J. Food Qual.* **2024**, *2024*, 6263938. [CrossRef]

5. Thomas, J.; Govender, N.; McCarthy, K.M.; Erasmus, L.K.; Doyle, T.J.; Allam, M.; Ismail, A.; Ramalwa, N.; Sekwadi, P.; Ntshoe, G.; et al. Outbreak of listeriosis in South Africa associated with processed meat. *N. Engl. J. Med.* **2020**, *382*, 632–643. [CrossRef] [PubMed]

6. Mead, P.S.; Dunne, E.F.; Graves, L.; Wiedmann, M.; Patrick, M.; Hunter, S.; Salehi, E.; Mostashari, F.; Craig, A.; Mshar, P.; et al. Nationwide outbreak of listeriosis due to contaminated meat. *Epidemiol. Infect.* **2006**, *134*, 744–751. [CrossRef] [PubMed]

7. Pichler, J.; Much, P.; Kasper, S.; Fretz, R.; Auer, B.; Kathan, J.; Mann, M.; Huhulescu, S.; Ruppitsch, W.; Pietzka, A. An outbreak of febrile gastroenteritis associated with jellied pork contaminated with *Listeria monocytogenes*. *Wien. Klin. Wochenschr.* **2009**, *121*, 149–156. [CrossRef] [PubMed]

8. Smith, B.; Larsson, J.T.; Lisby, M.; Müller, L.; Madsen, S.B.; Engberg, J.; Bangsborg, J.; Ethelberg, S.; Kemp, M. Outbreak of listeriosis caused by infected beef meat from a meals-on-wheels delivery in Denmark 2009. *Clin. Microbiol. Infect.* **2011**, *17*, 50–52. [CrossRef] [PubMed]

9. Awofisayo-Okuyelu, A.; Arunachalam, N.; Dallman, T.; Grant, K.; Aird, H.; McLauchlin, J.; Painset, A.; Amar, C. An outbreak of human listeriosis in England between 2010 and 2012 associated with the consumption of pork pies. *J. Food Prot.* **2016**, *79*, 732–740. [CrossRef]

0. Rivas, L.; Dupont, P.Y.; Wilson, M.; Rohleder, M.; Gilpin, B. An outbreak of multiple genotypes of *Listeria monocytogenes* in New Zealand linked to contaminated ready-to-eat meats—A retrospective analysis using whole-genome sequencing. *Lett. Appl. Microbiol.* **2019**, *69*, 392–398. [CrossRef]

1. Okpo, E.; Leith, J.; Smith-Palmer, A.; Bell, J.; Parks, D.; Browning, F.; Byers, L.; Corrigan, H.; Webster, D.; Karcher, A.M.; et al. An outbreak of an unusual strain of *Listeria monocytogenes* infection in North-East Scotland. *J. Infect. Public Health* **2015**, *8*, 612–618. [CrossRef] [PubMed]

2. Dahl, V.; Sundqvist, L.; Hedenström, I.; Löfdahl, M.; Alm, E.; Ringberg, H.; Lindblad, M.; Wallensten, A.; Thisted Lambertz, S.; Jernberg, C. A nationwide outbreak of listeriosis associated with cold-cuts, Sweden 2013–2014. *Infect. Ecol. Epidemiol.* **2017**, *7*, 1324232. [CrossRef] [PubMed]

3. Lachmann, R.; Halbedel, S.; Adler, M.; Becker, N.; Allerberger, F.; Holzer, A.; Boone, I.; Falkenhorst, G.; Kleta, S.; Al Dahouk, S. Nationwide outbreak of invasive listeriosis associated with consumption of meat products in health care facilities, Germany, 2014–2019. *Clin. Microbiol. Infect.* **2021**, *27*, 1035.e1–1035.e5. [CrossRef] [PubMed]

4. Duranti, A.; Sabbatucci, M.; Blasi, G.; Acciari, V.A.; Ancora, M.; Bella, A.; Busani, L.; Centorame, P.; Cammà, C.; Conti, F. A severe outbreak of listeriosis in central Italy with a rare pulsotype associated with processed pork products. *J. Med. Microbiol.* **2018**, *67*, 1351–1360. [CrossRef] [PubMed]

5. Maurella, C.; Gallina, S.; Ru, G.; Adriano, D.; Bellio, A.; Bianchi, D.M.; Chiavacci, L.; Crescio, M.I.; Croce, M.; D'Errico, V.; et al. Outbreak of febrile gastroenteritis caused by *Listeria monocytogenes* 1/2a in sliced cold beef ham, Italy, May 2016. *Eurosurveillance* **2018**, *23*, 17-00155. [CrossRef] [PubMed]

6. Althaus, D.; Jermini, M.; Giannini, P.; Martinetti, G.; Reinholz, D.; Nüesch-Inderbinen, M.; Lehner, A.; Stephan, R. Local outbreak of *Listeria monocytogenes* serotype 4b sequence type 6 due to contaminated meat pâté. *Foodborne Pathog. Dis.* **2017**, *14*, 219–222. [CrossRef]

7. European Centre for Disease Prevention and Control; European Food Safety Authority. Rapid Outbreak Assessment: Multi-Country Outbreak of Listeria monocytogenes Sequence Type 6 Infections Linked to Ready-to-Eat Meat Products. Available online: https://www.ecdc.europa.eu/en/publications-data/rapid-outbreak-assessment-multi-country-outbreak-listeria-monocytogenes-sequence (accessed on 11 February 2024).

8. Luque-Sastre, L.; Arroyo, C.; Fox, E.M.; McMahon, B.J.; Bai, L.; Li, F.; Fanning, S. Antimicrobial resistance in *Listeria* species. *Microbiol. Spectr.* **2018**, *6*, 237–259. [CrossRef] [PubMed]

9. Sarr, M.; Alou, M.T.; Padane, A.; Diouf, F.S.; Beye, M.; Sokhna, C.; Fenollar, F.; Mboup, S.; Raoult, D.; Million, M. A review of the literature of *Listeria monocytogenes* in Africa highlights breast milk as an overlooked human source. *Front. Microbiol.* **2023**, *14*, 1213953. [CrossRef] [PubMed]

0. Hof, H.; Nichterlein, T.; Kretschmar, M. Management of listeriosis. *Clin. Microbiol. Rev.* **1997**, *10*, 345–357. [CrossRef]

1. Lamont, R.F.; Sobel, J.; Mazaki-Tovi, S.; Kusanovic, J.P.; Vaisbuch, E.; Kim, S.K.; Uldbjerg, N.; Romero, R. Listeriosis in human pregnancy: A systematic review. *J. Perinat. Med.* **2011**, *39*, 227–236. [CrossRef]

2. Keet, R. *Categorization of* Listeria monocytogenes *from Food, Environmental, and Clinical Origin in the Western Cape (South Africa)*; Stellenbosch University: Western Cape, South Africa, 2020.

3. Keet, R.; Rip, D. *Listeria monocytogenes* isolates from Western Cape, South Africa exhibit resistance to multiple antibiotics and contradicts certain global resistance patterns. *AIMS Microbiol.* **2021**, *7*, 40–58. [CrossRef] [PubMed]

4. Janakiraman, V. Listeriosis in pregnancy: Diagnosis, treatment, and prevention. *Rev. Obstet. Gynecol.* **2008**, *1*, 179–185. [PubMed]

5. Dos Reis, J.O.; Vieira, B.S.; Cunha Neto, A.; Castro, V.S.; Figueiredo, E.E.S. Antimicrobial resistance of *Listeria monocytogenes* from animal foods to first- and second-line drugs in the treatment of listeriosis from 2008 to 2021: A systematic review and meta-analysis. *Can. J. Infect. Dis. Med. Microbiol.* **2022**, *2022*, 1351983. [CrossRef] [PubMed]

6. Ishihara, Y.; Akazawa, K. Treatment of *Listeria monocytogenes* bacteremia with oral levofloxacin in an immunocompromised patient. *IDCases* **2023**, *31*, e01680. [CrossRef]

7. Kayode, A.J.; Okoh, A.I. Antibiotic resistance profile of *Listeria monocytogenes* recovered from ready-to-eat foods surveyed in South Africa. *J. Food Prot.* **2022**, *85*, 1807–1814. [CrossRef]

58. Da Rocha, L.S.; Gunathilaka, G.U.; Zhang, Y. Antimicrobial-resistant *Listeria* species from retail meat in metro Detroit. *J. Food Prot.* **2012**, *75*, 2136–2141. [CrossRef]
59. Sosnowski, M.; Lachtara, B.; Wieczorek, K.; Osek, J. Antimicrobial resistance and genotypic characteristics of *Listeria monocytogenes* isolated from food in Poland. *Int. J. Food Microbiol.* **2019**, *289*, 1–6. [CrossRef]
60. Elsayed, M.E.; Abd El-Hamid, M.I.; El-Gedawy, A.; Bendary, M.M.; RM, E.L.; Alhomrani, M.; Alamri, A.S.; Alghamdi, S.A.; Arnout, M.; Binjawhar, D.N.; et al. New insights into *Listeria monocytogenes* antimicrobial resistance, virulence attributes and their prospective correlation. *Antibiotics* **2022**, *11*, 1447. [CrossRef]
61. Caruso, M.; Fraccalvieri, R.; Pasquali, F.; Santagada, G.; Latorre, L.M.; Difato, L.M.; Miccolupo, A.; Normanno, G.; Parisi, A. Antimicrobial susceptibility and multilocus sequence typing of *Listeria monocytogenes* isolated over 11 years from food, humans, and the environment in Italy. *Foodborne Pathog. Dis.* **2020**, *17*, 284–294. [CrossRef]
62. Hailu, W.; Helmy, Y.A.; Carney-Knisely, G.; Kauffman, M.; Fraga, D.; Rajashekara, G. Prevalence and antimicrobial resistance profiles of foodborne pathogens isolated from dairy cattle and poultry manure amended farms in Northeastern Ohio, the United States. *Antibiotics* **2021**, *10*, 1450. [CrossRef]
63. Lachtara, B.; Wieczorek, K.; Osek, J. Antimicrobial resistance of *Listeria monocytogenes* serogroups IIa and IVb from food and food-production environments in Poland. *J. Vet. Res.* **2023**, *67*, 373–379. [CrossRef] [PubMed]
64. Poyart-Salmeron, C.; Carlier, C.; Trieu-Cuot, P.; Courtieu, A.L.; Courvalin, P. Transferable plasmid-mediated antibiotic resistance in *Listeria monocytogenes*. *Lancet* **1990**, *335*, 1422–1426. [CrossRef] [PubMed]
65. Charpentier, E.; Courvalin, P. Antibiotic resistance in *Listeria* spp. *Antimicrob. Agents Chemother.* **1999**, *43*, 2103–2108. [CrossRef] [PubMed]
66. Willey, J.M.; Sherwood, L.M.; Woolverton, C.J. *Prescott's Microbiology*, 8th ed.; McGraw-Hill: New York, NY, USA, 2011; pp. 832–838.
67. Millanao, A.R.; Mora, A.Y.; Villagra, N.A.; Bucarey, S.A.; Hidalgo, A.A. Biological effects of quinolones: A family of broad-spectrum antimicrobial agents. *Molecules* **2021**, *26*, 7153. [CrossRef] [PubMed]
68. Lyon, S.A.; Berrang, M.E.; Fedorka-Cray, P.J.; Fletcher, D.L.; Meinersmann, R.J. Antimicrobial resistance of *Listeria monocytogenes* isolated from a poultry further processing plant. *Foodborne Pathog. Dis.* **2008**, *5*, 253–259. [CrossRef] [PubMed]
69. Morvan, A.; Moubareck, C.; Leclercq, A.; Hervé-Bazin, M.; Bremont, S.; Lecuit, M.; Courvalin, P.; Le Monnier, A. Antimicrobial resistance of *Listeria monocytogenes* strains isolated from humans in France. *Antimicrob. Agents Chemother.* **2010**, *54*, 2728–2731. [CrossRef] [PubMed]
70. Zakaria, A.I.; Sabala, R.F. Potential public health hazards related to consumption of poultry contaminated with antibiotic resistant *Listeria monocytogenes* in Egypt. *BMC Microbiol.* **2024**, *24*, 41. [CrossRef]
71. Hopkins, K.L.; Arnold, C.; Threlfall, E.J. Rapid detection of gyrA and parC mutations in quinolone-resistant *Salmonella enterica* using Pyrosequencing® technology. *J. Microbiol. Methods* **2007**, *68*, 163–171. [CrossRef]
72. Mata, M.; Baquero, F.; Perez-Diaz, J. A multidrug efflux transporter in *Listeria monocytogenes*. *FEMS Microbiol. Lett.* **2000**, *187*, 185–188. [CrossRef]
73. Godreuil, S.; Galimand, M.; Gerbaud, G.; Jacquet, C.; Courvalin, P. Efflux pump Lde is associated with fluoroquinolone resistance in *Listeria monocytogenes*. *Antimicrob. Agents Chemother.* **2003**, *47*, 704–708. [CrossRef]
74. Verraes, C.; Van Boxstael, S.; Van Meervenne, E.; Van Coillie, E.; Butaye, P.; Catry, B.; De Schaetzen, M.-A.; Van Huffel, X.; Imberechts, H.; Dierick, K. Antimicrobial resistance in the food chain: A review. *Int. J. Environ. Res. Public Health* **2013**, *10*, 2643–2669. [CrossRef]
75. Guerin, F.; Galimand, M.; Tuambilangana, F.; Courvalin, P.; Cattoir, V. Overexpression of the novel MATE fluoroquinolone efflux pump FepA in *Listeria monocytogenes* is driven by inactivation of its local repressor FepR. *PLoS ONE* **2014**, *9*, e106340. [CrossRef] [PubMed]
76. Courvalin, P. Antimicrobial drug resistance: "prediction is very difficult, especially about the future". *Emerg. Infect. Dis.* **2005**, *11*, 1503. [CrossRef] [PubMed]
77. Cernicchi, G.; Felicetti, T.; Sabatini, S. Microbial efflux pump inhibitors: A journey around quinoline and indole derivatives. *Molecules* **2021**, *26*, 6996. [CrossRef] [PubMed]
78. Troxler, R.; Von Graevenitz, A.; Funke, G.; Wiedemann, B.; Stock, I. Natural antibiotic susceptibility of *Listeria* species: L. grayi, L. innocua, L. ivanovii, L. monocytogenes, L. seeligeri and L. welshimeri strains. *Clin. Microbiol. Infect.* **2000**, *6*, 525–535.
79. Turner, J.; Muraoka, A.; Bedenbaugh, M.; Childress, B.; Pernot, L.; Wiencek, M.; Peterson, Y.K. The chemical relationship among beta-lactam antibiotics and potential impacts on reactivity and decomposition. *Front. Microbiol.* **2022**, *13*, 807955. [CrossRef] [PubMed]
80. Krawczyk-Balska, A.; Markiewicz, Z. The intrinsic cephalosporin resistome of *Listeria monocytogenes* in the context of stress response, gene regulation, pathogenesis and therapeutics. *J. Appl. Microbiol.* **2016**, *120*, 251–265. [CrossRef]
81. Mpundu, P.; Muma, J.B.; Mukubesa, A.N.; Kainga, H.; Mudenda, S.; Bumbangi, F.N.; Muleya, W.; Katemangwe, P.; Munyeme, M. Antibiotic resistance patterns of *Listeria* species isolated from broiler abattoirs in Lusaka, Zambia. *Antibiotics* **2022**, *11*, 591. [CrossRef] [PubMed]
82. Von Wintersdorff, C.J.; Penders, J.; Van Niekerk, J.M.; Mills, N.D.; Majumder, S.; Van Alphen, L.B.; Savelkoul, P.H.; Wolffs, P.F. Dissemination of antimicrobial resistance in microbial ecosystems through horizontal gene transfer. *Front. Microbiol.* **2016**, *7*, 173. [CrossRef]

3. Wiśniewski, P.; Zakrzewski, A.J.; Zadernowska, A.; Chajęcka-Wierzchowska, W. Antimicrobial resistance and virulence characterization of *Listeria monocytogenes* strains isolated from food and food processing environments. *Pathogens* **2022**, *11*, 1099. [CrossRef]
4. Michaelis, C.; Grohmann, E. Horizontal gene transfer of antibiotic resistance genes in biofilms. *Antibiotics* **2023**, *12*, 328. [CrossRef] [PubMed]
5. Ngema, S.S.; Khumalo, S.H.; Ojo, M.C.; Pooe, O.J.; Malilehe, T.S.; Basson, A.K.; Madoroba, E. Evaluation of antimicrobial activity by marine *Nocardiopsis dassonvillei* against foodborne *Listeria monocytogenes* and Shiga toxin-producing *Escherichia coli*. *Microorganisms* **2023**, *11*, 2539. [CrossRef] [PubMed]
6. Allen, K.J.; Wałecka-Zacharska, E.; Chen, J.C.; Katarzyna, K.-P.; Devlieghere, F.; Van Meervenne, E.; Osek, J.; Wieczorek, K.; Bania, J. *Listeria monocytogenes*–an examination of food chain factors potentially contributing to antimicrobial resistance. *Food Microbiol.* **2016**, *54*, 178–189. [CrossRef]
7. Srinivasan, V.; Nam, H.; Nguyen, L.; Tamilselvam, B.; Murinda, S.; Oliver, S. Prevalence of antimicrobial resistance genes in *Listeria monocytogenes* isolated from dairy farms. *Foodborne Pathog. Dis.* **2005**, *2*, 201–211. [CrossRef] [PubMed]
8. Saleh, S.O.; Hussien, A.A.; Youseef, A.G.; Younis, W.K.; Mubarak, A.G. Prevalence, antibiotic resistance, and phylogenetic analysis of *Listeria monocytogenes* isolated from various sources in Egypt: Fish, vegetables, and humans. *Iraqi J. Vet. Sci.* **2024**, *38*, 15–27. [CrossRef]
9. Shaw, K.; Rather, P.; Hare, R.; Miller, G. Molecular genetics of aminoglycoside resistance genes and familial relationships of the aminoglycoside-modifying enzymes. *Microbiol. Rev.* **1993**, *57*, 138–163. [CrossRef] [PubMed]
0. Moura, A.; Leclercq, A.; Vales, G.; Tessaud-Rita, N.; Bracq-Dieye, H.; Thouvenot, P.; Madec, Y.; Charlier, C.; Lecuit, M. Phenotypic and genotypic antimicrobial resistance of *Listeria monocytogenes*: An observational study in France. *Lancet Reg. Health Eur.* **2024**, *37*, 100800. [CrossRef] [PubMed]
1. Lancaster, H.; Roberts, A.P.; Bedi, R.; Wilson, M.; Mullany, P. Characterization of Tn *916* S, a Tn *916*-like element containing the tetracycline resistance determinant *tet*(S). *J. Bacteriol.* **2004**, *186*, 4395–4398. [CrossRef]
2. Charpentier, E.; Courvalin, P. Emergence of the trimethoprim resistance gene dfrD in *Listeria monocytogenes* BM4293. *Antimicrob. Agents Chemother.* **1997**, *41*, 1134–1136. [CrossRef]
3. Kayode, A.J.; Okoh, A.I. Antimicrobial-resistant *Listeria monocytogenes* in ready-to-eat foods: Implications for food safety and risk assessment. *Foods* **2023**, *12*, 1346. [CrossRef]
4. Huovinen, P.; Sundström, L.; Swedberg, G.; Sköld, O. Trimethoprim and sulfonamide resistance. *Antimicrob. Agents Chemother.* **1995**, *39*, 279–289. [CrossRef] [PubMed]
5. Schwarz, S.; Kehrenberg, C.; Doublet, B.; Cloeckaert, A. Molecular basis of bacterial resistance to chloramphenicol and florfenicol. *FEMS Microbiol. Rev.* **2004**, *28*, 519–542. [CrossRef] [PubMed]
6. Roberts, M.C.; Facinelli, B.; Giovanetti, E.; Varaldo, P.E. Transferable erythromycin resistance in *Listeria* spp. isolated from food. *Appl. Environ. Microbiol.* **1996**, *62*, 269–270. [CrossRef] [PubMed]
7. Granier, S.A.; Moubareck, C.; Colaneri, C.; Lemire, A.; Roussel, S.; Dao, T.-T.; Courvalin, P.; Brisabois, A. Antimicrobial resistance of *Listeria monocytogenes* isolates from food and the environment in France over a 10-year period. *Appl. Environ. Microbiol.* **2011**, *77*, 2788–2790. [CrossRef] [PubMed]
8. Doyle, M.P.; Loneragan, G.H.; Scott, H.M.; Singer, R.S. Antimicrobial resistance: Challenges and perspectives. *Compr. Rev. Food Sci. Food Saf.* **2013**, *12*, 234–248. [CrossRef]
9. Rostamian, M.; Kooti, S.; Mohammadi, B.; Salimi, Y.; Akya, A. A systematic review and meta-analysis of *Listeria monocytogenes* isolated from human and non-human sources: The antibiotic susceptibility aspect. *Diagn. Microbiol. Infect. Dis.* **2022**, *102*, 115634. [CrossRef]
00. Manyi-Loh, C.E.; Okoh, A.I.; Lues, R. Occurrence and multidrug resistance in strains of *Listeria monocytogenes* recovered from the anaerobic co-digestion sludge contained in a single stage steel biodigester: Implications for antimicrobial stewardship. *Microorganisms* **2023**, *11*, 725. [CrossRef]
01. Law, J.W.-F.; Letchumanan, V.; Tan, L.T.-H.; Ser, H.-L.; Goh, B.-H.; Lee, L.-H. The rising of "modern actinobacteria" era. *Prog. Microbes Mol. Biol.* **2020**, *3*, a0000064. [CrossRef]
02. Hui, M.L.; Tan, L.T.; Letchumanan, V.; He, Y.W.; Fang, C.M.; Chan, K.G.; Law, J.W.; Lee, L.H. The extremophilic actinobacteria: From microbes to medicine. *Antibiotics* **2021**, *10*, 682. [CrossRef]
03. Aly, M.M.; Bahamdain, L.A.; Aba, S.A. Unexplored extreme habitats as sources of novel and rare actinomycetes with enzyme and antimicrobial activities. *J. Pharm. Biol. Sci.* **2019**, *14*, 45–54.
04. Barka, E.A.; Vatsa, P.; Sanchez, L.; Gaveau-Vaillant, N.; Jacquard, C.; Klenk, H.-P.; Clément, C.; Ouhdouch, Y.; van Wezel, G.P. Taxonomy, physiology, and natural products of *Actinobacteria*. *Microbiol. Mol. Biol. Rev.* **2016**, *80*, 1–43. [CrossRef] [PubMed]
05. Willey, J.M.; Sandman, K.M.; Wood, D.H. *Prescott's Microbiology*, 11th ed.; McGraw-Hill Education: New York, NY, USA, 2020; pp. 538–545.
06. Li, F.; Liu, S.; Lu, Q.; Zheng, H.; Osterman, I.A.; Lukyanov, D.A.; Sergiev, P.V.; Dontsova, O.A.; Liu, S.; Ye, J. Studies on antibacterial activity and diversity of cultivable actinobacteria isolated from mangrove soil in Futian and Maoweihai of China. *Evid. Based Complement. Altern. Med.* **2019**, *2019*, 3476567. [CrossRef] [PubMed]
07. Abdel-Razek, A.S.; El-Naggar, M.E.; Allam, A.; Morsy, O.M.; Othman, S.I. Microbial natural products in drug discovery. *Processes* **2020**, *8*, 470. [CrossRef]

108. Pacios-Michelena, S.; Aguilar González, C.N.; Alvarez-Perez, O.B.; Rodriguez-Herrera, R.; Chávez-González, M.; Arredondo Valdés, R.; Ascacio Valdés, J.A.; Govea Salas, M.; Ilyina, A. Application of *Streptomyces* antimicrobial compounds for the control of phytopathogens. *Front. Sustain. Food Syst.* **2021**, *5*, 696518. [CrossRef]
109. Hifnawy, M.S.; Fouda, M.M.; Sayed, A.M.; Mohammed, R.; Hassan, H.M.; AbouZid, S.F.; Rateb, M.E.; Keller, A.; Adamek, M.; Ziemert, N.; et al. The genus *Micromonospora* as a model microorganism for bioactive natural product discovery. *RSC Adv.* **2020**, *10*, 20939–20959. [CrossRef] [PubMed]
110. Arifuzzaman, M.; Khatun, M.; Rahman, H. Isolation and screening of actinomycetes from Sundarbans soil for antibacterial activity. *Afr. J. Biotechnol.* **2010**, *9*, 4615–4619.
111. El-sersy, N.A.; Abou-Elela, G.M. Antagonistic effect of marine *Nocardia brasiliensis* against the fish pathogen *Vibrio damsela*: Application of Piackett-Burman experimental design to evaluate factors affecting the production of the antibacterial agent. *Int. J. Oceans Oceanogr.* **2006**, *1*, 141–150.
112. Naragani, K.; Munaganti, R.K.; Mangamuri, U.K.; Vijayalakshmi, M. Optimization of culture conditions for enhanced antimicrobial activity of *Rhodococcus erythropolis* VLK-12 isolated from South Coast of Andhra Pradesh, India. *Br. Microbiol. Res. J.* **2014**, *4*, 63–79. [CrossRef]
113. Stackebrandt, E.; Rainey, F.A.; Ward-Rainey, N.L. Proposal for a new hierarchic classification system, actinobacteria classis nov. *Int. J. Syst. Evol. Microbiol.* **1997**, *47*, 479–491. [CrossRef]
114. Bhairamkar, S.; Kadam, P.; Anjulal, H.; Joshi, A.; Chaudhari, R.; Bagul, D.; Javdekar, V.; Zinjarde, S. Comprehensive updates on the biological features and metabolic potential of the versatile extremophilic actinomycete *Nocardiopsis dassonvillei*. *Res. Microbiol.* **2023**, *in press*. [CrossRef] [PubMed]
115. Bennur, T.; Kumar, A.R.; Zinjarde, S.; Javdekar, V. *Nocardiopsis* species: Incidence, ecological roles and adaptations. *Microbiol. Res.* **2015**, *174*, 33–47. [CrossRef]
116. Ali, S.M.; Siddiqui, R.; Khan, N.A. Antimicrobial discovery from natural and unusual sources. *J. Pharm. Pharmacol.* **2018**, *70*, 1287–1300. [CrossRef] [PubMed]
117. Subramani, R.; Sipkema, D. Marine rare actinomycetes: A promising source of structurally diverse and unique novel natural products. *Mar. Drugs* **2019**, *17*, 249. [CrossRef] [PubMed]
118. Mohseni, M.; Norouzi, H.; Hamedi, J.; Roohi, A. Screening of antibacterial producing actinomycetes from sediments of the Caspian Sea. *Int. J. Mol. Cell. Med.* **2013**, *2*, 64. [PubMed]
119. Kamjam, M.; Sivalingam, P.; Deng, Z.; Hong, K. Deep sea actinomycetes and their secondary metabolites. *Front. Microbiol.* **2017**, *8*, 760. [CrossRef] [PubMed]
120. Yaradoddi, J.S.; Kontro, M.H.; Ganachari, S.V.; Banapurmath, N.R.; Oli, A.; Katti, A.S.; Sulochana, M. Actinobacteria in marine environments. In *Actinobacteria: Ecology, Diversity, Classification and Extensive Applications*; Yaradoddi, J.S., Kantro, M.H., Ganachari, S.V., Eds.; Springer: Singapore, 2022; pp. 21–38.
121. Zhang, L.; Xi, L.; Ruan, J.; Huang, Y. *Microbacterium marinum* sp. nov., isolated from deep-sea water. *Syst. Appl. Microbiol.* **2012**, *35*, 81–85. [PubMed]
122. Carroll, A.R.; Copp, B.R.; Davis, R.A.; Keyzers, R.A.; Prinsep, M.R. Marine natural products. *Nat. Prod. Rep.* **2023**, *40*, 275–325 [CrossRef] [PubMed]
123. Saha, M.; Jaisankar, P.; Das, S.; Sarkar, K.K.; Roy, S.; Besra, S.E.; Vedasiromani, J.R.; Ghosh, D.; Sana, B.; Mukherjee, J. Production and purification of a bioactive substance inhibiting multiple drug resistant bacteria and human leukemia cells from a salt-tolerant marine actinobacterium sp. isolated from the Bay of Bengal. *Biotechnol. Lett.* **2006**, *28*, 1083–1088. [CrossRef]
124. Arumugam, M.; Mitra, A.; Jaisankar, P.; Dasgupta, S.; Sen, T.; Gachhui, R.; Kumar Mukhopadhyay, U.; Mukherjee, J. Isolation of an unusual metabolite 2-allyloxyphenol from a marine actinobacterium, its biological activities and applications. *Appl. Microbiol. Biotechnol.* **2010**, *86*, 109–117. [CrossRef]
125. Siddharth, S.; Rai, V.R. Isolation and characterization of bioactive compounds with antibacterial, antioxidant and enzyme inhibitory activities from marine-derived rare actinobacteria, *Nocardiopsis* sp. SCA21. *Microb. Pathog.* **2019**, *137*, 103775. [CrossRef]
126. Gómez-Rodríguez, L.; Schultz, P.J.; Tamayo-Castillo, G.; Dotson, G.D.; Sherman, D.H.; Tripathi, A. Adipostatins E–J, new potent antimicrobials identified as inhibitors of coenzyme-A biosynthesis. *Tetrahedron Lett.* **2020**, *61*, 151469. [CrossRef] [PubMed]
127. El-Naggar, M.Y.; Barakat, K.M.; Aly-Eldeen, M.A.; Ghoneim, H.B.; Hamdan, A.M. Veratraldehyde as a food additive produced by the marine isolate *Streptomyces diastaticus* LC360811. *Egypt. J. Aquat. Biol. Fish* **2021**, *25*, 1–19. [CrossRef]
128. Yi, K.-X.; Xie, Q.-Y.; Ma, Q.-Y.; Yang, L.; Dai, H.-F.; Zhao, Y.-X.; Hao, Y.-E. Diverse ansamycin derivatives from the marine-derived *Streptomyces* sp. ZYX-F-97 and their antibacterial activities. *Fitoterapia* **2024**, *173*, 105814. [PubMed]
129. Uemura, M.; Kobayashi, K.; Sato, N.; Nagai, K.; Seki, R.; Kamio, M.; Fukuda, T.; Tsubouchi, T.; Tomoda, H.; Ohshiro, T.; et al. Haneummycin, a new 22-membered macrolide lactam antibiotic, produced by marine-derived *Streptomyces* sp. KM77-8. *J. Antibiot.* **2023**, *76*, 650–657. [CrossRef] [PubMed]
130. Schinke, C.; Martins, T.; Queiroz, S.C.; Melo, I.S.; Reyes, F.G. Antibacterial compounds from marine bacteria, 2010–2015. *J. Nat. Prod.* **2017**, *80*, 1215–1228. [CrossRef] [PubMed]
131. Wang, C.; Lu, Y.; Cao, S. Antimicrobial compounds from marine actinomycetes. *Arch. Pharm. Res.* **2020**, *43*, 677–704. [CrossRef] [PubMed]
132. Wang, C.; Du, W.; Lu, H.; Lan, J.; Liang, K.; Cao, S. A review: Halogenated compounds from marine Actinomycetes. *Molecules* **2021**, *26*, 2754. [CrossRef] [PubMed]

33. Liang, J.; She, J.; Fu, J.; Wang, J.; Ye, Y.; Yang, B.; Liu, Y.; Zhou, X.; Tao, H. Advances in natural products from the marine-sponge-associated microorganisms with antimicrobial activity in the last decade. *Mar. Drugs* **2023**, *21*, 236. [CrossRef] [PubMed]
34. Barbosa, F.; Pinto, E.; Kijjoa, A.; Pinto, M.; Sousa, E. Targeting antimicrobial drug resistance with marine natural products. *Int. J. Antimicrob. Agents* **2020**, *56*, 106005. [CrossRef]
35. World Health Organization. Prioritization of Pathogens to Guide Discovery, Research and Development of New Antibiotics for Drug-Resistant Bacterial Infections, Including Tuberculosis. Available online: https://www.who.int/publications/i/item/WHO-EMP-IAU-2017.12 (accessed on 16 August 2023).
36. Molinos, A.C.; Abriouel, H.; López, R.L.; Omar, N.B.; Valdivia, E.; Gálvez, A. Enhanced bactericidal activity of enterocin AS-48 in combination with essential oils, natural bioactive compounds and chemical preservatives against *Listeria monocytogenes* in ready-to-eat salad. *Food Chem. Toxicol.* **2009**, *47*, 2216–2223. [CrossRef]
37. Martínez, B.; Rodríguez, A.; Suárez, E. Antimicrobial peptides produced by bacteria: The bacteriocins. In *New Weapons to Control Bacterial Growth*; Villa, T.G., Vinas, M., Eds.; Springer: Cham, Switzerland, 2016; pp. 15–38.
38. Batiha, G.E.-S.; Hussein, D.E.; Algammal, A.M.; George, T.T.; Jeandet, P.; Al-Snafi, A.E.; Tiwari, A.; Pagnossa, J.P.; Lima, C.M.; Thorat, N.D. Application of natural antimicrobials in food preservation: Recent views. *Food Control* **2021**, *126*, 108066. [CrossRef]
39. Khorshidian, N.; Khanniri, E.; Yousefi, M. Antibacterial activity of pediocin and pediocin-producing bacteria against *Listeria monocytogenes* in meat products. *Front. Microbiol.* **2021**, *12*, 709959. [CrossRef]
40. Del Nobile, M.A.; Lucera, A.; Conte, A. Food applications of natural antimicrobial compounds. *Front. Microbiol.* **2012**, *3*, 28087.
41. Claverías, F.P.; Undabarrena, A.N.; González, M.; Cámara, B.P. Culturable diversity and antimicrobial activity of actinobacteria from marine sediments in Valparaíso bay, Chile. *Front. Microbiol.* **2015**, *6*, 121535. [CrossRef] [PubMed]
42. Shams, M.; Shahnavaz, B.; Ghazvini, K.; Valinasab, T. Screening of actinomycetes from Lipar area of Oman Sea to investigate the antibacterial compounds. *Avicenna J. Clin. Microbiol. Infect.* **2015**, *2*, 23621. [CrossRef]
43. Undabarrena, A.; Beltrametti, F.; Claverías, F.P.; Moore, E.R.; Cámara, B. Exploring the diversity and antimicrobial potential of marine actinobacteria from the comau fjord in Northern Patagonia, Chile. *Front. Microbiol.* **2016**, *7*, 210381. [CrossRef]
44. Undabarrena, A.; Ugalde, J.A.; Seeger, M.; Cámara, B. Genomic data mining of the marine actinobacteria *Streptomyces* sp. H-KF8 unveils insights into multi-stress related genes and metabolic pathways involved in antimicrobial synthesis. *PeerJ* **2017**, *5*, e2912. [PubMed]
45. Eythorsdottir, A.; Omarsdottir, S.; Einarsson, H. Antimicrobial activity of marine bacterial symbionts retrieved from shallow water hydrothermal vents. *Mar. Biotechnol.* **2016**, *18*, 293–300. [CrossRef] [PubMed]
46. Collins, F.W.; Walsh, C.J.; Gomez-Sala, B.; Guijarro-García, E.; Stokes, D.; Jakobsdóttir, K.B.; Kristjánsson, K.; Burns, F.; Cotter, P.D.; Rea, M.C. The microbiome of deep-sea fish reveals new microbial species and a sparsity of antibiotic resistance genes. *Gut Microbes* **2021**, *13*, 1921924. [CrossRef]
47. Uniacke-Lowe, S.; Collins, F.W.; Hill, C.; Ross, R.P. Bioactivity screening and genomic analysis reveals deep-sea fish microbiome isolates as sources of novel antimicrobials. *Mar. Drugs* **2023**, *21*, 444. [CrossRef]
48. Sarmiento-Vizcaíno, A.; González, V.; Braña, A.F.; Palacios, J.J.; Otero, L.; Fernández, J.; Molina, A.; Kulik, A.; Vázquez, F.; Acuña, J.L. Pharmacological potential of phylogenetically diverse actinobacteria isolated from deep-sea coral ecosystems of the submarine Avilés Canyon in the Cantabrian Sea. *Microb. Ecol.* **2017**, *73*, 338–352. [CrossRef] [PubMed]
49. Dangar, K.G.; Singh, S.P. Antimicrobial Activities and Antibiotic Resistance of *Nocardiopsis alba* Isolated from the Saline Habitats of Coastal Gujarat. In Proceedings of the National Conference on Innovations in Biological Sciences (NCIBS), Rajkot, India, 10 January 2020.
50. Atallah, B.M.; El-Mohsnawy, E.; El-Shouny, W.; Haroun, S. Identification and characterization of different potentially antibacterial compounds from a marine *Streptomyces* sp. Sp1. *J. Anim. Plant Sci.* **2023**, *33*, 166–173.
51. Sharma, N.; Khajuria, V.; Gupta, S.; Kumar, C.; Sharma, A.; Lone, N.A.; Paul, S.; Meena, S.R.; Ahmed, Z.; Satti, N.K. Dereplication based strategy for rapid identification and isolation of a novel anti-inflammatory flavonoid by LCMS/MS from *Colebrookea oppositifolia*. *ACS Omega* **2021**, *6*, 30241–30259. [CrossRef] [PubMed]

antibiotics

MDPI

Review

Overview of Bioactive Fungal Secondary Metabolites: Cytotoxic and Antimicrobial Compounds

Rafael Conrado, Tainah Colombo Gomes, Gabriella Sales Calaço Roque and Ana Olívia De Souza *

Development and Innovation Laboratory, Instituto Butantan, Avenida Vital Brasil, 1500, São Paulo 05503-900, SP, Brazil
* Correspondence: ana.souza@butantan.gov.br; Tel.: +55-11-3723-7034

Abstract: Microorganisms are known as important sources of natural compounds that have been studied and applied for different purposes in distinct areas. Specifically, in the pharmaceutical area, fungi have been explored mainly as sources of antibiotics, antiviral, anti-inflammatory, enzyme inhibitors, hypercholesteremic, antineoplastic/antitumor, immunomodulators, and immunosuppressants agents. However, historically, the high demand for new antimicrobial and antitumor agents has not been sufficiently attended by the drug discovery process, highlighting the relevance of intensifying studies to reach sustainable employment of the huge world biodiversity, including the microorganisms. Therefore, this review describes the main approaches and tools applied in the search for bioactive secondary metabolites, as well as presents several examples of compounds produced by different fungi species with proven pharmacological effects and additional examples of fungal cytotoxic and antimicrobial molecules. The review does not cover all fungal secondary metabolites already described; however, it presents some reports that can be useful at any phase of the drug discovery process, mainly for pharmaceutical applications.

Keywords: fungi; secondary metabolites; cytotoxicity; antimicrobial

1. Introduction

Natural products can be obtained from plants, animals, microorganisms, or marine organisms, which are present in different environments, such as in the aquatic, terrestrial, or even in the air. Natural products have been relevant sources for drug discovery and the development of medicines since ancient times [1–3].

Newman and Cragg [1] showed that from January 1981 to September 2019, at least 1881 new drugs were approved worldwide for the treatment of all types of diseases, and among them, 75.38% (1418) were derived from natural products.

According to the Web of Science, from January 1989 to August 2022 (Figure 1), there were 486,109 publications related to the keyword "natural products", 14,364 related to "natural products fungi", and 9615 addressing "natural products secondary metabolites". Figure 1 shows the huge increase in publications related to natural products since 1989, and in addition, an increase of approximately 4-fold in the number of publications related to "secondary metabolite natural products" in the last 10 years can be observed.

Secondary metabolites (SMs), also known as natural products, are biomolecules with a low molecular weight that, although not essential for the development of its producer, can increase their tolerance to different types of environmental stresses and hostile conditions, and consequently, their survival rate [1–3].

Bioactive SMs are produced mainly by microorganisms, which can be isolated from soil microbiota, marine environments, or extreme environments. The most promising are those obtained from underexplored niches, including plant-associated microbes called endophytes, which live in symbiotic relationships with host plants in their internal tissues and structures. Endophytes can be isolated from herbaceous plants, grasses, trees, shrubs, mosses, ferns, and aquatic plants [4].

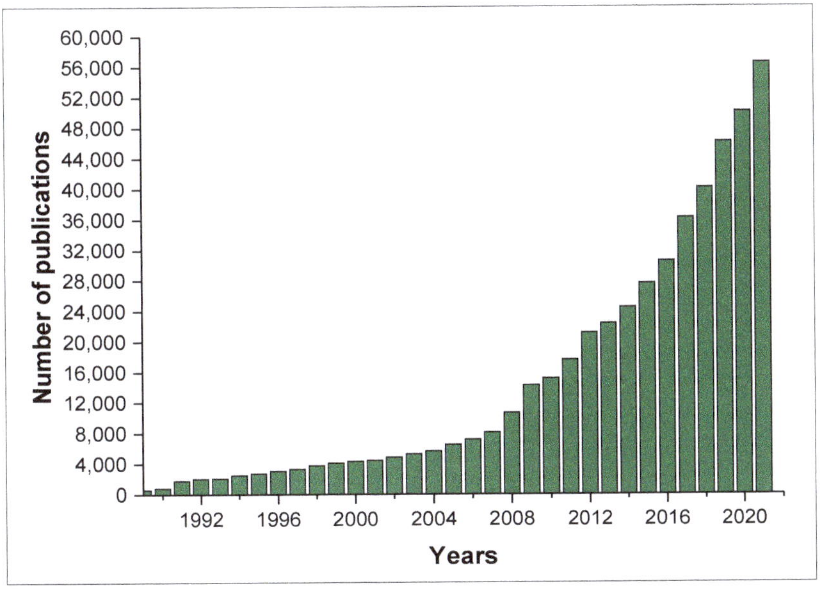

Figure 1. Overview of publications profile containing in their topic the keyword "natural product", from 1989 to 2021. Source: Web of Science database (accessed on 15 August 2022).

Recent research has demonstrated that among fungi species, the endophytes are a great source of bioactive SMs, and approximately 20,000 compounds were already isolated from these microorganisms [5–10]. The diversity of fungal endophytes is wide mainly due to the different environments to which they are exposed, also considering the seasonal variations, host plants' genotype, and physiology [11]. These microorganisms have functional diversity and, in association with plants, can produce SMs that protect the host from insects or other microorganisms' attacks [12].

SMs are used by microorganisms as chemical signs for communication, to defend the habitat, or to inhibit the growth of competitors [13] and differ from their primary metabolite equivalents by the timing of their biosynthesis and dispensability for fungal growth [14]. The evolution of SMs biosynthesis by microorganisms is mainly due to their exposure to complex ecosystems.

Some SMs have afforded society-changing benefits; however, some compounds are associated with serious problems and negative impacts, such as mycotoxins, due to their widespread occurrence as food contaminants for humans and livestock, as well as mold-contaminated indoor environments [15]. Even so, previous data shows that among the 500,000 SMs described, around 70,000 compounds are obtained from microorganisms, and roughly 33,500 are bioactive compounds, of which 47% are of fungal origin (15,600) [16,17].

2. Obtaining Microorganism Strains for Biosynthesis of Bioactive Secondary Metabolites

Figure 2 briefly illustrates a few steps from the isolation of a fungi strain to the development of a pharmaceutical product. The process begins with the isolation of the fungus strain from diverse environmental sources (e.g., water, plants, and soil), and the culture of the microorganisms. The product of the culture is processed for extraction and purification of the SM. Subsequently, biological assays and clinical trials are performed to define the possibility of future pharmaceutical applications of the SM (Figure 2).

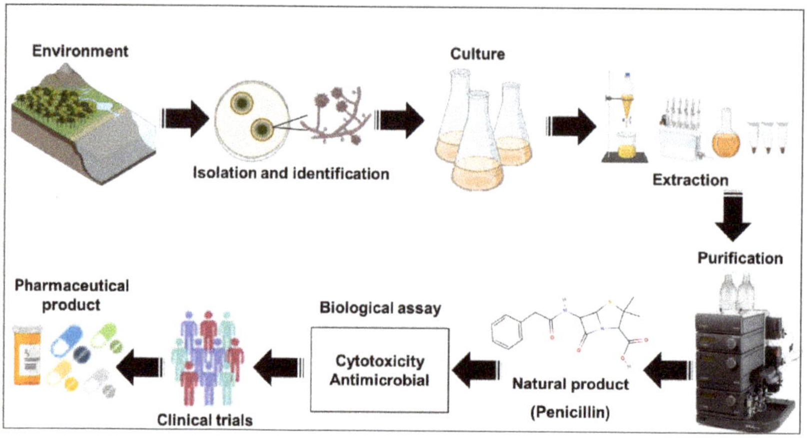

Figure 2. Representation of fungal strain isolation and development of medicines containing natural products as the bioactive component. Penicillin is being used just as an example.

The limitations for the discovery of new SMs from microorganisms begin in the handling and isolation of new strains. The isolation process of microorganisms, including that for endophytes, is well described; however, there is some challenge in isolating new fungi species, mainly due to the current techniques applied for the cultivation and strains purification that usually have remained the same in the last years. For obtaining new species, it is mandatory to design new approaches that are able to simulate the condition of the natural habitat from which the microorganisms were isolated. In this aspect, it is essential to overcome challenges related to growing conditions such as pH, temperature, nutrients, and preservation in the laboratory.

3. Obtaining Fungal Secondary Metabolites and Structural Characterization

Once the limitations for microorganisms isolation have been overcome, the next step is to proceed with its culture, followed by extraction, purification, and physicochemical characterization of the SMs. These steps are very well described and widely applied in natural product research.

Fungal culture for SMs biosynthesis can be performed in either solid-state fermentation (SSF) or submerged fermentation (SmF). The first one, on a solid substrate, can produce a high concentration of metabolites. In this system, the low water concentration favors oxygen transfer for the growth of the microorganisms. Among solid substrates, rice straw, rice hull, and sugarcane bagasse are common [18]. The SmF occurs in aqueous liquid nutrient media, and it is the most widely used due to the ease of control of fermentation parameters such as pH, temperature, dissolved oxygen, and types of culture media [18].

A good example is the lovastatin production that has been attempted in both SSF and SmF, where using SmF, the lovastatin production is about five times higher [19].

In addition to different strategies for the microorganism culture, the profile of the obtained metabolites also depends on the extraction technique and solvent applied for it. Appropriate organic solvents such as chloroform, dichloromethane, hexane, or ethyl acetate are the most commonly applied for the extraction of the metabolites from both culture methods. However, it is important to optimize the process to obtain a full profile of SMs. The most applied techniques are liquid–liquid extraction (LLE) and ultrasonic-assisted extraction (UAE). Through the LLE, the metabolites from the aqueous phase are transferred to the organic solvents and obtained by phase separation; in the UAE, ultrasound waves increase the extraction efficiency by disrupting cells for the effective release of the compounds to the solvents [20].

For SMs extraction, an effective and recent innovative approach is the Solid-Phase Extraction Embedded Dialysis (SPEED technology), which is based on a physical separation between the microbial biomass and the resin used for in situ solid-phase extraction [21]. This technology consists of a two-layer barrier with an external nylon filter cloth (NFC) and an internal dialysis tube (DT) containing the resin beads. The molecular cut of the tube allows only molecules with appropriate molecular weight can flow from the culture broth to the resin. This selectivity makes SPEED technology a very interesting tool for the study of specialized microbial metabolites.

In addition to the development of genome-based technologies, the employment of modern omics tools and dereplication strategies can potentially increase the efficiency of the discovery of new molecules from microorganisms making the process approach wider and more effective [22]. To avoid the rediscovery of known molecules, it is important to identify the obtained compounds and to prioritize those potential strains for new compound biosynthesis. In this regard, the dereplication using a metabolomics profile based on mass spectrometry has been useful in selecting these strains, and, in addition to that, the data can be integrated into a Global Natural Product Social Molecular Networking (GNPS) [23]. Nuclear magnetic resonance (^{1}H, ^{13}C NMR) spectroscopy [24,25], 2D NMR [26], and mass spectrometry (MS, MS/MS) [27,28] fingerprints are widely used, and well-established techniques for the dereplication of natural products, and in addition to that, Fourier transform infrared spectroscopy (FTIR) was proposed by Grkovic et al., [29] to be equally valuable in the dereplication workflow. FTIR application is supported by modern instruments, which have a relatively small footprint, are sensitive, and provide a large data set across the spectroscopic range, suitable for library generation and pattern matching [29].

The dereplication combining different analytical tools enables rapid identification of known compounds and introduces knowledge of active principle structural class early in the isolation workflow [24].

After the extraction, the fractionating and purification of compounds present in a crude extract are performed mainly by chromatographic techniques, including thin layer chromatography (TLC), liquid chromatography (LC), high-performance liquid chromatography (HPLC) [30–32], and supercritical fluid chromatography (SFC). The HPLC analyses can be operated by distinct modes categorized mainly by the type of stationary phase, being (i) normal phase, (ii) reversed phase (RP-HPLC), (iii) hydrophilic interaction liquid chromatography (HILIC), and (iv) ion chromatography [32]. The interaction of the sample with the stationary phase depends on the relative solubilities, surface absorption or adsorption, ion exchange, and steric effects [32].

Normal-phase HPLC and RP-HPLC differ from each other according to the stationary and mobile phases used. In normal HPLC, the separation of non-polar compounds occurs at the stationary phase that is composed of hydrophilic groups, followed by the polar ones in the sequence. In RP-HPLC, lipophilic groups form the stationary phase, and the polar compounds are eluted first [30–32].

RP-HPLC has a wide capacity for analyzing different compounds due to its broad applicability and reproducibility. The hydrophobic properties of the stationary phase allow the reverse HPLC use for separation of mixtures of hydrophilic (i.e., polar), hydrophobic (i.e., non-polar), ionic and/or ionizable molecules resulting in the purification of their components, according to the employed procedure [31]. The chromatographic selectivity of RP-HPLC can be changed through modifications in the mobile phase characteristics, and excellent resolution can be achieved, even for very closely related structures present in the mixture.

The HPLC, ultra-performance liquid chromatography (UPLC), and ultra-high PLC (UHPLC) can be temperature controlled and are usually coupled with different detectors. The analytes can be detected by UV/Vis, photodiode array (PDA), and/or mass spectrometry (MS) [32]. The UV/Vis detector can be operated in dual wavelength mode, whereas the PDA can measure the peaks in a wavelength-scanning mode resulting in a total absorbance spectrum of a single peak. The main advantages of PDA detection are the resolution and wavelength range allowing the selection of the best wavelength for analysis. In addition to that, for specific naturally fluorescent compounds, the detection can be performed by fluorescence (FLD). Moreover, although not regularly used, the chromatographs can be coupled with detectors system through electrical conductivity, electrochemical, refractive index (RI), light scattering, evaporative light scattering, infrared (IR), transport, aerosol-based, and chiral and pulsed amperometric [32].

FT-IR is an auxiliary tool for molecular identification through the transitions between vibrational states that are mainly related to vibrations centered on functional groups that, in combination, results in a specific spectrum fingerprint unique to each compound and allows it to be distinguished from other molecules present in database libraries, such as SpectraBase, NIST, Scifinder, etc. [33].

Beyond that, with the development of methodologies for analytical tools, several platforms were applied, such as HPLC–MS/MS, UPLC coupled to high-resolution mass spectrometry (UPLC-HRMS), or UHPLC–MS/MS. The MS detector is able to an unequivocal mass identification providing structural information of the analytes, and all these platforms are widely used for physicochemical characterization of the molecular structure of pure compounds [34,35].

Along with the ability of physical separation of compounds, the HPLC–MS/MS and UPLC-HRMS are employed to identify molecular structures by analysis of their fragmentation derived from mass spectrometry. Due to their robustness, these analytical techniques are extensively used for qualitative and quantitative analysis [31,32,34,35].

An overview of the concepts and applications of the different types of chromatographic techniques was recently published, including the most recent technological advances and achievements that improved its contribution to the science and technological process in several areas, such as in agricultural and food science, environmental sciences, chemical and petrochemical industries, health sciences and medicine, pharmacology and biotechnology [32].

X-ray diffraction (XRD) and NMR spectroscopies are analytical tools for the characterization of molecular structures of pure compounds at the atomic level [35,36].

Regardless of some limitations, such as time consumption and the requirement of large amounts of crystalline samples, the use of XRD has increased in the pharmaceutical area and can be used to determine structures at the atomic level, mainly for biological macromolecules [35].

Although XRD and NMR can be complementary techniques, NMR has the advantage of being automatable, allowing high throughput for large-scale sample analysis, requiring less sample, and being able to preserve it [35–37]. With less than a milligram of sample, in addition to the molecular structure characterization, the NMR allows to determine the content and purity level of compounds and can be applied to characterize the molecular conformation in solution. At the molecular level, data regarding the physical properties of compounds can be obtained by NMR, such as conformational exchange, solubility, phase changes, and diffusion [36]. More detailed descriptions of the NMR fundaments and application in drug discovery and development can be found in previous reports [35–37].

Although not applied for stereochemical information, for structural characterization of lipids and other volatile compounds, gas chromatography/mass spectrometry (GC-MS) is the most recommended analysis [38]. Although the development of pharmaceutical drugs from volatile organic compounds (VOC) is not viable, VOCs can act as autoinducers, defend against competing species and play essential roles in attracting pollinators for spreading fungal spores.

GC/MS analytical methods are very useful for application in pharmaceutical development to detect unknown molecules and impurities in the drug product.

Bacteria, Eukaryota, and Archaea can produce and respond to chemical signals called autoinducers (AIs) that are extracellular small diffusible molecules [39]. These compounds have low molecular mass (100–300 Daltons, up to C20), generally with a low boiling point and high vapor pressure [40]. Several of these small VOCs play important roles in signaling in intra- and inter-species to kingdom and domain interactions, survival, or virulence. This cell-to-cell communication through AIs is known as quorum sensing (QS) and occurs mainly in the presence of high microbial cell population density [41].

In the presence of a sufficient concentration of AIs in the extracellular medium, the activation of a cognate response regulator can occur [42]. For example, in a community of fungi and bacteria in soil, the bacterial population can be combated by fungi for nutrition, space, or pathogenicity by the action of mycotoxins and QS inhibitors (enzymes and other chemicals) [43].

Previous studies showed that fungal QS can regulate important functions such as pathogenesis, morphogenesis, and filamentation by the action of QS molecules such as lipids (oxylipins), peptides (pheromones), alcohols derived from aromatic amino acids (tyrosol, farnesol, tryptophol, and 1-phenylethanol), acetaldehydes and some VOCs [44–46].

The VOCs 3-octanol, 3-octanone, and 1-octen-3-ol induced conidiation in *Trichoderma* spp. in darkness culture [47]; however, at concentrations above 500 μM, the growth and conidia formation were suppressed [48].

Among 474 VOCs across 221 bacterial and fungal genera analyzed in 2020, 61 showed phylogenetic signals across bacterial domain and fungal Kingdom, 11 volatiles were phylogenetically conserved in genera from both bacteria and fungi (e.g., geosmin), whereas 17 were phylogenetically conserved in the fungal Kingdom (e.g., aristolochene) [49]. The study showed evidence that microbial VOCs can be studied by integrating metabolomic and phylogenetic data.

In general, VOCs are heterocyclic compounds, sesquiterpenoids, long-chain fatty acids, and aromatics. Among them, the compounds 1-octen-3-ol, 3-octanone, and 2-pentylfuran are considered common fungal VOCs.

The AIs molecules are not generally strain-specific, and a huge diversity of them has been reported in fungi. As an example, the acyclic sesquiterpene alcohol farnesol is involved in the filamentation of the pathogenic polymorphic yeast *Candida albicans* reducing its ability for the yeast-to-hyphal switch, being able to inhibit hyphae formation [50], as well as tryptophol and 1-phenylethanol [51].

Farnesol also affects the growth of other yeasts and fungi, including *Candida tropicalis* and *Candida parapsilosis*, *Sacharomyces cerevisiae*, *Aspergillus fumigatus*, and *Aspergillus nidulans* [52]. Moreover, in *S. cerevisiae* the morphogenesis is regulated during nitrogen starvation by the aromatic alcohols, 1-phenylethanol, and tryptophol that act as quorum sensing molecules (QSMs) [45].

Still in *C. albicans*, the aromatic alcohol tyrosol acts as a QSM controlling the growth, morphogenesis and biofilm formation [45]. Moreover, in the yeast form, *Histoplasma capsulatum* produces some specific cell wall polysaccharides α-(1,3)-glucan, required for its virulence [53].

Since different organisms employ molecular mechanisms to block the QS perception and related functions, quorum quencher (QQ) molecules are being prospected as antimicrobial compounds to target the QS and inhibit its function [54–56].

In the last years, more than 2000 VOCs produced by around 1000 bacterial and fungi species have been identified and systematically organized in the mVOC 2.0 database [57,58]. The database consists of VOCs collected from ca. 300 original publications and allow searching for specific microbial VOCs by name, molecular formula, molecular weight, chemical classification, Pubchem ID, bacterial, or fungal species [59]. The search using a microorganism species will indicate all the VOCs produced by that specific species.

In drug discovery, the main analytical tools have advantages and disadvantages regarding sensitivity and costs, and in general, are applied conjointly, in order to conclude a structural characterization of organic compounds [3,32–38,60,61].

Liquid-chromatography-high-resolution mass spectrometry (LC-HRMS) and NMR are advanced, highly sensitive, and selective analytical tools. Used in metabolomic studies, the NMR and MS techniques are able to detect samples in the range of 1 μM and 10 nM, respectively [37]. These tools offer the advantage of comparing the obtained data with wide spectral databases, such as MS-Finder, METLIN, National Institute of Science and Technology (NIST), Dictionary of Natural Products (DNP), AntiBase, DrugBank, FooDB, Chemical Entities of Biological Interest (ChEBI), and others [37,60].

As it is known, the purification of bioactive compounds involves several bioassay-guided steps and usually takes some time. To accelerate natural-product-based drug discovery, many natural product research groups are working with prefractionated natural product libraries [61]. For this purpose, solid-phase extraction (SPE) is a useful technique that has been successfully applied for metabolite fractionating [24]. This methodology is based on the separation of the crude extract into several fractions in C18 or C8 cartridges and water/methanol step gradients. The obtained fractions can be evaluated for biological activity in cell-based and/or cell-free assays [62].

In this regard, the National Cancer Institute (NCI) launched the Program for Natural Products Discovery (NPNPD) with the goal of generating prefractionated samples (up to 1,000,000) for modern high-throughput screening technologies (HTS) and developing integrated analytical resources for rapid isolation and structure elucidation of biologically active natural products [63]. The NPNPD platform provides 45 mg of a compound with a purity of ~97% in only 2 days, while by using traditional methods, the procedure would take 2–3 weeks. The NPNPD proposal has the potential to significantly reduce natural-product-based active principle identification timelines, cost of screening, and enable faster outcomes for assaying and identifying natural products in the HTS platform [63].

In the NPNPD platform [63], natural products extracted from plants, marine invertebrates, and microbial are fractionated through a high-throughput and automated SPE system into seven fractions of decreasing polarity. The available NCI library is an important source of new drugs and drug leads that can be accessed, free of charge, by a single principal investigator of a single project or by research groups or screening centers under completing and signing of a Material Transfer Agreement (MTA). Launched in 2019, the NPNPD library already has 326,000 natural product fractions, and the project purpose is to add new fraction sets at each consecutive year until reaching 1,000,000 fractions. Detailed information on the NPNPD platform is available at the National Cancer Institute Website (https://dtp.cancer.gov/organization/npb/npnpd_prefractionated_library.htm, accessed on 25 October 2022).

4. Aspects of Secondary Metabolism Pathways in Microorganisms

In microorganisms, complex biosynthetic pathways depend on gene regulation, enzymes, substrates, cofactors, and intermediates to form end products [13,14].

The SMs are represented mainly by polyketides, nonribosomal peptides (NRPs), ribosomal peptides, terpenes, shikimate-derived, and compounds that emerged from hybrid pathways [13,14,64–68]. The classification of these molecules is related to the primary metabolites from which they are derived and built through building blocks, forming molecules that are more complex.

The cellular regulatory elements and mechanisms involved in the biosynthetic pathways of SMs by microorganisms were previously very well discussed [13,14,66–68]. Particularly in fungi, the SMs are chemical classes classified according to their starter substrates (acyl-CoA, amino acids, nucleotides, carbohydrates, etc.), which are commonly incorporated in the final structure by specific enzymes, such as polyketide synthases (PKSs), nonribosomal peptide synthetases (NRPSs), dimethylallyl tryptophan synthetases (DMAT), geranylgeranyl diphosphate synthetases (GGPS), prenyltransferases, and terpene cyclases (TC) [13,14,65–68].

Polyketides are part of the most diverse class of SMs and are formed through the polymerization of acyl-coA and malonyl-coA units [69–74] by the action of the enzyme complexes called PKS, briefly represented in Figure 3. Polyketides are formed by three main β-ketoacyl synthase domains, responsible for the Claisen condensation and acyl transferase, which performs the transfer of the new building block to the nascent molecule and the acyl carrier protein domain, which acts as an anchor, binding covalently.

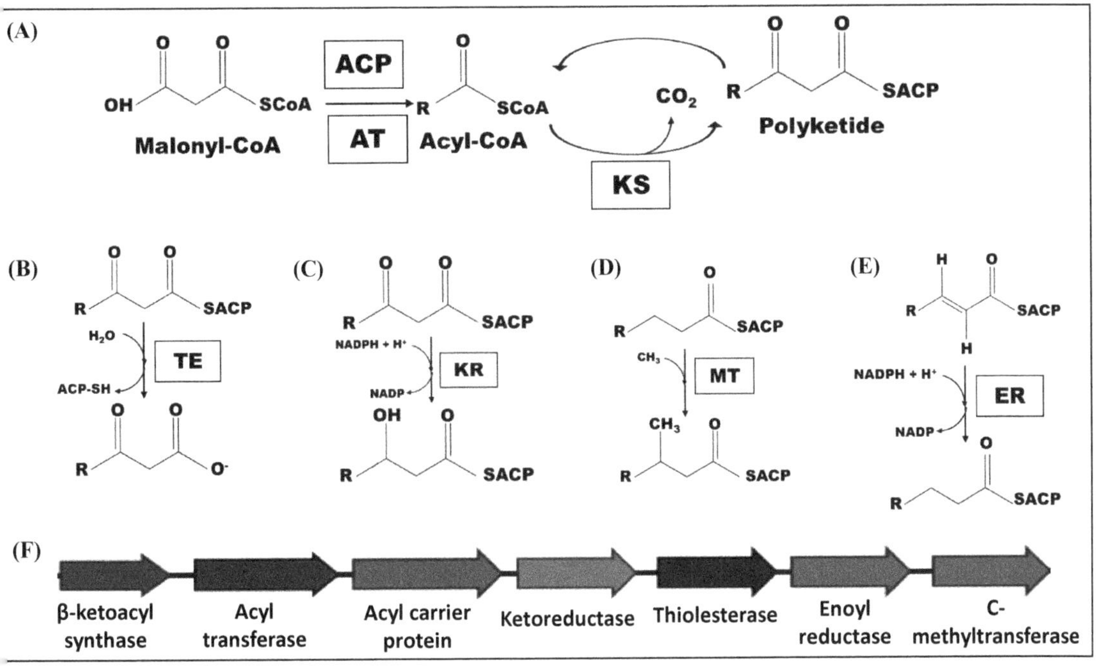

Figure 3. (**A**) Scheme of polyketide biosynthesis by polyketide synthetase. AT—Acyl transferase; ACP—Acyl carrier protein; KS—β-ketoacyl synthase; R—Carbon Chain. (**B–E**) Scheme of tailoring enzymes action on polyketide chain. TE—Thiolesterase; KR—Ketoreductase; MT—C-methiltrasferase; ER—Enoylreductase, respectively. (**F**) Genetic representation of gene cluster of polyketide synthetase and tailoring enzymes.

Modifications in the carbon skeleton of polyketides are carried out by tailoring enzymes, among which the main domains are thiolesterase, C-methyltransferase, enoyl reductase, and ketoreductase. These enzymes are responsible for the hydrolysis of thiol groups, the addition of a methyl group at the α position, and the reduction of the double bond of the enoyl-ACP group and the β-ketoacyl domains, respectively [71–74].

Fungi produce the greatest diversity of NRPs, among which are the β-lactams antibiotics, such as penicillin and cephalosporin [75,76]. The biosynthesis of these compounds is similar to that of polyketides. The NRPSs enzymatic complex also has three main domains. The adenylation domain (A), which through ATP will adenylate amino acids; the peptidyl carrier protein domain (T), responsible for the binding of a thioester group to the nascent chain; and finally, the condensation domain (C), which catalyzes the peptide bonds in the nascent chain and the amino acids [77,78]. After the beginning of the synthesis, structural changes such as epimerization, n-methylation, cyclization, acylation, glycosylation, hydroxylation, and halogenation occur [79].

An emerging class of SMs is small peptides, known as ribosomally synthesized and posttranslationally modified peptides (RiPPs) [80,81]. These compounds produced by fungi can be applied as antimicrobial and antiviral [82]. The RiPPs biosynthesis via the ribosome is differentiated from NRPS through post-translational modifications such as cyclization, hydroxylation, modifications in amino acid residues, and formation of new bonds [83,84].

Terpenoids are derived from the organic compounds obtained via the mevalonate pathway, isopentenyl diphosphate (IPP), and dimethylallyl diphosphate (DMAPP). The terpene biosynthesis pathway comprises condensation steps catalyzed by enzymatic complexes called terpene synthase, which will synthesize monoterpenes using the precursor geranyl disphophates (C10), sesquiterpenes from the precursor farnesyl diphosphate (C15), and diterpenes from geranylgeranyl diphosphate (C20). Compounds with C30 or C40 are formed by the precursors' squalene and phytoene, respectively [85,86]. These linear chains undergo modifications via enzyme complexes, prenyl transferase, terpene cyclases, and other terpene synthases, resulting in dephosphorylations, condensation, and cyclization of the hydrocarbon backbone [78].

As mentioned above, it is well known that genes encoding the enzymes for the biosynthesis and transport of SMs and those for regulating pathway-specific are encoded in the genome, forming biosynthetic gene clusters (BGCs) [14]; however, under standard laboratory conditions, most BGCs remain silent or are expressed only weakly [87]. To awaken or activate the expression of cryptic synthetic pathways and increase the biosynthesis of SMs by fungi, some strategies such as One-Strain Many-Compounds (OSMAC), Microbial Co-culture, Chemical Epigenetic Modification, and Molecular-Based approaches are being carried out [4,14]. These strategies were previously very well discussed, and all of them have strengths and weaknesses [13,14].

Epigenetic modifications in BGCs comprehend chromatin remodeling by histone post-translational modifications, DNA methylation, and RNA interference. Therefore, in this direction, small-molecule epigenetic modifiers are able to suppress or activate the associated enzymes, resulting in the regulation of the expression of specific genes that encode the SMs biosynthetic pathways in fungi. Among them is the DNA methyltransferase inhibitors (DNMT) 5-azacytidine, 5-aza-2′-deoxycytidine (decitabine), hydralazine hydrochloride, and the histone deacetylase (HDAC) inhibitors suberoylanilide hydroxixamic acid (vorinostat or SAHA), suberoyl bishydroxamic acid (SBHA), trichostatin A, trapoxin B, sodium butyrate, valproic acid, and nicotinamide [4]. These inhibitors are effective in low concentrations or in association. Regulation of genetic expression is an approach that relies on genome mining of the target strain to awaken cryptic BGCs using bioinformatics tools for the regulation of the SMs biosynthesis [65].

In fungi from distinct sources, BGCs play an important role in the production of chemical skeletons and structures with wide possibilities of bioactivities. Most of these gene clusters encode SMs; however, they are silent under laboratory conditions. To activate the biosynthetic potential of these genes, many methods or tools at the genome, proteome, transcriptome, or metabolome have been developed [14,65]. For this purpose, microorganism co-culture is an effective strategy in which, cell–cell contact, the chemical signal between the cells, or even biotransformation of one metabolite by another microorganism are possible mechanisms able to increase the production or activate the synthesis of novel SMs [88].

5. Bioactive Fungal Secondary Metabolites and Derivatives Successfully Applied by the Pharmaceutical and Agrochemical Companies

In microorganisms, the SMs are produced as a strategy to enhance their growth or even to improve their defenses or survival ability; however, SMs produced by fungi had their importance highlighted mainly after the isolation of penicillin **(1)** by Fleming, from the fungus *Penicillium chrysogenum* [89], lately re-identified as *Penicillium rubens* [90].

The interaction of SMs with biological receptors can be favored due to the high diversity and specificity of these molecules, and Figures 4 and 5 illustrate a few examples of SMs and their derivatives successfully applied by pharmaceuticals and agrochemicals industries.

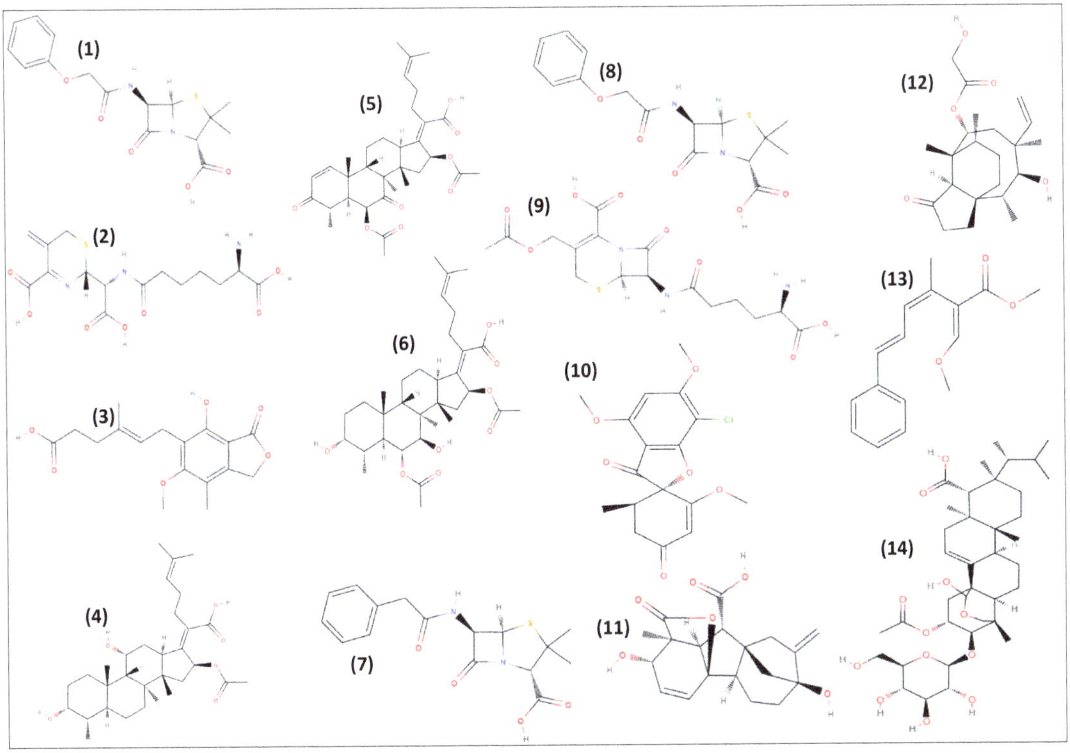

Figure 4. Examples of secondary metabolites from fungi and some of their derivatives successfully applied by pharmaceuticals and agrochemicals industries. **(1)** penicillin, **(2)** cephalosporin, **(3)** mycophenolic acid, **(4)** fusidic acid, **(5)** helvolic acid, **(6)** cephalosporin P1, **(7)** penicillin G, **(8)** penicillin V, **(9)** cephalosporin C, **(10)** griseofulvin, **(11)** gibberellic acid, **(12)** pleuromutilin, **(13)** strobilurins A, and **(14)** oudemansins. Structures were obtained from https://molview.org/ (accessed on 2 November 2022).

In the last years, due to studies on fungal biology, the importance of fungal SMs has been more widely recognized, and their pivotal participation in the survival, defense, and adaptation against biotic and abiotic conditions is well described. Over time, several different SMs were discovered, such as statins, cephalosporin **(2)**, and mycophenolic acid **(3)**, thus expanding and encouraging the studies for the discovery of more bioactive compounds by pharmaceuticals and agrochemicals companies [5–10,15,91].

Fusidic acid **(4)** was described in 1962 and is produced by the fungus *Fusidium coccineum* [92]. Commercialized as Fucithalmic, Fucidin, Usidin, Fucicort, Fucibet, or Taksta this antibiotic is indicated for treatments of topical or eye infections caused by Gram-positive bacteria, including methicillin-resistant *Staphylococcus aureus* acting by inhibiting translocation during protein synthesis [93]. Recent studies are in progress to the discovery and synthesis of more active analogs of fusidic acid **(4)**, including compounds with anticancer activity [94]. Interestingly, fusidic acid **(4)**, helvolic acid **(5)**, and cephalosporin P1 **(6)** are fusidane-type antibiotics that share a common early stage biosynthetic route catalyzed by six conserved enzymes, which are then divergent to generate fusidic acid **(4)**, helvolic acid **(5)**, and cephalosporin P1 **(6)**, respectively, under the action of different post-modification enzymes [95].

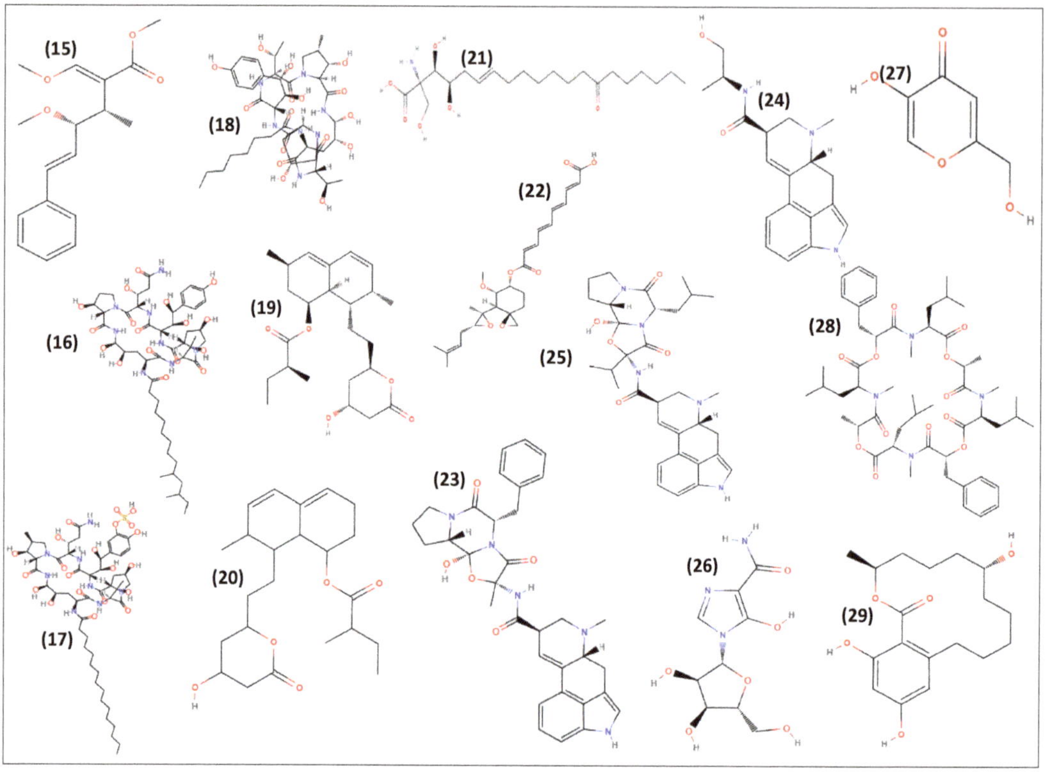

Figure 5. Examples of secondary metabolites from fungi and some of their derivatives successfully applied by pharmaceuticals and agrochemicals industries: **(15)** pneumocandin B0, **(16)** FR901379, **(17)** echinocandin B, **(18)** enfumafungin, **(19)** lovastatin, **(20)** compactin (ML-236B), **(21)** myriocin (ISP-I), **(22)** fumagillin, **(23)** ergotamine, **(24)** ergometrine, **(25)** ergocryptine, **(26)** mizoribine, **(27)** kojic acid, **(28)** PF1022A, and **(29)** α-Zearalanol (α-ZAL). Structures were obtained from https://molview.org/ (accessed on 2 November 2022).

Penicillins **(1)** and cephalosporin **(2)** are known antibacterial compounds that interfere with cell wall peptidoglycan biosynthesis and are used in the treatment of Gram-positive and some Gram-negative bacterial infections. Penicillins G **(7)** and V **(8)** are NRPs obtained from *P. rubens* and resulted in the development of benzylpenicillin and phenoxymethylpenicillin. The semisynthetic penicillins (piperacillin, amoxicillin, ampicillin) are derived from the acylation of 6-aminopenicillanic acid derived from penicillin G **(7)**. Cephalosporin C **(9)** is also an NRP produced by *Acremonium chrysogenum*, and from its structural core 7-amino cephalosporanic acid several derivatives such as cephalothin, cephalexin, cephradine, and cefadroxil were obtained [15].

A (Antibiotic S 7481F1, Ramihyphin A, CsA) is also an NRP produced by SmF of the aerobic fungus *Tolypocladium inflatum* [96], and it is available in the drug market as Sandimmune, Neoral, Restasis, Gengraf or as a generic drug. Cyclosporin A acts by binding to cyclophilin A, resulting in the inhibition of calcineurin, which activates the transcription of interleukin 2. This drug has been used as immunomodulatory to prevent organ transplantation and tissue graft rejection [97].

The fungi *Penicillium griseofulvum*, *Penicillium aethiopicum*, and *Penicillium coprophilum* are sources of the halogenated polyketide griseofulvin **(10)**, which is present in the commercial products Fulcine, Fulsovin, and Grisovin. This antifungal agent inhibits mitosis by binding to tubulin and preventing tubulin polymerization and has been used in the treatment of fungal infections of the skin, hair, and nails [98].

The diterpene gibberellic acid **(11)** is produced by the fermentation of the fungus *Fusarium fujikuroi*. This hormone is commonly used for plant growth through plant tissue culture and applications for certain high-value crops [99].

Another interesting SM is the diterpene pleuromutilin **(12)**, produced by the fungi *Clitopilus passeckerianus* and *Clitopilus* spp. This molecule is a precursor of the partial chemical synthesis of the topic antibiotic retapamulin (Altabax) and of tiamulin and valnemulin (Econor). The two latter compounds are for animal use and have been used for topical treatment of Gram-positive bacterial infections in cattle, acting by targeting the peptyl drug transferase center of the bacterial ribosome [100].

Strobilurins A-D **(A = 13)** and oudemansins **(14)** are polyketides structurally related to an initial unit of benzoyl CoA and are produced by *Strobilurus tenacellus*. Applying Quantitative Structural Activity Relationship (QSAR) on the structures of the natural strobilurins, many pesticide companies discovered several synthetic analogs that are more efficacious and stable fungicides, including the synthesis of fenamidone, azoxystrobin, kresoxim methyl, fluoxastrobin, trifloxystrobin, pyraclostrobin—carbamate, picoxystrobin, and dimoxystrobin [101]. Part of these compounds is registered worldwide for agrochemical use, and others are in the process of registration. This class of fungicides, for phytopathogenic fungi acts on the mitochondrial cytochrome b, which interferes with the respiratory electron transport in fungi.

Pneumocandin B0 **(15)** is an NRP acylated to polyketide and produced by *Glarea lozoyensis*, being the precursor of the partial chemical synthesis of the antifungal echinocandin, caspofungin (MK-0991) [102].

The fungi *Coleophoma cylindrospora* or *Coleophoma empetri* F-11899 produces the NRP acylated to fatty acid, coded as FR901379 **(16)**. This sulfated echinocandin is a precursor of the partial chemical synthesis of the echinocandin, micafungin, commercially sold as Mycamine and Fungiguard [103]. Recently, the BGCs involved in the biosynthesis of FR901379 **(16)** were analyzed and an efficient clustered regularly interspaced short palindromic repeats/Cas9-based gene editing tool for the industrial production strain C. *empetri* SIPI1284 was established [104].

The fungi *Aspergillus pachycristatus* and *A. nidulans* from the *Nidulants* section produce a molecule identified as echinocandin B **(17)**. This compound is classified as a fatty acid-acylated NRP, which is a semi-synthetic echinocandin synthesized from the fermentation product of *Aspergillus* sp. anidulifungin, which has been marketed as Eraxis [105].

The triterpene glycoside, enfumafugin was isolated from *Hormonema carpetanum*, an endophytic isolated from leaves of *Juniperus communis* and showed high antifungal activity *in vitro* against *Candida* sp. and *A. fumigatus*, and moderate efficacy in an *in vivo* mouse model of disseminated candidiasis [106]. Modifications in the structure of this compound to improve the oral bioavailability and pharmacokinetic properties led to the development of a semi-synthetic derivative named ibrexafungerp (IBX, MK-3118, SCY-078), which is currently in phase III clinical trials [107].

As well as caspofungin, micafungin, and echinocandin B **(17)**, enfumafungin **(18)** inhibits 1,3-beta-D-glucan synthesis, a major structural component of the fungal cell wall, which is not present in mammalian cells, consequently avoiding toxicity problems, and is used in the treatment of systemic fungal infections.

Lovastatin **(19)** (mevinolin, monacolin K, L-154803, MK-803, ML-530B) is a bioactive polyketide commercially sold as Mevacor. Its semisynthetic product, simvastatin, is on the market as Zocor and Simvador. This statin blocks 3-hydroxy-3-methylglutaryl-coenzyme A reductase, and prevents the conversion of mevalonate into cholesterol, being the most frequently used drug in the treatment of hypercholesterolemia to reduce the risk of cardiovascular diseases and to manage abnormal lipid levels by inhibiting the endogenous production of cholesterol in the liver [108]. Additional studies have indicated several important applications for lovastatin **(20)** as an antimicrobial and agent for treatments of cancers and bone diseases [19,108].

Lovastatin **(19)** biosynthesis has been reported in several fungal species, including *Aspergillus terreus*, *Monascus purpureus*, *Penicillium citrinum*, *Paecilomyces viridis*, *Penicillium purpurogenum*, *Pleurotus* sp., and *Trichoderma viride*; however, *A. terreus* is the species used for industrial production. Furthermore, to elucidate the lovastatin **(19)** biosynthesis, most genomic, and transcriptomic studies were performed in *A. terreus* ATCC 20542. Studies regarding the biosynthetic pathway showed that lovastatin **(19)** is synthesized from two-chain reactions using acetyl and malonyl-CoA and methionine as substrates through the biosynthetic pathway PKS. Two key multifunctional enzymes are involved in biosynthesis, which are encoded by specific genes organized in clusters on the fungal genome [19,108].

In the group of statins, compactin (ML-236B) **(20)** produced by the fungi *P. citrinum* and *Penicillum solitum* also reduces the risk of cardiovascular diseases and is used as a substrate for the biotransformation or partial chemical synthesis of mevastatin. Commercially, it is available as Pravachol, Selektine, or Lipostat. This statin inhibits 3-hydroxy-3-methylglutaryl-coenzyme A reductase and is widely used for the treatment of hypercholesterolemia by preventing the conversion of mevalonate into cholesterol [109].

Myriocin (ISP-I) **(21)** is an amino acid lipid produced by the fungus *Isaria sinclairii*, which structural model is used as a template for the synthesis of fingolimod, commercially available as Gilenya. Used in the treatment of multiple sclerosis through the activation of its structure, this drug connects with extracellular G protein-coupled receptors, preventing the release of lymphocytes that enter the central nervous system [110].

Mycophenolic acid **(3)** is an immunosuppressive drug used to prevent organ transplant rejection. Synthesized by the fungus *Penicillium brevicompactum*, this compound, classified as a meroterpenoid, is used to produce mycophenolate mofetil and mycophenolate sodium, which are commercially available as Cellcept and Myfortic. This drug blocks inosine monophosphate dehydrogenase (IMPDH), consequently causing a decrease in DNA synthesis [91].

Another meroterpenoid is the compound fumagillin **(22)**, which is produced by the fungus *A. fumigatus* and is commercially available as dicyclohexyl ammonium fumagillin **(22)** (Fumidil B®). This compound is a methionine aminopeptidase2 (MetAP2) inhibitor used to control Nosema disease in bees caused by *Nosema apis* [111].

The alkaloids ergotamine **(23)**, ergometrine **(24)**, and ergocryptine **(25)** are prenylated NRPs obtained as SMs from the fungi *Claviceps purpurea*, *Claviceps fusiformis*, and *Claviceps paspali*.

Ergotamine **(23)** acts as a vasoconstrictor, being used in migraine attacks, and is commercially available as ergotamine **(23)** tartrate (Ergomar and Migril), dihydroergotamine mesylate (Migranal), dihydroergotamine (DHE 45), or associated with caffeine (Cafergot, Cafetrate). This alpha-1 selective adrenergic agonist vasoconstrictor is an agonist of 5-OH tryptamine (5-HT2) receptor 1B, norepinephrine, dopamine, and serotonin receptors. This ergot alkaloid is used as an anti-migraine agent, also associated with belladonna and phenobarbital, for the relief of menopausal hot flashes [112].

The synthetic derivative ergometrine **(24)** maleate is commercially available as Ergotrate and Ergovin, and the derivative methylergometrine can be combined with oxytocin (Syntomethrin) for veterinary use. Both ergometrine **(24)** and methylergometrine are used for the treatment of postpartum hemorrhage. Ergometrine **(24)** is a serotonin antagonist, activating or blocking 5-HT2 receptors and used as the precursor for partial chemical synthesis of lysergic acid diethylamide, LSD-25, and LSD (Delysid) [113].

The chemical structure of ergocriptine **(25)** is the core for the partial chemical synthesis of bromocriptine (Parlodel), an agonist of dopamine D2 and various serotonin and adrenergic receptors used in the treatment of reproductive disorders (galactorrhea, prolactin-dependent mammary carcinoma, amenorrhea, acromegaly, or anovulation) or complications such as osteoporosis. Bromocriptine is also an option for the treatment of Parkinson's disease [114].

The imidazole nucleoside mizoribine **(26)** is produced by the ascomycete *Penicillium brefeldianum* [115] and is the active component of Bredinin™ OD25 and Bredinin™ OD50, marketed since 2016 by the company Asahi Kasei Pharma as an immunosuppressant for renal disorders and transplant.

The fungi *Aspergillus oryzae*, *Aspergillus tamarii*, and *Aspergillus flavus* are producers of the kojic acid **(27)**, classified as a pyrone, derived from glucose. In industry, it is often formulated as kojic dipalmitate to increase stability and shelf life. As an inhibitor of tyrosinase, the kojic acid **(28)** suppresses melanogenesis and is used as an antioxidant in cosmetic products to lighten skin color and treat abnormal hyperpigmentation [116].

PF1022A **(28)** is an NRP produced by the fungus *Rosellinia* sp. and its derivative, bisparamorphonyl, is produced commercially as Emodepside. This form disrupts the function of the target neuromuscular nematode by the interactions with the latrophilin, such as HC110-R G protein-coupled and SLO-1 potassium channels. This NRP can be associated with the anthelmintic drug praziquantel to treat roundworms, hookworms, and tapeworms in cats [117].

α-Zearalanol **(29)** (α-ZAL) is a polyketide produced by *Fusarium* spp., including *Fusarium culmorum*, *Fusarium equiseti*, *Fusarium sporotrichioides*, and *Fusarium semitectum*. This estrogenic mycotoxin is obtained from the fermentation of zearalenone by a highly reducing and non-reducing PKS, followed by hydrogenation to form α-zearalanol **(29)**. The synthetic version of α-ZAL, zeranol, is marketed as a growth promotor in the United States and Canadian livestock industries, whereas experimental animal models have typically demonstrated impaired growth following prenatal exposure [118].

In addition to the fungal SMs with proven biological effects, there are several others in the study by different groups in many countries for application as cytotoxic or antimicrobial agents, and a few of them are discussed in the next topics.

6. Studies of Fungal Secondary Metabolites as Cytotoxic and Antimicrobial Agents

6.1. Cytotoxic Compounds

Cancer has been the leading cause of death and an important barrier to increasing life expectancy around the world [119]. According to the World Health Organization [120], in 2018, cancer was the second leading cause of death globally. It was responsible for at least one in every six deaths, accounting for at least 9.6 million deaths. An estimate for 2020 indicates that there were 19.3 million new cases of cancer and 10.0 million deaths worldwide [121]. In Brazil, it is estimated that between 2020 and 2022, there will be around 625,000 new cases of cancer, the most frequent being non-melanoma skin cancer, followed by breast, prostate, colon, rectum, lung, and stomach cancer [122].

Despite the progress in research for cancer treatments, which has improved the life expectancy and cure of many patients, the complications from the adverse effects due to chemotherapy and radiotherapy are disastrous. In addition, there are cases of chemoresistance and cancer recurrences.

Currently, there is a great demand for new oncological therapies that reduce or do not cause severe adverse effects to patients, providing an improvement in quality of life. In this context, several studies seek new antitumor molecules from natural sources, attempting to reduce treatment costs, increase specificity, and decrease the side effects [123].

Among all anticancer drugs described from 1946 to 2019, in a total of 321, about 71% (230) are based on or derived from natural products [1]. According to Keller [65], from 1993 to 2001, among the 1500 fungal SMs, 50% were described as antimicrobial and antitumor compounds.

Various tumor cell lines from human and animal sources have been useful and applied for *in vitro* cytotoxicity and *in vivo* studies of the anticancer potential of natural products, whether by the academy or by pharmaceutical companies.

The main human tumor cells applied are from lung carcinoma (A549, HL251, 95-D, NCI-H460), melanoma (A375-S2, SK-MEL-28), glioblastoma (SF268, U-251, U373), caucasian esophageal squamous carcinoma (OE21), oligodendroglioma (Hs683), colorectal carcinoma

(HCT 116, HT29), prostate adenocarcinoma (PC-3, PC-3M), leukemia (HLK 210, Jurkat), acute promyelocytic leukemia (HL-60), chronic myelogenous leukemia—CML (K562), breast carcinoma or adenocarcinoma (BT-20, BT-474, MDA-MB-435, MDA-MB-468, MCF7, MCF7-dox T47D, MDA-MB-231, SK BR-3), cervical carcinoma (HeLa), hepatocellular carcinoma (HepG2, MHCC97H, BEL-7402), gastric carcinoma (MGC-803), ovarian carcinoma (HO-8910), and neuroblastoma (SH-SY5Y). Among mouse cells are melanoma (B16-F10), lymphoma (L5178Y), and lymphocytic leukemia (L1210) [5,124]. Many other specific cell lines are used or were mentioned throughout this review discussion.

Several fungal SMs are reported as revealing potential anticancer activity through different mechanisms of action, such as apoptosis, angiogenesis, cell cycle regulation, metastasis, and signal transduction cascades [124]. Table 1 presents several examples of cytotoxic fungal SMs already described.

In the last years, various fungal bioactive SMs have been described [5–10,125], such as paclitaxel. Paclitaxel was initially discovered from the medicinal plant *Taxus brevifolia*, and its cytotoxicity and apoptotic effect were studied *in vitro* on different cancer cell lines such as BT220 (breast), H116 (colon), Int407 (intestine), HL251 (lung), and HLK210 (leukemia) and MCF-7 (breast cancer). Paclitaxel, commercially called Taxol®, was approved by the United States Food and Drug Administration (FDA) Agency in 1992, and is clinically applied in the treatment of breast, throat, neck, lung, and ovary cancer [126]. To attend to the demand, paclitaxel is produced by a fermentation process of *Taxus* sp. leaves in aqueous media, and despite the huge progress in the biotechnological approaches, *Taxus* spp. cell culture productivity remains a big challenge [127].

Table 1. Examples of cytotoxic secondary metabolites produced by fungi species.

Metabolites	Fungi Species	References
Bislongiquinolide (trichotetronine), dihydrotrichodimerol	*T. citrinoviride*	[124]
Paclitaxel	*B. robillardoides* *L. theobromae* *Paraconiothyrium* sp. *P. microspore* *P. neglecta* *P. versicolor* *T. andreanae*	[128–133]
Camptothecin, podophyllotoxin, vinblastine, vincristine	*T. radicus*	[134]
Altenuene, stemphyperylenol	*Botryosphaeriadothidea*	[135]
Alterperylenol, altertoxin II	*Alternaria* sp. A744	[136]
Alterperylenol, stemphyperylenol	*Alternaria* sp.	[137]
(-)-(10E,15S)-10(11)–dehydrocurvularin, (-)-(10E,15S)-6-Chloro-10(11)-dehydrocurvularin	*Alternaria* sp. AST0039	[138]
Phomoxanthone A	*P. longicolla*	[139]
Embellicines A and B	*E. eureka*	[140]
Pyrrocidin A	*N. ramulariae* Wollenw KS-246	[141]
Aphidicolin	*N. sphaerica*	[142]
Steroid (22E,24R)-8,14-epoxyergosta-4,22-diene-3,6-dione	*P. immersa*	[142]
12′-Hydroxyroridin E, 13′,14′-hydroxymytoxin B, 14′-hydroxymytoxin B, vertisporin, mytoxin C, roridin E, 2′,3′-epoxymyrotecin A, myrothecin A, myrotoxin A	*M. roridum*	[143]
Daldinone I	*Annulohypoxylon* sp.	[144]
Cladosporol A	*C. cladosporioides*	[145]

Table 1. *Cont.*

Metabolites	Fungi Species	References
Engyodontiumones H, AGI-B4	*E. album* DFFSCS021	[145]
Rosoloactone	*T. roseum*	[146]
Penipacide A and E	*P. paneum*	[147]
Beauvericin	*Fusarium oxysporum* EPH2R$_{AA}$	[148]
Bikaverin	*F. oxysporum* CEC1S	[148]
Beauvericin	*F. oxysporum*	[149]
Penichryfurans A and B	*P. chrysogenum*	[150]
Brefeldin A (BFA)	*P. brefeldianum*	[151]
Porritoxin	*A. porri*	[152]
Amides (AI-77-B, AI-77-F, Sg17-1-4)	*A. tenuis* sg17-1	[153]
Dehydroaltenusin	*A. tenuis*	[154]
Alternariol, alternariol 5-*O*-sulfate, alternariol 5-*O*-methyl ether, altenusin, desmethykaltenusin	*Alternaria* sp.	[154]
Alternol	*A. alternata var. monosporus*	[155]
Cajanol (5-hydroxy-3-(4-hydroxy-2-methoxyphenyl)-7-methoxychroman-4-one)	*H. lixii*	[156]
Sorbicillinoids and bisorbicillinoids	*P. chrysogenum* *Penicillium* sp. *P. terrestre* *Phialocephala* sp. *S. album* *Trichoderma* sp. *Trichothecium* sp.	[157–168]
Sorbicillinoids: tricoreeseione A and B, tricodermolide B, 13-hydroxytricodermolide, 24-hydroxy-tricodimerol, 15-hydroxybisvertinol Analogs of sorbicillinoids: trichodimerol, 24-hydroxy-bisvertinol, bisvertinol	*T. reesei* HN-2016-018	[169]
Deacetylcytophasin C, zygosporin D	*C. taii*	[170]
Ylarone A, (-)5-methylmelein	*X. psidii*	[171]
Diapolic acid AB, xylarolide, fomolide	*D. terebinthifolli*	[172]
Pestalrone B	*P. karstenii*	[173]
Siccayne [2-(3-methyl-3-buten-1-ynyl) hydro]	*P. fici*	[173]
Chaetoglobosins A, B, D, E, F, Fex, 20-dihydrochaetoglobosin A	*C. globosum*	[174,175]
Pensulfonoxy, Pensulfonamide	*P. aculeatum*	[176]
Polonidine A, fructigenine A	*P. polonicum* TY12	[177]
Giluterrin	*A. terreus* P63	[178]
Arvoredol	*Penicillium* sp. F37	[179]
Anhydrofusarubin, fusarubin	*Cladosporium* sp.	[180]
Xanthocillins X and Y1	*P. chrysogenum* CCTCC M 2020019	[181]

Over time, in an attempt to find a good source of paclitaxel, fungi species have been studied for this purpose. First, paclitaxel was identified as an SM of the endophytic fungus *Taxomyces andreanae*, isolated from the plant with the same name [128]. Later, it was also identified as an SM of at least 35 endophytic fungi [12].

Endophytes isolated from *Taxus wallachiana* in the Himalayas can produce paclitaxel and other chemicals with valuable antitumor activity, and among them, the fungi *Pestalotiopsis microspora* [129]. The endophytes *Bartalinia robillardoides* isolated from the medicinal plant *Aegle marmelos* [130] and *Lasiodiplodia theobromae* from *Morinda citrifolia* [131] are other examples of paclitaxel biosynthesis.

Among the fungi species able to produce paclitaxel, *Pestalotiopsis neglecta* and *Pestalotiopsis versicolor,* which were isolated as endophytic from the Japanese Yew tree, *Taxus cuspidata,* produced paclitaxel with a yield of 375 and 478 μg/L, respectively. This yield is 9560 times higher than what was previously reported as obtained for the fungus *T. andreanae* [132].

The co-culture of *Paraconiothyrium* sp., a paclitaxel-producing fungus with *Alternaria* sp., increased the paclitaxel production by 3-fold, and when the *Phomopsis* sp. was added to the coculture, the yield increased to around 7.8-fold [133].

The compounds camptothecin, vincristine, vinblastine, and podophyllotoxin, produced by plants, and their semisynthetic derivatives, are clinically used for cancer treatments (e.g., ovarian, breast, prostate, lung cancers, and leukemias). However, these compounds were also described as fungal SMs of *Talaromyces radicus* isolated from *Catharanthus roseus,* in India, with good yields of vincristine (670 μg/L) in modified M2 medium and of vinblastine (70 μg/L) in potato dextrose broth medium [136]. Similarly, the endophytic fungus *F. oxysporum*, isolated from the leaves of the same plant, also in India, produced vincristine and vinblastine, resulting in a yield of 67 and 76 mg mg/L, respectively [182].

Stemphyperylenol and altenuene were isolated from the endophytic *B. dothidea* of *Melia azedarach* and exhibited cytotoxicity against HCT116 cells with an IC50 value of 3.13 μM, while the positive control etoposide showed IC50 of 2.13 μM [135].

The compounds alterperylenol and altertoxin II were described as SMs of the endophytic fungus *Alternaria* sp. A744 derived from *Morinda officinalis*, and both compounds significantly inhibited the proliferation of the tumor cell lines MCF-7, HepG-2, NCI-H460, and SF-268, with IC50 values in the range of 1.91–9.67 μM [136]. Later, Zhao et al. [137] also reported the isolation of the two perylenequinones, stemphyperylenol, and alterperylenol, from *Alternaria* sp. endophyte from marine plants in China.

The *Alternaria* sp. AST0039, derived from *Astragalus lentiginosus* collected from a roadside area in central Arizona, in the USA, provided the resorcylic acid lactones curvularins, (-)-(10E,15S)-6-chloro-10(11)-dehydrocurvularin, and (-)-(10E,15S)-10(11)–dehydrocurvularin, and both compounds exhibited cytotoxicity to the tumor cells NCI-H460 (human non-small cell lung cancer), SF-268 (human CNS glioma), MCF-7 (human breast cancer), and PC-3M (metastatic human prostate cancer adenocarcinoma), with IC50 lower than 5.0 μM. Additionally, (−)-(10E,15S)-6-chloro-10(11)-dehydrocurvularin inhibited the growth of MDA-MB-231 (human metastatic breast adenocarcinoma) cells with IC50 of 2.95 μM) [138].

Phomoxanthone A is an SM produced by the endophytic fungus *Phomopsis longicolla* isolated from the mangrove plant *Sonneratia caseolaris* (*Lythraceae*) collected in southern China [139]. This tetrahydroxanthone atropisomer derived was cytotoxic with IC50 of 0.7, 0.8, and 5.2 μM on human esophagus Kyse510, human tongue Cal27, and human ovarian carcinoma A2780 cell lines sensitive to cisplatin, and IC50 of 0.9, 0.8 and 0.9 5.6 μM on these same cell lines, however cisplatin-resistant, respectively. Additional data showed that this compound was selective to tumor cells, showing IC50 of 61.2 μM on healthy human peripheral blood mononuclear cells (PBMCs) and of 0.1 μM on the human Burkitt's lymphoma cell line DG75 and human T cell lymphoma cell line Jurkat [139]. Besides the pro-apoptotic activity, phomoxanthone A activated murine T lymphocytes, NK cells, and macrophages, suggesting an activation of the immune system. The association of this dual effect to combat the cancer cells could help in fighting resistance during chemotherapy.

The fungus *Embellisia eureka*, endophytic from the plant *Cladanthus arabicus*, collected in Morocco, produced the SMs pyrrolidinone alkaloids, embellicines A and B, and the bioactivity screening of these congeners indicated pronounced cytotoxicity against human chronic myeloid leukemia cells (K562) with IC50 lower than 10 μM. Embelicin B was the most cytotoxic, with IC50 of 0.92, 0.32, 0.25, and 0.21 μM after 8, 24, 48, and 72 h of treatment, respectively [140].

Uesugi et al. [141] performed a study with pyrrocidin A, a close derivative of embellicines A and B, which was isolated from the fungus *Neonectria ramulariae* Wollenw KS-246, derived from a dead twig (plant species not reported) collected in Japan. Pyrrocidin A, originally described as a natural product with antibiotic activity, exhibited cytotoxic properties against human acute promyelocytic leukemia HL60 cells with IC50 of 0.12 μM.

The diterpene aphidicolin and the steroid (22E,24R)-8,14-epoxyergosta-4,22-diene-3,6-dione were described as SMs from the endophytes *Nigrospora sphaerica* and *Papulaspora immersa*, respectively, isolated from the plant *Smallanthus sonchifolius*. These compounds were also cytotoxic to HL60 cells with IC50 of 0.27 and 1.65 μM, respectively [142].

The fungus *Myrothecium roridum*, isolated from roots of *Ajuga decumbens* collected in Fuzhou, Fujian, China, provided nine trichothecenes derivatives, which were evaluated for cytotoxicity on the human hepatocellular carcinoma cells A549, MCF-7, HepG2, and SMMC-7721 [143]. The compounds 2′,3′-epoxymyrotecin A, myrothecin A, and 12′-hydroxyroridin E were the less cytotoxic, with IC50 values in the range of 1.25–95 μM on these cell lines; however, the other six derivatives (13′,14′-hydroxymytoxin B, myrotoxin A, 14′- hydroxymytoxin B, vertisporin, mytoxin C, roridin E) were more effective, with IC50 from 1.42 to 75 nM. It is known that the 12,13-epoxy ring is the structural feature relevant to the remarkable differences in toxicity, which explains the much lower cytotoxicity of the trichothecenes 2′,3′-epoxymyrotecin A and myrothecin A.

Among the nine SMs (daldinone H, daldinone I, daldinone J, daldinone C, hypoxylonol C, daldinone B, 3,4-dihydro-3,4,6,8-trihydroxy-l(2H)-naphthalenone, (R)-scytalone, and 1-hydroxy-8-methoxynaphthalene) of the endophyte *Annulohypoxylon* sp. isolated from the mangrove plant *Rhizophora racemosa* collected in Cameroon, only daldinone I inhibited the growth of human leukemia and lymphoma cell lines, Jurkat J16 and Burkitt's lymphoma (Ramos) with IC50 of 14.1 and 6.6 mM, respectively [144]. The cytotoxicity mechanism was described as being by induction of intrinsic apoptosis and by blocking autophagy, a potential pro-survival pathway for cancer cells.

Chromatographic processing of the crude extract obtained from the culture of the endophytic fungus *Cladosporium cladosporioides* derived from *Datura innoxia* (Jammu, India) provided the tetralone derivative, cladosporol A [145]. The cytotoxicity assay against the NCI60 panel of human cancer cells revealed its antiproliferative effect on various cancer cell lines with IC50 values from 8.7 to 15.6 μM, being more effective against the MCF-7 cell line of breast cancer [145]. The data showed that cladosporol A triggered apoptotic as well as autophagic death of human breast cancer (MCF-7) cells.

Natural diterpenoids have been reported to induce tumor cell apoptosis. In this class, the compound rosoloactone was isolated from the endophytic fungus *Trichothecium roseum*, derived from *Ginkgo biloba* L. in Shandong Province, China. Rosoloactone displayed significant cytotoxicity *in vitro*, inducing anti-proliferative and pro-apoptotic effects in human cervical cancer HeLa cells, which were associated with endoplasmic reticulum stress (ERS) and mitochondrial damage [146]. An accumulation of misfolded or unfolded proteins in the ER lumen was observed, causing excessive ERS, as well as mitochondrial damage followed by the release of cytochrome c into the cytosol, activation of caspase-3 and -9, with subsequent activation of mitochondria-mediated apoptosis. There was a simultaneous and marked ROS production [146].

Engyodontium album DFFSCS021, isolated from a marine sediment sample collected in the South China Sea, produces 19 polyketides, including engyodontiumones A–J, 2-methoxyl cordyol C, sydowinin A-B, pinselin, aspergillusone B, AGI-B4, diorcinol, cordyol C, and hydroxysydonic acid, which were screened for cytotoxicity, antibacterial, and antilarval activities. Among them, engyodontiumones H and AGI-B4 were cytotoxic on the human histiocytic lymphoma U937 cell line, with IC50 values of 4.9 and 8.8 μM, respectively [183].

The anthranilic acid derivatives penipacide A and E produced by *Penicillium paneum*, also isolated from marine sediment in the South China Sea, exhibited cytotoxicity on the RKO cell line (colon carcinoma), with IC50 values of 8.4 and 9.7 μM, respectively [147].

The pharmacological plants from the Brazilian biodiversity *Ageratum myriadenia*, *Palicourea tetraphylla*, *Piptadenia adiantoides*, and *Trixis vauthieri* were used as a source of 121 fungal endophytes, which were identified as belonging to the genera Alternaria, Arthrinium, Cochliobolus, Colletotrichum, Penicillium, Fusarium, and Gibberella. The fungi strains were applied for fermentation, and crude organic extracts obtaining using ethyl acetate [184]. Among the extracts, 17 (14%) were cytotoxic for at least one human cancer cell line, with IC50 values ranging from >0.2 to 25 μg/mL, and selected for future compound isolation. Later, Alves et al. [185] demonstrated that the organic extract (EtOAc/MeOH 1:1 solvent) obtained using the culture of the endophytic fungus *Dichotomophthora* sp. was cytotoxic on cell strains of colon adenocarcinoma (HCT-166) and prostate cancer (PC3), inhibiting, respectively, 100% and 96.41% of the cellular growth.

Beauvericin and bikaverin produced by the strains of *F. oxysporum* $EPH2R_{AA}$ and CEC1S endophytes from the plants collected from the Sonoran desert *Ephedra fasciculate* and *Cylindropuntia echinocarpus*, respectively, were active showing selective toxicity toward NCI-H460 and MIA Pa Ca-2, respectively [148].

Similarly, a strain of *F. oxysporum* isolated from the bark of *Cinnamomum kanehirae* from Jiaoban Mountain, Taiwan Province, along with other SMs provided the cyclodepsipeptide beauvericin that showed cytotoxicity against PC-3, PANC-1, and A549, with IC50 ranging from 10.4 to 49.5 1.6μM [149].

Among Penicillium species able to produce cytotoxic SMs is the endophytic fungus *P. chrysogenum*, from the marine medical red alga *Grateloupia turuturu* [150]. Through *in vitro* assay, it was possible to demonstrate the cytotoxicity activity of the N-acetyl–glucosamine derivatives, penichryfurans A and B against A549, HeLa, and HepG2 cell lines. Penichryfurans A was the most effective on the HepG2 cell line with an IC50 of 9 μM.

In addition to mizoribine **(27)** (Bredinin™ OD25 and Bredinin™ OD50) and other SMs, *P. brefeldianum* produces brefeldin A (BFA), a lactone with antifungal, antiviral, and antitumor activity. BFA is able to trigger apoptosis in tumor cells, and besides its biological properties, its structural features have aroused the interest of medicinal chemistry in developing numerous analogs to improve its bioavailability and its antiproliferative effect [151].

Lou et al. [186] reviewed metabolites from *Alternaria* sp. and their bioactivities, showing several SMs with cytotoxic effects on Hela, KB, MDA-MB-435 breast cancer, and L5178Y mouse lymphoma cells. Among them, *Alternaria alternata* JS0515 was described as an endophyte from *Vitex rotundifolia* rhizomes, and in the chemical composition of its metabolism products were found phenolics, pyranones, quinones, steroids, terpenoids, and nitrogen-containing compounds, some of which showed phytotoxic, cytotoxic, antifungal, and antimicrobial activities [186–189].

The SM porritoxin from *Alternaria porri* showed antitumor-promoting activity [152], and amides (AI-77-B, AI-77-F, Sg17-1-4) from the marine fungus *Alternaria tenuis* sg17-1 were cytotoxic on human malignant A375-S2 and human cervical cancer Hela cells in the range from 0.02 to 0.4 mM [153]. Alternariol and its derivatives from *Alternaria* sp., including alternariol 5-*O*-sulfate, alternariol 5-*O*-methyl ether, altenusin, and desmethykaltenusin were described as cytotoxic on mouse lymphoma cells (L5178Y), with IC50 of 1.7 to 7.8 μg/mL [154]. Alternariol was described as topoisomerase I and II poison, which can interfere with the impairment of DNA integrity in human colon carcinoma cells [190]. Alternariol and its 9-methyl ether induced cytochrome P450 1A1 and apoptosis in murine hepatoma cells dependent on the aryl hydrocarbon receptor [191]. Dehydroaltenusin obtained from *A. tenuis* showed specific inhibition of eukaryotic DNA polymerase α to show its high cytotoxicity on tumor cells [192].

Liu et al. [155] demonstrated for the first time that alternol, a compound purified from the metabolism of the fungus *A. alternata var. monosporus*, from yew bark (*T. brevifolia*), inhibits proliferation and induces apoptosis in mouse lymphocytic leukemia L1210 cells. Alternol induced activation of caspase-3 and caspase-9, but not caspase-8. In addition, there was a significant increase in ROS production, which may also play a role in apoptosis. Based on this study, *A. alternata* was suggested as an interesting source for its ability to produce the cytotoxic compound alternol.

A study carried out by Fernandes et al. [193] demonstrated the cytotoxic potential of SMs from the endophytic fungus *A. alternata* isolated from coffee leaves of *Coffea arabica* L. Through *in vitro* assay, the organic crude extract obtained using dichloromethane showed cytotoxicity with IC50 of 400 μg/mL on human cervical cancer cells (HeLa).

In another study, Bhimba et al. [194] showed that the organic extract obtained from the culture of the fungus *Hypocrea lixii*, isolated from leaves of the mangrove *Rhizophora mucronata*, *Avicennia officinalis*, and *Avicenna marina*, displayed cytotoxic activity on human hepatocellular carcinoma (HepG2) and breast cancer (MCF-7) cells. This extract also presented antimicrobial effects on *Bacillus subtilis*, *S. aureus*, *Escherichia coli*, *Proteus vulgaris*, and mainly on *Pseudomonas aeruginosa*. Following that, a study published by Zhao et al. [156] demonstrated that the endophytic fungus *H. lixii*, isolated from the medicinal plant pigeon pea (*Cajanus cajan* [L.] Millsp.), in China, produces the SM cajanol, which showed cytotoxicity on human lung carcinoma cells (A549).

Cajanol, 5-hydroxy-3-(4-hydroxy-2-methoxyphenyl)-7-methoxychroman-4-one, is an isoflavone, and previous studies showed its antiplasmodial activity [195] and inhibition of prostate-specific antigen secretion in LNCaP cells [196]. Despite its pharmacological potential, cajanol occurrence is rare. Among 135 endophytic fungi isolated from roots, stems, and leaves of pigeon pea, only three strains of *H. lixii* were able to produce cajanol [156].

Sorbicillinoids (also called vertinoids) are hexaketide metabolites, first isolated in 1948 as an impurity in penicillin **(1)** [197]. This family of compounds has more than 100 analogs already described, having a specific carbon skeleton of sorbicillin, in which the cyclization has taken place on the carboxylate terminus [198–200]. According to their basic structural features, sorbicillinoids are classified as monomeric, dimeric, trimeric, polycyclic, and vertinolides [198–200]. Due to their unique structures, sorbicillinoids have received particular attention for drug development in the pharmaceutical and agrochemical areas.

The structures of 91 sorbicillinoids were reviewed, including those from 62 new sorbicillinoids obtained from fungi between 2016 and 2021, as well as their bioactivities as anti-inflammatory, antiviral, radical scavenging, cytotoxicity, antimicrobial, anti-inflammatory, phytotoxic, and α-glucosidase inhibitory [198].

These classes of SMs have been isolated from terrestrial or marine fungi. Among the sorbicillinoids producers, the species *Aspergillus* sp., *P. citrinum*, *Penicillium notatum*, *Emericella* sp., and *Phaeoacremonium* sp. belong to the Trichocomaceae Family. In the class of Sordariomycetes, the sorbicillinoid producers are *Trichoderma* sp., *T. viride*, *Trichoderma citrinoviride*, *Trichoderma longibrachiatum*, *Verticillium intertextum*, *Acremonium* sp., *Acremonium strictum*, *Phaeoacremonium* sp., *Clonostachys rosea*, and *Scytalidium* sp. Among them, cytotoxic sorbicillinoids were isolated from the terrestrial fungus [201], and also from the marine-derived fungi *Trichoderma* sp., including bisorbicillinoids [157–159], *Penicillium terrestre* [160,161], *Penicillium* sp. [162], *Phialocephala* sp. [163–166], *Trichothecium* sp. [167], and *P. chrysogenum* [168]. All the above sorbicillinoid-producing species are from the Pezizomycotina subphylum from the Ascomycota phylum.

With interest in discovering new natural compounds with biological or therapeutic applications, researchers have also focused on studies of fungi of the *Trichoderma* genus to isolate new compounds. Rehman et al. [169] isolated and identified six sorbicillinoids (tricoreeseione A and B, tricodermolide B, 13-hydroxytricodermolide, 24-hydroxy-tricodimerol and 15-hydroxybisvertinol), and three analogs (trichodimerol, 24-hydroxy-bisvertinol, and bisvertinol) from *Trichoderma reesei* (HN-2016-018), derived from the fresh internal tissue of an unidentified sponge collected in China. Among them, 24-hydroxy-trichodimerol exhibited great potential for anticancer drugs, being cytotoxic against A549 lung carcinoma cell lines, MCF-7, and HCT116 cell lines with IC50 of 5.1, 9.5, and 13.7 μM, respectively [169].

Similarly, bislongiquinolide (trichotetronine) and dihydrotrichodimerol were isolated from *T. citrinoviride* and inhibited the proliferation of human cancer cell lines (U373 glioblastoma, Hs683 oligodendroglioma, A549, OE21 esophageal cancer, and SKMEL-28 melanoma), and B16F10 mouse melanoma cancer cell lines significantly through cytostatic and not cytotoxic activity [124]. Literature data show that dihydrotrichodimerol acts by activation of peroxisome proliferator-activated receptor-ϒ (PPAR-ϒ), which has an important role in cancer cell biology, being able to suppress the production of tumor necrosis factor-R (TNF-R) and nitric oxide in LPS-stimulated RAW264.7 cells [202].

The endophytes *F. oxysporum* CFE177 and *A. oryzae* CFE108 were isolated from the Brazilian medicinal plant *Combretum leprosum*, and their SMs present in the crude extract obtained from their culture showed cytotoxicity on the HeLa, ECV304, B16F10, J744, P388, Jurkat and k562 cell strains [203].

The *Cordyceps taii* species is normally found as a parasite in insect or arthropod larvae and produces the compounds deacetylcytophasin C and zygosporin D, which by the sulforhodamine B (SRB) method were cytotoxic for the 95-D human lung cancer cells. The cytotoxicity was more effective than that of the antitumor cisplatin [170].

Arora et al. [171] demonstrated that ylarone A and (-)5-methylmelein metabolites produced by the fungus *Xylaria psidii*, isolated from the leaf of the medicinal plant *A. marmelos*, have a cytotoxic effect against pancreatic cancer cells (MIA PaCa-2). These metabolites interrupt the cell cycle in the sub-G1 phase and induce cell death by activating apoptotic mechanisms.

The endophytic fungus *Diaporthe terebinthifolli*, isolated from the rhizome of *Glycyrrhiza glabra*, a native of India, produces bioactive compounds such as diapolic acid AB, xylarolide, and fomolide. These compounds were assayed for *in vitro* cytotoxicity using the cancer cell lines MIA PaCa-2 (pancreatic cancer), A549 (human lung adenocarcinoma), HCT-116 (human colon cancer), and T47D (human breast cancer). The xylarolide inhibited the growth of T47D cells with IC50 of 7 μM, through induction of apoptosis, and, in association with doxorubicin, inhibited the formation of tumor cell colonies [172].

In 2012, pestalrone B, an oxysporone derivative from the endophytic fungus *Pestalotiopsis karstenii* isolated from stems of *Camellia sasanqua*, showed significant cytotoxicity against the HeLa, HepG2, and U-251 cancer cell lines with IC50 of 12.6, 31.7, and 5.4 μg/mL, respectively [173]. In 2013, the compound siccayne [2-(3-methyl-3-buten-1-ynyl) hydro], a quinone isolated from the endophytic fungus *Pestalotiopsis fici* was cytotoxic against the human cancer cell lines HeLa and HT29, with IC50 of 48.2 and 33.9 μM, respectively [204].

Mady et al. [205] identified the molecules meleagrin, roquefortine C, and dehydrohistidyltryptophenyl-diketopiperazine (DHTD) produced by the endophytic fungus *P. chrysogenum* isolated from the leaf of the olive tree *Olea europea*. Roquefortine C and DHTD showed moderate antiproliferative activity against human breast cancer cells of the lines MDA-MB-231, MDA-468, BT-474, SK BR-3, MCF7, and MCF7-dox, while the compound meleagrin showed high inhibition in the growth of these same cell lines. In addition, the three compounds showed minimal toxicity to the MCF-10A non-tumorigenic human mammary epithelial cells [205], indicating its specificity for tumor cells.

A study published by Momesso et al. [174] showed that the fungus *Chaetomium globosum*, isolated from the leaves of *Viguiera robusta*, produced the chaetoglobosins B, D, and E. Chaetoglobosins are typically cytotoxic molecules, and the chaetoglobosin B derivative showed 89.55% and 57.10% inhibition of the growth of the tumor cells, Jurkat (leukemia) and B16F10 (melanoma), at 0.1 mg/mL.

Li et al. [175] also studied the cytotoxicity of SMs from the fungus *C. globosum*, isolated from *G. biloba*, and chaetoglobosins A, F, Fex, and 20-dihydrochaetoglobosin A showed IC50 values ranging from 3.15 to 8.44 μM on HCT116 cells, compared to the control drug, etoposide (IC50= 2.13 μM).

In a study performed by Hawas et al. [176], the fungus identified as *Penicillium aculeatum* was isolated from the marine red alga *Laurencia obtusa* and produced the sulfonyl metabolites pensulfonoxy and pensulfonamide. Preliminaries assays showed that these compounds were more cytotoxic than the antitumor paclitaxel. Pensulfonamide was cytotoxic to MCF-7 cells and pensulfonoxy to HCT-116 cells with IC50 of 2.18 and 5.23 μM, respectively.

Penicillium polonicum TY12 is a terrestrial endophytic fungus from the roots of *Aconitum vilmorinianum* from which a new indole-type monoterpene, polonidine A, and the already described compounds cyclopenol, verrucosidin, fructigenine A, 3-*O*-methylviridicatin, and aurantiomides C were isolated [177]. The cytotoxicity of these compounds was tested on the cell lines MHCC97H (human hepatoma), BT549 (human breast cancer), H1299 (lung cancer), SW620 (colon cancer), T98G (human glioma), and on the A549 (human lung cancer cell). Polonidine A exhibited moderate cytotoxicity on MHCC97H, BT549, and SW620 cell lines with IC50 from 6.1 to 20 μg/mL, while fructigenine A was less effective with IC50 of 20 μg/mL on T98G cells.

From the culture of the fungus *A. terreus* P63, endophyte of the grass *Axonopus leptostachyus*, was isolated the prenylated indole alkaloid, giluterrin, with an unprecedented carbon skeleton, which showed an antiproliferative effect on prostate (PC-3) and kidney (786-0) cancer cell lines [178].

6.2. Antimicrobial Compounds

The research for new antimicrobial molecules derivating from fungi metabolism is promising mainly due to the diversity of microorganisms not yet studied, which can contribute to the development of more specific drugs, with lower cost and effectiveness, even against drug-resistant pathogens.

Microbial resistance is considered a global health problem, which compromises the effectiveness of antibiotics, making the treatment of common infections unfeasible. Resistance occurs when microorganisms undergo genetic mutation due to intense exposure to antimicrobial drugs, and these microorganisms are considered superbugs. During the mutation phenomenon, bacteria, as well as fungi and yeasts, are able to develop strategies to prevent the antimicrobial effect of antibiotics, and are able to grow and multiplicate, promoting the emergence of drug-resistant strains.

Despite the severity of microbial resistance, the development of new antibiotics in the last decades has been rare. Although the incentives from the traditional market were not enough in the last years, it can hardly be able to solve the innovation gap in this area, especially in the context of restricted use of these drugs and sustainable discovery and development of new antibiotics [206]. Table 2 presents several examples of antimicrobial fungal SMs described in the last years.

Table 2. Examples of antimicrobial secondary metabolites produced by fungi.

Metabolites	Fungi Species	References
Oxalicine C, penicierythritol A	*P. chrysogenum* XNM-12	[7]
Stemphyperylenol	*B. dothidea*	[135]
Alterperylenol, stemphyperylenol, 7-acetyl-1,3,6-trihydroxyanthracene-9,10-dione, (11*S*)-1,3,6-trihydroxy-7-(1-hydroxyethyl)anthracene-9,10-dione, and 7-acetyl-1,3,6-trihydroxyanthracene-9,10-dione	*Alternaria* sp.	[136]
Pensulfonoxy, pensulfonamide	*P. aculeatum*	[176]
Cyclopenol, verrucosidin and fructigenine A, polonide A	*P. polonicum* TY12	[177]
Arvoredol	*Penicillium* sp. F37	[179]
Fusarubin	*Cladosporium* sp.	[180]
Engyodontiumones H, aspergillusone B, AGI-B4	*E. album* DFFSCS021	[183]

Table 2. *Cont.*

Metabolites	Fungi Species	References
1,3-dihydroxy-5-(12-hydroxyheptadecyl)benzene, 1,3-dihydroxy-5-(12 sulfoxyheptadecyl)benzene, and penixillarins B and C	Coculture of *P. crustosum* PRB-2 and *Xylaria* sp. HDN13-249A	[207]
10-formyl andrastone A, 10-dymethylated andrastone A, 15-deacetylated citreohybridone E, citreohybridonol, andrastins A and B	*P. vulpinum*	[208]
Ergosterol peroxide, monocerin, annularin I and J, ravenelin	*E. rostratum*	[209,210]
Emericelactone A-D, emericellins A and B	*Emericella* sp. XL 029	[211,212]
Irpexlacte A, furan derivatives irpexlacte B (5-(2α-hydroxypentyl) furan-2-carbaldehyde), irpexlacte C (5-(1α-hydroxypentyl) furan-2-carbaldehyde), irpexlacte D (5-(5-(2-hydroxypropanoyl) furan-2-yl) pentan-2-one)	*I. lacteus* DR10-1	[213]
Fusariumins C and D	*F. oxysporum* ZZP-R1	[214]
Fusarithioamide B	*F. chlamydosporium*	[215]
Anhydrofusarubin, fusarubin, 3-deoxyfusarubin, bostrycoidin, ergosterol, 3,5,9-trihydroxyergosta-7,22-diene-6-one, 4-hydroxybenzaldehyde	*F. solani*	[216]
Beauvericin	*F. oxysporum* EPH2R$_{AA}$ and CEC1S *F. oxysporum* 5-19	[148,217]
4-hydroxycinnamic acid derivative, methyl 2-{(E)-2-[4-(formyloxy)phenyl]ethenyl}-4-methyl-3-oxopentanoate	*Pyronema* sp. A2-1 *Pyronema* sp. D1-2	[218]
p-Coumaric acid	*A. alternata*	[219]

Since the discovery of penicillin **(1)**, Penicillium species have been known and widely studied as important sources of new antimicrobial compounds. Among them, the *Penicillium* sp. F37 isolated from the marine sponge *Axinella corrugata* and closely related to *Penicillium maximae* (Figure 6) produced a novel chlorinated polyketide with 6,7-dihydro-4(5H)-benzofuranone moiety, which was characterized by spectroscopic methods as arvoredol [179]. This polyketide prevented biofilm formation by *Staphylococcus epidermidis*, and was also active in inhibiting the colorectal carcinoma HCT116 cells proliferation at 7.9 µg/mL.

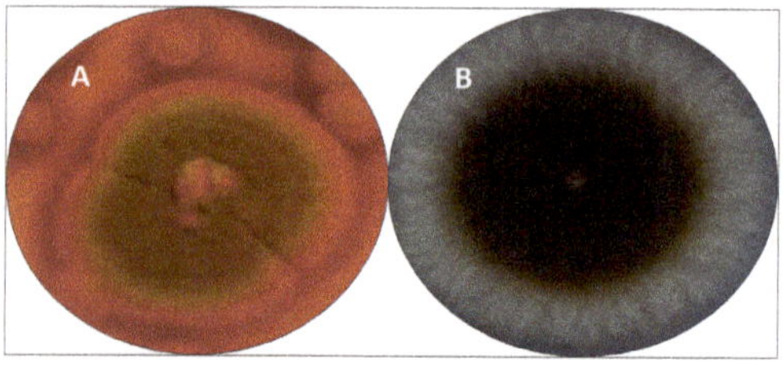

Figure 6. Illustration of the fungi species (**A**) *P. maximae* and (**B**) *E. rostratum* isolated by our group from the Brazilian biome, Caatinga.

The fungus identified as *P. aculeatum*, isolated from the marine red alga *L. obtuse*, is able to produce the sulfonyl metabolites pensulfonoxy and pensulfonamide. The crude ethyl acetate extract obtained from the culture product of this fungus inhibited the growth of *E. coli*, while the purified compound pensulfonamide inhibited the growth of *C. albicans* [176].

In another study with the fungus *P. chrysogenum* XNM-12 isolated from the seaweed *Leathesia nana*, the compounds, oxalicin C, penicierythritols A and B, and the monoterpenes decaturins B, C, D, and F were isolated [7]. The antimicrobial activity of the compounds was tested against *E. coli*, *Micrococcus luteus*, *P. aeruginosa* and on the phytopathogens *Ralstonia solanacearum*, *Valsa mali*, *A. alternata*, *Botrytis cinerea*, *F. oxysporum*, and *Penicillium digitatum*. Oxalicine C and penicierythritol A showed a moderate antibacterial effect on the bacterium *R. solanacearum* with MICs values of 8 and 4 μg/mL, respectively. Penicierythritol A possesses moderate antifungal properties against *A. alternata*, showing an MIC value of 8 μg/mL [7].

Sophora tonkinensis is a plant commonly applied in traditional Chinese medicine for the treatment of various diseases. From the roots of this plant, Qin et al. [208] isolated the fungus *Penicillium vulpinum*, which produced seven SMs derived from andrastin. These SMs were identified as 10-formyl andrastone A, 10-dymethylated andrastone A, 15-deacetylated citreohybridone E, citreohybridonol, and as andrastins A, B, and G. Except for andrastin G, the antibacterial activity was evaluated for all the others compounds that showed moderated effect with MIC ranging from 6.25 to 25 μg/mL on *Bacillus megaterium*, *Bacillus paratyphosus*, *Enterobacter aerogenes*, *E. coli*, and *S. aureus* [208].

Among the SMs obtained from the terrestrial fungus *P. polonicum* TY12, the compounds cyclopenol, verrucosidin, and fructigenine A showed a weak antibacterial effect with MIC of 643 μg/mL on *S. aureus* [177]. In spite of that, polonide A was the most effective, with MIC of 4 and 32 μg/mL on *S. aureus* and *B. subtilis*, respectively.

From the Brazilian biodiversity, the fungus *Exserohilum rostratum* (Figure 6) was isolated from *Bauhinia guianensis* [209] and *Phanera splendens* [210], which are typical Amazonian plants used in popular medicine to combat infections. Among the SMs produced by this species are the polyketide monocerin, ergosterol peroxide, annularins, and ravenelin, which presented moderated antimicrobial activity on *B. subtilis*, *S. aureus*, *E. coli*, *P. aeruginosa*, and *B. subtilis* and *Salmonella typhimurium* [209,210].

In 2018, from the endophytic fungus *Emericella* sp. XL 029, associated with the leaves of *Panax notoginseng*, four new polyketides (emericelactone A-D) were isolated and showed moderate antimicrobial activity against the phytopathogenic fungi *Verticillium dahliae Kleb*, *Rhizoctonia solani* and *Gibberella saubinetii*, and on the human pathogenic bacteria *Micrococcus lysodeikticus* and *Salmonella typhi* [211]. Emericelactone A, an optically active brown oil, contained on its structure an unprecedented linear pentaene substructure ending in an oxabicyclo[2.2.1]heptane moiety, while emericelactone B-D was characterized as epimers with an unprecedented linear triene structure with two cyclic moieties of an oxabicyclo[2.2.1]heptane and a cyclopentan-1-one. This study was the first one reporting the presence of emericelactones as SMs in Emericella species, suggesting that these polyketides might be acting as phytoalexins on the *P. notoginseng*, protecting it against pathogen attack. In the same year, another study with this same endophyte, demonstrated the isolation of two novel sesquiterpenoids, emericellins A and B, and both compounds showed moderate antimicrobial activity against *B. subtilis*, *Bacillus cereus*, and *E. coli* and on the fungi *V. dahliae* Kleb, *Helminthosporium maydis*, and *B. dothidea* [212].

The above-described compounds produced by *Alternaria* sp., alterperylenol, stemphyperylenol, 7-acetyl-1,3,6-trihydroxyanthracene-9,10-dione, (11*S*)-1,3,6-trihydroxy-7-(1-hydroxyethyl)anthracene-9,10-dione, and 7-acetyl-1,3,6-trihydroxyanthracene-9,10-dione showed an antimicrobial effect on phytopathogens [136]. Alterperylenol showed prominent antibacterial activity through membrane hyperpolarization without evidence of destruction of cell membrane integrity against *Clavibacter michiganensis* with MIC of 1.95 μg/mL, while the positive control streptomycin showed MIC of 3.90 μg/mL. 7-acetyl-1,3,6-trihydroxyanthracene-9,10-dione showed potent antibacterial activity toward the phytopathogenic bacteria *Pseudomonas syringae* pv. *lachrymans*, *Acidovorax avenae*, and *Erwinia carotovora* with MIC values of 3.91, 3.91, and 7.81 μg/mL, respectively, whereas (11*S*)-1,3,6-trihydroxy-7-(1-hydroxyethyl)anthracene-9,10-dione showed weak antibacterial effect against the same bacterial strains with MIC in the range of 7.81 to 15.6 μg/mL.

Alterperylenol showed antifungal activity with MIC values of 7.81 and 125 μg/mL against *Pestallozzia theae* and *Alternaria brassicicola*, respectively. 7-acetyl-1,3,6-

trihydroxyanthracene-9,10-dione and (11*S*)-1,3,6-trihydroxy-7-(1-hydroxyethyl)anthracene-9,10-dione displayed moderate antifungal activity on *P. theae* with MIC value of 31.3 μg/mL (carbendazim MIC: 7.81 μg/mL) and stemphyperylenol displayed potent antifungal activity against *P. theae* and *A. brassicicola* with MIC values equal to those of carbendazim (carbendazim MIC: 7.81 μg/mL) [136].

Produced by the endophytic *B. dothidea* of *M. azedarach*, stemphyperylenol inhibited the phytopathogens *Alternaria solani* with MIC of 1.57 μM, being comparable to the commonly used fungicide carbendazim, and pycnophorin significantly inhibited the growth of *B. subtilis* and *S. aureus* with MIC of 25 μM [135].

The SMs engyodontiumones H, aspergillusone B, and AGI-B4 from *E. album* DFF-SCS021 showed mild antibacterial activity against *E. coli* and *B. subtilis*, and aspergillusone B presented potent antilarval activity against barnacle *Balanus amphitrite* larval [183].

Four bioactive compounds were isolated from the fungus *Irpex lacteus* DR10-1; an endophytic from the water-tolerant plant *Distylium chinense*. The compounds were identified as the tremulane sesquiterpene irpexlacte A, and the furan derivatives irpexlacte B (5-(2α-hydroxypentyl) furan-2-carbaldehyde), irpexlacte C (5-(1α-hydroxypentyl) furan-2-carbaldehyde), and irpexlacte D (5-(5-(2-hydroxypropanoyl) furan-2-yl) pentan-2-one). These metabolites exhibited moderate antimicrobial activity against *P. aeruginosa* with MIC of 23.8 to 35.4 μM [213]. It is interesting that irpexlacte A and B presented remarkable antioxidant activity, and this effect may be related to the survival of the plant *D. chinense* under waterlogging stress, due to the release of ROS.

Although the Fusarium genus is known to harbor pathogenic species for plants, animals, and humans, several compounds isolated from species belonging to this genus are pointed out as antimicrobial agents [149,203,214–217,220,221]. In this regard, the fungus *F. oxysporum* CFE177, an endophyte from the Brazilian medicinal plant *C. leprosum*, produced SMs not yet identified, which showed an antifungal effect on important clinical species of *C. albicans*, *Candida krusei*, *Candida glabrata*, *Candida guillermondi*, *C. parapsilosis*, *C. tropicalis*, *Cryptococcus neoformans*, and on the filamentous fungi *Trichophyton rubrum* [203]. From this same plant, the endophyte *A. oryzae* CFE108 also produced antimicrobial SMs [203].

The SM fusarithioamide B was biosynthesized by the endophytic fungus *Fusarium chlamydosporium* isolated from the leaves of the plant *Anvillea garcinii* (Burm.f.) DC. On the bacteria *S. aureus*, *B. cereus*, and *E. coli*, the fusarithioamide B was more effective than the antimicrobial agent ciprofloxacin showing MIC from 2.5 to 3.7 μg/mL [215].

The cyclodepsipeptide, beauvericin isolated from *F. oxysporum* 5-19 endophytic on *Edgeworthia chrysantha* Linn showed an antimicrobial effect (MIC = 3.91 μM) on *S. aureus* similar to that of amoxicillin [217]. Furthermore, when produced by *F. oxysporum* EPH2R$_{AA}$ and CEC1S, beauvericin exhibited antibacterial activity towards methicillin-resistant *S. aureus* (MIC = 3.125 μg/mL) and *Bacillus subtilis* (MIC = 3.125 μg/mL) [148].

The chemical investigation of SMs produced by *Fusarium solani* endophyte from *Casia alata* resulted in the isolation of the three naphthoquinones (anhydrofusarubin, fusarubin, and 3-deoxyfusarubin), one aza-anthraquinone (bostrycoidin), two sterols (ergosterol and 3,5,9-trihydroxyergosta-7,22-diene-6-one) and 4-hydroxybenzaldehyde. Beyond cytotoxicity, the napthoquinones and aza-anthraquinones exhibited antimicrobial and antioxidant effects; however, napthoquinone, aza-anthraquinone (bostrycoidin) was more bioactive [216]. Fusarubin was also found in *Cladosporium* sp. described as endophytes from the leaves of *Rauwolfia serpentina* [180]. Lately, the cytotoxicity of fusarubin and anhydrofusarubin was demonstrated through the inhibition of cell proliferation and an increase in apoptosis mediated by p21/p53- pathway [180].

The endophyte *F. oxysporum* ZZP-R1, from the coastal plant *Rumex madaio* Makino, used in traditional Chinese medicine, produced the two terpenes, fusariumins C and D, which showed an antibacterial effect on *S. aureus* with MIC of 6.25 μM and 25.0 μM [162]; however, both compounds were not effective on *E. coli* and *C. albicans*.

The 4-hydroxycinnamic acid derivative, methyl 2-{E-2-[4-(formyloxy)phenyl]ethenyl}-4-methyl-3-oxopentanoate, was isolated from the endophyte *Pyronema* sp. (A2-1 and D1-2) and expressed moderate antimycobacterial effect on *Mycobacterium marinum* [218].

The antimicrobial potential of the metabolites penixillarins A-C, and the biogenetically related 1,3-dihydroxy-5-(12-hydroxyheptadecyl)benzene and 1,3-dihydroxy-5-(12-sulfoxyheptadecyl)benzene obtained from the coculture of *P. crustosum* PRB-2 and *Xylaria* sp. HDN13-249A was evaluated on *Mycobacterium phlei, B. subtilis, Vibrio parahaemolyticus, E. coli, P. aeruginosa, C. albicans,* and *P. vulgaris* [207]. Among them, only the compounds 1,3-dihydroxy-5-(12-hydroxyheptadecyl)benzene, 1,3-dihydroxy-5-(12 sulfoxyheptadecyl)benzene, and penixillarins B and C were effective on *M. phlei, B. subtilis,* and *V. parahaemolyticus* with MIC in the range from 6.25 to 100 μM [207].

The ethyl acetate (EA) extract and the *p*-coumaric acid obtained from the endophytic fungus *A. alternata* isolated from *C. roseus* leaves showed concentration-dependent broad-spectrum antimicrobial activity against *E. coli, Klebsiella pneumoniae, P. vulgaris, P. aeruginosa, S. typhimurium, S. aureus,* and *Streptococcus faecalis,* and on the yeast *C. albicans* and *C. neoformans* [222]. These EA extracts and the *p*-coumaric acid were also effective as antifungal on the phytopathogens *A. brassicicola,* Alternaria geophila, A. flavus (aflatoxigenic strain), *A. fumigatus,* Aspergillus ochraceus, A. tamarii, A. terreus, Curvularia tetramera, F. oxysporum, Fusarium lateritium, F. equiseti, Fusarium udum, Fusarium verticillioides (fumonisinogenic strain), *P. citrinum,* and *Penicillium expansum,* showing MIC from 7.8 to 250 μg/mL. The authors suggested that the EA extract and *p*-coumaric acid could be alternatives to be explored as protection for preventing microbial deterioration and mycotoxins accumulation in food and feedstuffs storage.

In 2016, Shin et al. [219] performed a screening for the discovery and characterization of antimicrobial compounds from 342 entomopathogenic fungi strains in Korea, and showed that the obtained SMs completely suppressed the growth of *B. cereus* and *B. cinerea.* Although the compounds were not purified and physically chemically characterized, the results demonstrated that entomopathogenic fungi could be a promising source of antimicrobial compounds.

6.3. Cytotoxic and Antimicrobial Compounds Produced by Extremophile Fungi

Microorganisms that thrive under extremely adverse environmental conditions are classified as extremophiles [223], and possess mechanisms developed to adapt to these conditions through the regulation of the expression of specific genes in their genome. The extremophiles' classification takes into account the environmental condition of the habitat in which they can grow and survive. The parameters for the classification are usually based on extremes of temperature (thermophiles, hyperthermophiles, psychrophiles), pH (acidophiles, alkaliphiles), pressure (piezophiles), salinity (halophiles), nutrient availability (oligotrophic), desiccation (xerophilesdry), metal concentration (metallophiles), radiation (radiophiles), and oxygen (microaerophiles). Moreover, there are polyextremophiles adapted to more than one extreme physicochemical factor.

The survival strategy of extremophiles can result in the biosynthesis of enzymes and SMs that can have applications in distinct areas such as environmental, industrial, and medical [224,225]. In light of that, extremophiles are considered an excellent source of new bioactive molecules [226]; however, currently, few biomolecules obtained from extremophiles have had successful biotechnological industrial applications. Besides that, due to several different reasons, extremophile microorganisms are not deeply studied concerning the chemistry of natural compounds and their potential as bioactive agents.

Literature data have shown that the majority of fungi species are able to tolerate extreme conditions of temperature, keeping the production of different thermostable enzymes, or SMs, useful in several distinct areas. Previously, the main enzymes and SMs produced by extremophile fungi were discussed by our group [227], including aspects related to mechanisms involved in their biosynthesis. Regarding this issue, Baranova et al. [228] summarized antibiotics described from 2018–2019, which resulted from the biosynthesis of extremophilic micromycetes, and classified these SMs according to the structure and biosynthetic origin.

In this review, Table 3 shows examples of cytotoxic and antimicrobial SMs produced by extremophilic fungi species. As an example, the species *Emericellopsis alkalina* is an extremophilic fungus found in waters with high levels of salinity and alkalinity, and these environmental conditions can contribute to the production of a great diversity of biological compounds, such as a peptaibol called emericellipsin A, which, unlike other peptaibols, has a smaller structure with nine residues [229]. Emericellipsin A showed cytotoxicity on the cell line HepG2 and HeLa *in vitro*, with IC50 of 2.8 and <0.5 μM, respectively. In addition, this compound had less toxicity on fibroblast cells than the antitumor drug doxorubicin, showing its specificity for tumor cells.

In addition to previous studies [209,210], the fungus *E. rostratum* was isolated by our group from the endemic plant *Croton blanquetianus* from the Caatinga, a Brazilian biome with 844,000 km^2, along with a semi-arid climate considered extreme, and its SMs, monocerin, and annularin I, showed moderated antimicrobial activity and absence of toxicity on human vein endothelial cells (HUVECs), inducting cell proliferation [unpublished data]. This species has intense melanin-based pigmentation, being resistant to stresses induced by UV and ionizing radiation, desiccation and heat, toxic metals, and organic pollutants [230].

A total of 52 fungal strains were isolated from Lake Magadi, in the East African Rift valley and phylogenetically classified as belonging to 18 different genera, represented mainly by Aspergillus, Penicillium, Cladosporium, Phoma, and Acremonium [231]. According to the conditions of growth (elevated temperature, pH, and salts), the isolates were of haloalkaliphilic nature, and different strains produced enzymes such as chitinases, proteases, amylases, cellulases, lipases, and pectinases. The isolate coded as 11M was identified as *P. chrysogenum* and produced antimicrobial SMs present in cell-free extracts and crude extracts which inhibited the growth of *B. subtilis, E. coli, P. aeruginosa, Salmonella* spp., *Shigella* spp., *C. albicans* and of the phytopathogens *Schizophyllum commune, Epicoccum sorghinum, A. fumigatus, Cladosporium halotolerans, Phoma destructive*, and *Didymella glomerata* [231]. The study was important to show that diversified culture conditions are an effective strategy to recover more phylotypes from extremes environment.

Three glacial ice fragments from the Antarctica Peninsula, of approximately 20 kg each, were fragmented in small pieces, melted and, after filtration of 3 L in a membrane of 0.45 μM, the retained material was inoculated in Sabouraud agar and minimal medium containing chloramphenicol and incubated at 10 °C for 60 days [232]. Following the process, 66 strains were isolated and classified into 27 taxa of 14 genera. The dominant species were *Penicillium palitans, Penicillium* sp., *Thelebolus balaustiformis, Glaciozyma antarctica, Rhodotorula mucilaginosa*, and *Rhodotorula dairenensis*. Crude extracts obtained from the culture of *P. chrysogenum, P. palitans*, and *Penicillium* spp. displayed leishmanicidal, trypanocidal, and herbicidal activities, with *P. chrysogenum* extract presenting the broadest and highest biological effect.

The strain of *P. chrysogenum* CCTCC M 2020019 was isolated from the feces of Adelie penguins in the Antarctic and its SMs, xanthocillins X and Y1, showed intense cytotoxicity against SF-268, MCF-7, HepG2 and A549 cancer cell lines with IC50 values ranging from 0.26 to 5.04 μM [233]. Xanthocillin X inhibited the growth of the Gram-negative pathogens *Acinetobacter baumannii, K. pneumoniae*, and *P. aeruginosa* growth with MIC values of 0.125 μg/mL [233]. Gubiani et al. [181] described that the soil-derived fungus *Aspergillus* sp SDC28, isolated from a Brazilian cave, produced the anthraquinones versicolorin C and versiconol, which were cytotoxic on human ovarian cancer cells (OVCAR3) with IC50 of 0.24 and of 1.07 μM, respectively, while the IC50 for taxol was 7 ± 2 nM. Beyond the cytotoxic effect, the study was the first one to report the spectroscopic characterization, including the absolute configuration of both anthraquinones, attesting to the relevance in drug discovery using extremophile fungi.

Table 3. Examples of cytotoxic and antimicrobial secondary metabolites produced by extremophilic fungi species.

Metabolites	Fungi Species	References
Xanthocillins X (cytotoxicity and antimicrobial) and Y1 (antimicrobial)	*P. chrysogenum* CCTCC M 2020019	[181]
Monocerin and annularin I (antimicrobial)	*E. rostratum*	[209,210]
Emericellipsin A (cytotoxicity)	*E. alkalina*	[229]
Versicolorin C and versiconol (cytotoxicity)	*Aspergillus* sp. SDC28	[234]

As a strategy to increase the production of bioactive compounds by different species of microorganisms, coculture was used for the culture of the deep-seawater derived fungus from Antarctica, *Penicillium crustosum* PRB-2, and of the mangrove fungus *Xylaria* sp. HDN13-249A [207]. Six SMs classified as polyketides were obtained and named penixillarins A-D, 1,3-dihydroxy-5-(12-hydroxyheptadecyl)benzene, and 1,3-dihydroxy-5-(12-sulfoxyheptadecyl) benzene. Through UPLC-MS analysis, it was shown that penixillarins A and B were produced by the interaction of both fungi species. The compounds penixillarin C and 1,3-dihydroxy-5-(12-sulfoxyheptadecyl)benzene were produced only by *Xylaria* sp. HDN13-249; however, in a much higher concentration under the cocultivation conditions. The cytotoxicity of all six metabolites was evaluated on 12 cell lines (HL-60, K562, BEL-7402, HCT-116, A549, HeLa, L-02, MGC-803, HO8910, SH-SY5Y, PC-3, and MDA-MB-231); nevertheless, no compound was effective until the concentration of 30 μM [207].

7. General Considerations

According to Wang et al. [234], from 1998 to 2019, 272 antimicrobial compounds were obtained from marine fungal strains isolated from sediments, sponges, algae, mangroves, and corals. China was the country responsible for the discovery of the majority of these compounds, which were obtained mainly from *Aspergillus* sp. followed by *Penicillium* sp. In spite of the fact that 207 compounds showed antibacterial activity and 68 were antifungal agents, none of them are in clinical trials.

In a wider period, between 1981 and 2019, among 164 new antimicrobial drugs approved, 36 were of biological origin, of which 89% were prophylactic agents. In the same period, although fungal diseases were frequently reported, only two antifungal drugs were approved, both of synthetic origin [1].

In addition to the indiscriminate use of antibiotics by the population around the world, microorganisms' resistance to antimicrobials is also related to the high demand for animal protein for human consumption due to the regular use of antimicrobials in the modern production of this food. According to Van Boeckel et al. [235], the global average consumption of antimicrobials is from 45 to 172 mg/kg of animal produced per year, considering cattle, chicken, and pigs. In this sector, globally, the use of antimicrobials tends to increase by 67% in developed countries between 2010 and 2030, while in countries with low *Human Development Index* (HDI), such as Brazil, Russia, India, China, and South Africa, this rate may increase, reaching 99% [235].

Historically, it is clear that drug discovery is still not enough to attend to the high demand for new antimicrobial and antitumor agents, evidencing the relevance of intensifying studies in order to have a sustainable use of the huge world biodiversity, as well as of microorganisms.

Although the production of SMs by fungi is not always economically interesting due to the low yield from the fermentation process, potential species can be selected and genetically engineered to enhance the production of specific bioactive compounds through sustainable processes. Furthermore, chemical processes can be developed by medicinal chemistry for the production of such compounds with high yield, or their structure can be modified to reach specific antitumor effects.

8. Conclusions

The studies involving fungal SMs have been increased due to their relevance to biotechnology, medicine, and the pharmaceutical industry. Several SMs possess different pharmacological properties, and in an effort to discover anticancer and antimicrobial compounds and to attend to the high demand, research groups from all around the world are intensely exploring this possibility. In this review, various molecules from different chemical classes produced by fungi metabolism were presented, as well as their antimicrobial or cytotoxic effect. The SMs can be useful to improve the current treatments of infections and cancer, and their effects can go far beyond than described in this review. Some reported SMs are from endophytes isolated mainly from medicinal plants, which, in addition to their inherent effect, are also an interesting source of microorganism species and, consequently, from new molecules. Several SMs were assayed only *in vitro*, and their effects were not promising to encourage new assays; however, even these data are important for future screening and studies.

The relevance of fungi as a source of new molecules for the drug discovery process is evident; however, it is important to explore the compounds already described for obtaining new derivatives or analogs that can be more effective. As was discussed, fungi species producers of bioactive and promising SMs can also be engineered through molecular-based approaches and tools to increase their production, reduce adverse effects or improve their pharmacological properties. The main purpose is obtaining specific treatments with reduced or absence of severe side effects and low cost. Considering the wide and rich biodiversity around the entire world and all the available technology and knowledge accumulated over the years, this is a viable approach able to improve the quality of human life.

Funding: This research received no external funding.

Acknowledgments: This work was supported by grant #2014-19152-8, #2017-26524-7, #2020/04799-7 from São Paulo Research Foundation (FAPESP) and Fundação Butantan.

Conflicts of Interest: The authors declare no conflict of interest.

References

1. Newman, D.J.; Cragg, G.M. Natural products as sources of new drugs over the nearly four decades from 01/1981 to 09/2019. *J. Nat. Prod.* **2020**, *83*, 770–803. [CrossRef] [PubMed]
2. Newman, D.J.; Cragg, G.M. Plant endophytes and epiphytes: Burgeoning sources of known and "unknown" cytotoxic and antibiotic agents? *Planta Med.* **2020**, *86*, 891–905. [CrossRef] [PubMed]
3. Přichystal, J.; Schug, K.A.; Lemr, K.; Novák, J.; Havlíček, V. Structural analysis of natural products. *Anal. Chem.* **2016**, *88*, 10338–10346. [CrossRef] [PubMed]
4. Gakuubi, M.M.; Munusamy, M.; Liang, Z.-X.; Ng, S.B. Fungal endophytes: A promising frontier for discovery of novel bioactive compounds. *J. Fungi* **2021**, *7*, 786. [CrossRef] [PubMed]
5. Hridoy, M.; Gorapi, M.Z.H.; Noor, S.; Chowdhury, N.S.; Rahman, M.M.; Muscari, I.; Masia, F.; Adorisio, S.; Delfino, D.V.; Mazid, M.A. Putative anticancer compounds from plant-derived endophytic fungi: A review. *Molecules* **2022**, *27*, 296. [CrossRef] [PubMed]
6. Ancheeva, E.; Daletos, G.; Proksch, P. Bioactive secondary metabolites from endophytic fungi. *Curr. Med. Chem.* **2020**, *27*, 1836–1854. [CrossRef]
7. Xu, K.; Wei, X.L.; Xue, L.; Zhang, Z.F.; Zhang, P. Antimicrobial meroterpenoids and erythritol derivatives isolated from the marine-algal-derived endophytic fungus *Penicillium chrysogenum* XNM-12. *Mar. Drugs* **2020**, *18*, 578. [CrossRef]
8. Meena, H.; Hnamte, S.; Siddhardha, B. Secondary metabolites from endophytic fungi: Chemical diversity and application. In *Advances in Endophytic Fungal Research*; Singh, B., Ed.; Fungal Biology; Springer: Cham, Switzerland, 2019; Volume 1, pp. 145–169. [CrossRef]
9. Soares, D.A.; Rosa, L.H.; Silva, J.M.F.; Pimenta, R.S. A review of bioactive compounds produced by endophytic fungi associated with medicinal plants. *Bol. Mus. Para. Emílio Goeldi Cienc. Nat.* **2017**, *12*, 331–352. [CrossRef]

0. Nisa, H.; Kamili, A.N.; Nawchoo, I.A.; Shafi, S.; Shameem, N.; Bandh, S.A. Fungal endophytes as prolific source of phytochemicals and other bioactive natural products: A review. *Microb. Pathog.* **2015**, *82*, 50–59. [CrossRef]
1. Kim, C.K.; Eo, J.K.; Eom, A.H. Diversity and seasonal variation of endophytic fungi isolated from three conifers in Mt. Taehwa, Korea. *Mycobiology* **2013**, *41*, 82–85. [CrossRef]
2. Swamy, M.K.; Tuyelee, D.; Nandy, S.; Mukherjee, A.; Pandey, D.K.; Dey, A. Endophytes for the production of anticancer drug, paclitaxel. In *Paclitaxel: Sources, Chemistry, Anticancer Actions, and Current Biotechnology*; Swamy, M.K., Pullaiah, T., Chen, Z.-S., Eds.; Academic Press: Cambridge, MA, USA, 2022; pp. 203–228. [CrossRef]
3. Brakhage, A.A.; Schroeckh, V. Fungal secondary metabolites—Strategies to activate silent gene clusters. *Fungal Genet. Biol.* **2010**, *48*, 15–22. [CrossRef]
4. Pfannenstiel, R.T.; Keller, N.P. On top of biosynthetic gene clusters: How epigenetic machinery influences secondary metabolism in fungi. *Biotechnol. Adv.* **2019**, *37*, 107345. [CrossRef]
5. Bills, G.F.; Gloer, J.B. Biologically active secondary metabolites from the fungi. *Microbiol. Spect.* **2016**, *4*, FUNK-0009-2016. [CrossRef]
6. Bérdy, J. Thoughts and facts about antibiotics: Where we are now and where we are heading. *J. Antibiot.* **2012**, *65*, 385–395. [CrossRef]
7. Nett, M.; Ikeda, H.; Moore, B.S. Genomic basis for natural product biosynthetic diversity in the actinomycetes. *Nat. Prod. Rep.* **2009**, *26*, 1362–1384. [CrossRef]
8. Srivastava, N.; Srivastava, M.; Ramteke, P.W.; Mishra, P.K. Solid-state fermentation strategy for microbial metabolites production: An overview. In *New and Future Developments in Microbial Biotechnology and Bioengineering*; Gupta, V.J., Pandey, A., Eds.; Elsevier: Amsterdam, The Netherlands, 2019; Volume 1, pp. 345–354. [CrossRef]
9. Mulder, K.C.; Mulinari, F.; Franco, O.L.; Soares, M.S.; Magalhães, B.S.; Parachin, N.S. Lovastatin production: From molecular basis to industrial process optimization. *Biotechnol. Adv.* **2015**, *1*, 648–665. [CrossRef]
0. Zahari, N.A.A.R.; Chong, G.H.; Abdullah, L.C.; Chua, B.L. Ultrasonic-assisted extraction (UAE) process on thymol concentration from *Plectranthus Amboinicus* leaves: Kinetic modeling and optimization. *Processes* **2020**, *8*, 322. [CrossRef]
1. Mai, P.-Y.; Le Goff, G.; Poupon, E.; Lopes, P.; Moppert, X.; Costa, B.; Beniddir, M.A.; Ouazzani, J. Solid-Phase extraction embedded dialysis (speed), an innovative procedure for the investigation of microbial specialized metabolites. *Mar. Drugs* **2021**, *19*, 371. [CrossRef]
2. Wolfender, J.-L.; Litaudon, M.; Touboul, D.; Queiroz, E.F. Innovative omics-based approaches for prioritisation and targeted isolation of natural products—new strategies for drug discovery. *Nat. Prod. Rep.* **2019**, *36*, 855–868. [CrossRef]
3. Fox Ramos, A.E.; Evanno, L.; Poupon, E.; Champy, P.; Beniddir, M.A. Natural products targeting strategies involving molecular networking: Different manners, one goal. *Nat. Prod. Rep.* **2019**, *36*, 960–980. [CrossRef]
4. Grkovic, T.; Pouwer, R.H.; Vial, M.-L.; Gambini, L.; Noel, A.; Hooper, J.N.A.; Wood, S.A.; Mellick, G.D.; Quinn, R.J. NMR fingerprints of the drug-like natural-product space identify Iotrochotazine A: A chemical probe to study parkinson's disease. *Angew. Chem. Int. Ed.* **2014**, *53*, 6070–6074. [CrossRef] [PubMed]
5. Lang, G.; Mayhudin, N.A.; Mitova, M.I.; Sun, L.; Van der Sar, S.; Blunt, J.W.; Cole, A.L.J.; Ellis, G.; Laatsch, H.; Munro, M.H.G. Evolving trends in the dereplication of natural product extracts: New methodology for rapid, small-scale investigation of natural product extracts. *J. Nat. Prod.* **2008**, *71*, 1595–1599. [CrossRef] [PubMed]
6. Zhang, C.; Idelbayev, Y.; Roberts, N.; Tao, Y.; Nannapaneni, Y.; Duggan, B.M.; Min, J.; Lin, E.C.; Gerwick, E.C.; Cottrell, G.W.; et al. Small molecule accurate recognition technology (SMART) to enhance natural products research. *Sci. Rep.* **2017**, *7*, 14243. [CrossRef] [PubMed]
7. Wang, M.; Carver, J.J.; Phelan, V.V.; Sanchez, L.M.; Garg, N.; Peng, Y.; Nguyen, D.D.; Watrous, J.; Kapono, C.A.; Luzzatto-Knaan, T.; et al. Sharing and community curation of mass spectrometry data with Global Natural Products Social Molecular Networking. *Nat. Biotechnol.* **2016**, *34*, 828–837. [CrossRef] [PubMed]
8. Ho, C.S.; Lam, C.W.K.; Chan, M.H.M.; Cheung, R.C.K.; Law, L.K.; Lit, L.C.W.; Ng, K.F.; Suen, M.W.M.; Tai, H.L. Electrospray ionisation mass spectrometry: Principles and clinical applications. *Clin. Biochem. Rev.* **2003**, *24*, 3–12.
9. Grkovic, T.; Akee, R.K.; Thornburg, C.C.; Spencer, K.; Trinh, S.K.; Britt, J.R.; Harris, M.J.; Evans, J.R.; Kang, U.; Ensel, S.; et al. National Cancer Institute (NCI) Program for Natural Products Discovery: Rapid isolation and identification of biologically active natural products from the nci prefractionated library. *ACS Chem. Biol.* **2020**, *15*, 1104–1114. [CrossRef]
0. Tsao, R.; Yang, R.; Young, J.C.; Zhu, H. Polyphenolic profiles in eight apple cultivars using high-performance liquid chromatography (HPLC). *J. Agric. Food Chem.* **2003**, *51*, 6347–6353. [CrossRef]
1. Aguilar, M.I. Reversed-phase high-performance liquid chromatography. In *HPLC of Peptides and Proteins*; Methods in Molecular Biology; Springer: Totowa, NJ, USA, 2004; p. 251. [CrossRef]
2. Câmara, J.S.; Martins, C.; Pereira, J.A.M.; Perestrelo, R.; Rocha, S.M. Chromatographic-based platforms as new avenues for scientific progress and sustainability. *Molecules* **2022**, *27*, 5267. [CrossRef]
3. Coates, J. Interpretation of Infrared Spectra, A Practical Approach. In *Encyclopedia of Analytical Chemistry*; John Wiley & Sons, Ltd.: Chichester, UK, 2006; pp. 1–23. [CrossRef]
4. Nagarajan, K.; Ibrahim, B.; Ahmad Bawadikji, A.; Lim, J.-W.; Tong, W.-Y.; Leong, C.-R.; Khaw, K.Y.; Tan, W.-N. Recent developments in metabolomics studies of endophytic fungi. *J. Fungi* **2022**, *8*, 28. [CrossRef]

35. Krishnan, V.V.; Rupp, B. Macromolecular structure determination: Comparison of X-ray crystallography and NMR spectroscopy. In *eLS*; John Wiley & Sons: Hoboken, NJ, USA, 2012. [CrossRef]
36. Zloh, M. NMR spectroscopy in drug discovery and development: Evaluation of physico-chemical properties. *ADMET DMPK* **2019**, *7*, 242–251. [CrossRef]
37. Emwas, A.-H.; Roy, R.; McKay, R.T.; Tenori, L.; Saccenti, E.; Gowda, G.A.N.; Raftery, D.; Alahmari, F.; Jaremko, L.; Jaremko, M. NMR spectroscopy for metabolomics research. *Metabolites* **2019**, *9*, 123. [CrossRef]
38. Bartle, K.D.; Myers, P. History of gas chromatography. *TrAC—Trends Analyt. Chem.* **2002**, *21*, 547–557. [CrossRef]
39. Zapata, L.S.; Tabarez, M.R.; Alvarez, J.C.; Escobar, V.V. Reviewing microbial behaviors in ecosystems leading to a natural quorum quenching occurrence. Biological and applied sciences. *Braz. Arch. Biol. Technol.* **2017**, *60*, 1–12. [CrossRef]
40. Schulz, S.; Dickschat, J.S. Bacterial volatiles: The smell of small organisms. *Nat. Prod. Rep.* **2007**, *24*, 814–842. [CrossRef]
41. Singh, B.N.; Prateeksha, U.D.K.; Singh, B.R.; Defoirdt, T.; Gupta, V.K.; de Souza, A.O.; Singh, H.B.; Barreira, J.C.; Ferreira, I.C.; Vahabi, K. Bactericidal, quorum quenching and anti-biofilm nanofactories: A new niche for nanotechnologists. *Crit. Rev. Biotechnol.* **2017**, *37*, 525–540. [CrossRef]
42. Hogan, D.A. Talking to themselves: Autoregulation and quorum sensing in fungi. *Eukaryot. Cell* **2006**, *5*, 613–619. [CrossRef]
43. Frisvad, J.C.; Andersen, B.; Thrane, U. The use of secondary metabolite profiling in chemotaxonomy of filamentous fungi. *Mycol. Res.* **2008**, *112*, 231–240. [CrossRef]
44. Cottier, F.; Mühlschlegel, F.A. Communication in fungi. *Int. J. Microbiol.* **2011**, *2012*, 351832. [CrossRef]
45. Albuquerque, P.; Casadevall, A. Quorum sensing in fungi—A review. *Med. Mycol.* **2012**, *50*, 337–345. [CrossRef]
46. Polke, M.; Hube, B.; Jacobsen, I.D. *Candida* survival strategies. *Adv. Appl. Microbiol.* **2015**, *91*, 139–235. [CrossRef]
47. Nemcovic, M.; Jakubíková, L.; Víden, I.; Farkas, V. Induction of conidiation by endogenous volatile compounds in *Trichoderma* spp. *FEMS Microbiol. Lett.* **2008**, *284*, 231–236. [CrossRef] [PubMed]
48. Chitarra, G.S.; Abee, T.; Rombouts, F.M.; Dijksterhuis, J. 1-Octen-3-ol inhibits conidia germination of *Penicillium paneum* despite of mild effects on membrane permeability, respiration, intracellular pH, and changes the protein composition. *FEMS Microbiol. Ecol.* **2005**, *54*, 67–75. [CrossRef] [PubMed]
49. Elmassry, M.M.; Farag, M.A.; Preissner, R.; Gohlke, B.O.; Piechulla, B.; Lemfack, M.C. Sixty-one volatiles have phylogenetic signals across bacterial domain and fungal kingdom. *Front. Microbiol.* **2020**, *11*, 557253. [CrossRef] [PubMed]
50. Hornby, J.M.; Jensen, E.C.; Lisec, A.D.; Tasto, J.J.; Jahnke, B.; Shoemaker, R.; Dussault, P.; Nickerson, K.W. Quorum sensing in the dimorphic fungus *Candida albicans* is mediated by farnesol. *Appl. Environ. Microbiol.* **2001**, *67*, 2982–2992. [CrossRef] [PubMed]
51. Kruppa, M. Quorum sensing and *Candida albicans*. *Mycoses* **2009**, *52*, 1–10. [CrossRef]
52. Weber, K.; Schulz, B.; Ruhnke, M. The quorum-sensing molecule E,E-farnesol—its variable secretion and its impact on the growth and metabolism of *Candida species*. *Yeast* **2010**, *27*, 727–739. [CrossRef]
53. Sprague, G.F.; Winans, S.C. Eukaryotes learn how to count: Quorum sensing by yeast. *Genes Dev.* **2006**, *20*, 1045–1049. [CrossRef]
54. Bassler, B.L. How bacteria talk to each other: Regulation of gene expression by quorum sensing. *Curr. Opin. Microbiol.* **1999**, *2*, 582–587. [CrossRef]
55. Rasmussen, T.B.; Skindersoe, M.E.; Bjarnsholt, T.; Phipps, R.K.; Christensen, K.B.; Jensen, P.O.; Andersen, J.B.; Koch, B.; Larsen, T.O.; Hentzer, M.; et al. Identity and effects of quorum-sensing inhibitors produced by *Penicillium species*. *Microbiology* **2005**, *151*, 1325–1340. [CrossRef]
56. Padder, S.A.; Prasad, R.; Shah, A.H. Quorum sensing: A less known mode of communication among fungi. *Microbiol. Res.* **2018**, *210*, 51–58. [CrossRef]
57. Lemfack, M.C.; Nickel, J.; Dunkel, M.; Preissner, R.; Piechulla, B. mVOC: A database of microbial volatiles. *Nucleic Acids Res.* **2014**, *42*, 744–748. [CrossRef]
58. Lemfack, M.C.; Gohlke, B.-O.; Toguem, S.M.T.; Preissner, S.; Piechulla, B.; Preissner, R. mVOC 2.0: A database of microbial volatiles. *Nucleic Acids Res.* **2018**, *46*, 1261–1265. [CrossRef]
59. Microbial Volatile Organic Compound Database. Available online: https://bioinformatics.charite.de/mvoc/index.php?site=home (accessed on 6 November 2022).
60. Mishra, S.; Sahu, P.K.; Agarwal, V.; Singh, N. Exploiting endophytic microbes as micro-factories for plant secondary metabolite production. *Appl. Microbiol. Biotechnol.* **2021**, *105*, 6579–6596. [CrossRef]
61. Vinaixa, M.; Schymanski, E.L.; Neumann, S.; Navarro, M.; Salek, R.M.; Yanes, O. Mass spectral databases for LC/MS- and GC/MS-based metabolomics: State of the field and future prospects. *Trends Anal. Chem.* **2016**, *78*, 23–35. [CrossRef]
62. Harvey, A.L.; Edrada-Ebel, R.; Quinn, R.J. The reemergence of natural products for drug discovery in the genomics era. *Nat. Rev. Drug Discov.* **2015**, *14*, 111–129. [CrossRef]
63. Thornburg, C.C.; Britt, J.R.; Evans, J.R.; Akee, R.K.; Whitt, J.A.; Trinh, S.K.; Harris, M.J.; Thompson, J.R.; Ewing, T.L.; Shipley, S.M. NCI program for natural product discovery: A publicly-accessible library of natural product fractions for high-throughput screening. *ACS Chem. Biol.* **2018**, *13*, 2484–2497. [CrossRef]
64. Herrmann, K.M.; Weaver, L.M. The shikimate pathway. *Annu. Rev. Plant Physiol. Plant Mol. Biol.* **1999**, *50*, 473–503. [CrossRef]
65. Keller, N.P. Fungal secondary metabolism: Regulation, function and drug discovery. *Nat. Rev. Microbiol.* **2018**, *17*, 167–180. [CrossRef]
66. Nielsen, J.C.; Prigent, S.; Grijseels, S.; Workman, M.; Ji, B.; Nielsen, J. Comparative transcriptome analysis shows conserved metabolic regulation during production of secondary metabolites in filamentous fungi. *mSystems* **2019**, *4*, e00012-19. [CrossRef]

67. El Hajj Assaf, C.; Zetina-Serrano, C.; Tahtah, N.; Khoury, A.E.; Atoui, A.; Oswald, I.P.; Puel, O.; Lorber, S. Regulation of secondary metabolism in the *Penicillium* Genus. *Int. J. Mol. Sci.* **2020**, *21*, 9462. [CrossRef]
68. Deepika, V.B.; Murali, T.S.; Satyamoorthy, K. Modulation of genetic clusters for synthesis of bioactive molecules in fungal endophytes: A review. *Microbiol. Res.* **2016**, *182*, 125–140. [CrossRef] [PubMed]
69. Liou, G.F.; Khosla, C. Building-block selectivity of polyketide synthases. *Curr. Opin. Chem. Biol.* **2003**, *7*, 279–284. [CrossRef]
70. Weissman, K.J. Introduction to polyketide biosynthesis. *Methods Enzymol.* **2009**, *459*, 3–16. [CrossRef] [PubMed]
71. Qadri, M. Polyketides: Bioactive Secondary Metabolites, Biosynthesis, and Their Modulation. In *Endophyte Biology: Recent Findings from the Kashmir Himalayas*, 1st ed.; Zahoor, A.W., Masroor, Q., Palak, A., Khalid, R.H., Eds.; Apple Academic Press: New York, NY, USA, 2022; pp. 1–346. [CrossRef]
72. Cox, R.; Skellam, E.; Williams, K. Biosynthesis of fungal polyketides. In *The Mycota: Physiology and Genetics*, 2nd ed.; Anke, T., Schüffler, A., Eds.; Springer: Cham, Switzerland, 2018; Volume 15, pp. 385–412. [CrossRef]
73. Cox, R.J.; Simpson, T.J. Fungal type I polyketide synthases. *Methods Enzymol.* **2009**, *459*, 49–78. [CrossRef] [PubMed]
74. Fujii, I. Functional analysis of fungal polyketide biosynthesis genes. *J. Antibiot. Res.* **2010**, *63*, 207–218. [CrossRef]
75. Martínez-Núñez, M.A.; López, V.E.L.Y. Nonribosomal peptides synthetases and their applications in industry. *Sustain. Chem. Process.* **2016**, *13*, 1–8. [CrossRef]
76. Martín, J.F.; García-Estrada, C.; Zeilinger, S. *Biosynthesis and Molecular Genetics of Fungal Secondary Metabolites*, 1st ed.; Fungal Biology; Springer: New York, NY, USA, 2014; pp. 1–351. [CrossRef]
77. Zhang, L.; Wang, C.; Chen, K.; Zhong, W.; Xu, Y.; Molnár, I. Engineering the biosynthesis of fungal nonribosomal peptides. *Nat. Prod. Rep*, 2022; *Ahead of print*. [CrossRef]
78. Bills, G.; Li, Y.; Chen, L.; Yue, Q.; Niu, X.M.; An, Z. New insights into the echinocandins and other fungal non-ribosomal peptides and peptaibiotics. *Nat. Prod. Rep.* **2014**, *31*, 1348–1375. [CrossRef]
79. Qiao, K.; Zhou, H.; Xu, W.; Zhang, W.; Garg, N.; Tang, Y. A fungal nonribosomal peptide synthetase module that can synthesize thiopyrazines. *Org. Lett.* **2011**, *13*, 1758–1761. [CrossRef]
80. Cao, L.; Do, T.; Link, A.J. Mechanisms of action of ribosomally synthesized and posttranslationally modified peptides (RiPPs). *J. Ind. Microbiol. Biotechnol.* **2021**, *48*, kuab005. [CrossRef]
81. Arnison, P.G.; Bibb, M.J.; Bierbaum, G.; Bowers, A.A.; Bugni, T.S.; Bulaj, G.; Julio, A.C.; Dominic, J.C.; Gregory, L.C.; Jon, C. Ribosomally synthesized and post-translationally modified peptide natural products: Overview and recommendations for a universal nomenclature. *Nat. Prod. Rep.* **2013**, *30*, 108–160. [CrossRef]
82. Fu, Y.; Jaarsma, A.H.; Kuipers, O.P. Antiviral activities and applications of ribosomally synthesized and post-translationally modified peptides (RiPPs). *Cell. Mol. Life Sci.* **2021**, *78*, 3921–3940. [CrossRef]
83. Wu, C.; van der Donk, W.A. Engineering of new-to-nature ribosomally synthesized and post-translationally modified peptide natural products. *Curr. Opin. Chem. Biol.* **2021**, *69*, 221–231. [CrossRef]
84. Hetrick, K.J.; van der Donk, W.A. Ribosomally synthesized and post-translationally modified peptide natural product discovery in the genomic era. *Curr. Opin. Chem. Biol.* **2017**, *38*, 36–44. [CrossRef]
85. Yin, Z.; Dickschat, J.S. Engineering fungal terpene biosynthesis. *Nat. Prod. Rep*, 2022; *Ahead of print*. [CrossRef]
86. Wawrzyn, G.T.; Bloch, S.E.; Schmidt-Dannert, C. Discovery and characterization of terpenoid biosynthetic pathways of fungi. *Meth. Enzymol.* **2012**, *515*, 83–105. [CrossRef]
87. Schüller, A.; Wolansky, L.; Berger, H.; Studt, L.; Gacek-Matthews, A.; Sulyok, M.; Strauss, J. A novel fungal gene regulation system based on inducible VPR-dCas9 and nucleosome map-guided sgRNA positioning. *Appl. Microbiol. Biotechnol.* **2020**, *104*, 9801–9822. [CrossRef]
88. Goers, L.; Freemont, P.; Polizzi, K.M. Co-culture systems and technologies: Taking synthetic biology to the next level. *J. R. Soc. Interface* **2014**, *11*, 20140065. [CrossRef]
89. Fleming, A. On the antibacterial action of cultures of a *Penicillium*, with special reference to their use in the isolation of *B. influenza*. *Br. J. Exp. Pathol.* **1929**, *10*, 226–236. [CrossRef]
90. Houbraken, J.; Frisvad, J.C.; Samson, R.A. Fleming's penicillin producing strain is not *Penicillium chrysogenum* but *P. rubens*. *IMA Fungus* **2011**, *2*, 87–95. [CrossRef]
91. Patel, G.; Patil, M.D.; Soni, S.; Chisti, Y.; Banerjee, U.C. Production of mycophenolic acid by *Penicillium brevicompactum* using solid state fermentation. *Appl. Biochem. Biotechnol.* **2017**, *182*, 97–109. [CrossRef]
92. Godtfredsen, W.O.; Jahnsen, S.; Lorck, H.; Roholt, K.; Tybring, L. Fusidic acid: A new antibiotic. *Nature* **1962**, *4819*, 987. [CrossRef]
93. Fernandes, P. Fusidic acid: A bacterial elongation factor inhibitor for the oral treatment of acute and chronic *staphylococcal* infections. *Cold Spring Harb. Perspect. Med.* **2016**, *6*, a025437. [CrossRef] [PubMed]
94. Ni, J.; Guo, M.; Cao, Y.; Lei, L.; Liu, K.; Wang, B.; Lu, F.; Zhai, R.; Gao, X.; Yan, C.; et al. Discovery, synthesis of novel fusidic acid derivatives possessed amino-terminal groups at the 3-hydroxyl position with anticancer activity. *Eur. J. Med. Chem.* **2019**, *162*, 122–131. [CrossRef] [PubMed]
95. Hu, D.; Gao, H.; Yao, X.S. Biosynthesis of triterpenoid natural products. In *Comprehensive Natural Products III*, 3rd ed.; Hung-Wen, B.L., Tadhg, P.B., Eds.; Elsevier: Alpharetta, GA, USA, 2020; pp. 1–37.
96. Koble, H.; Traber, R. Directed biosynthesis of cyclosporins. *Eur. J. Appl. Microbiol. Biotechnol.* **1982**, *14*, 237–240. [CrossRef]
97. Agathos, S.N.; Parekh, R. Enhancement of cyclosporin production in a *Tolypocladium inflatum* strain after epichlorohydrin treatment. *J. Biotechnol.* **1990**, *13*, 73–81. [CrossRef]

98. Petersen, A.B.; Rønnest, M.H.; Larsen, T.O.; Clausen, M.H. The chemistry of griseofulvin. *Chem. Rev.* **2014**, *114*, 12088–12107. [CrossRef] [PubMed]
99. Desjardins, A.E.; Manandhar, H.K.; Plattner, R.D.; Manandhar, G.G.; Poling, S.M.; Maragos, C.M. *Fusarium* species from *Nepalese rice* and production of mycotoxins and gibberellic acid by selected species. *Microbiol. Appl. Environ.* **2000**, *66*, 1020–1025. [CrossRef]
100. Bailey, A.M.; Alberti, F.; Kilaru, S.; Collins, C.M.; de Mattos-Shipley, K.; Hartley, A.J.; Hayes, P.; Griffin, A.; Lazarus, C.M.; Cox, R.J.; et al. Identification and manipulation of the pleuromutilin gene cluster from *Clitopilus passeckerianus* for increased rapid antibiotic production. *Sci. Rep.* **2016**, *6*, 25202. [CrossRef]
101. Balba, H. Review of strobilurin fungicide chemicals. *J. Environ. Sci. Health B* **2007**, *42*, 441–451. [CrossRef]
102. Chen, L.; Yue, Q.; Li, Y.; Niu, X.; Xiang, M.; Wang, W.; Bills, G.F.; Liu, X.; An, Z. Engineering of *Glarea lozoyensis* for exclusive production of the pneumocandin B0 precursor of the antifungal drug caspofungin acetate. *Appl. Environ. Microbiol.* **2015**, *81*, 1550–1558. [CrossRef]
103. Men, P.; Wang, M.; Li, J.; Geng, C.; Huang, X.; Lu, X. Establishing an efficient genetic manipulation system for sulfated echinocandin producing fungus *Coleophoma empetri*. *Front. Microbiol.* **2021**, *12*, 734780. [CrossRef]
104. Wei, T.-Y.; Zheng, Y.; Wan, M.; Yang, S.; Tang, J.; Wu, Y.; Li, J.; Chen, S.-X. Analysis of FR901379 biosynthetic genes in Coleophoma empetri by CRISPR/Cas9-based genomic manipulation. *ACS Chem. Biol.* **2022**, *17*, 8, 2130–2141. [CrossRef]
105. Norris, T.; VanAlsten, J.; Hubbs, S.; Ewing, M.; Cai, W.; Bordner, J.; Jensen, O.G. Commercialization and Late-Stage Development of a Semisynthetic Antifungal API: Anidulafungin/D-Fructose (Eraxis). *Org. Process Res. Dev.* **2008**, *12*, 447–455. [CrossRef]
106. Peláez, F.; Cabello, A.; Platas, G.; Díez, M.T.; Del Val, A.G.; Basilio, A.; Martán, I.; Vicente, F.; Bills, G.E.; Giacobbe, R.A.; et al. The discovery of enfumafungin, a novel antifungal compound produced by an endophytic Hormonema species biological activity and taxonomy of the producing organisms. *Syst. Appl. Microbiol.* **2000**, *23*, 333–343. [CrossRef]
107. Apgar, J.M.; Wilkening, R.R.; Parker, D.L., Jr.; Meng, D.; Wildonger, K.J.; Sperbeck, D.; Greenlee, M.L.; Balkovec, J.M.; Flattery, A.M.; Abruzzo, G.K.; et al. Ibrexafungerp: An orally active beta-1,3-glucan synthesis inhibitor. *Bioorg. Med. Chem. Lett.* **2020**, *32*, 127661. [CrossRef]
108. Xie, L.G.; Zhu, G.D.; Shang, J.J.; Chen, X.M.; Zhang, C.T.; Ji, X.L.; Zhang, Q.; Wei, Y.L. An overview on the biological activity and anti-cancer mechanism of lovastatin. *Cell Signal.* **2021**, *87*, 110122. [CrossRef]
109. Baba, S.; Abe, Y.; Suzuki, T.; Ono, C.; Iwamoto, K.; Nihira, T.; Hosobuchi, M. Improvement of compactin (ML-236B) production by genetic engineering in compactin high-producing *Penicillium citrinum*. *Appl. Microbiol. Biotechnol.* **2009**, *83*, 697–704. [CrossRef]
110. Fujita, T.; Matsumoto, N.; Uchida, S.; Kohno, T.; Shimizu, T.; Hirose, R.; Yanada, K.; Kurio, W.; Watabe, K. Antibody against a novel, myriocin (ISP-I)-based immunosuppressant, FTY720. *Bioorg. Med. Chem. Lett.* **2000**, *10*, 337–339. [CrossRef]
111. Guruceaga, X.; Perez-Cuesta, U.; Abad-Diaz de Cerio, A.; Gonzalez, O.; Alonso, R.M.; Hernando, F.L.; Ramirez-Garcia, A.; Rementeria, A. Fumagillin, a mycotoxin of *Aspergillus fumigatus*: Biosynthesis, biological activities, detection, and applications. *Toxins* **2019**, *12*, 7. [CrossRef]
112. Amici, A.M.; Minghetti, A.; Scotti, T.; Spalla, C.; Tognoli, L. Production of ergotamine by a strain of *Claviceps purpurea* (Fr.) Tul. *Experientia* **1966**, *22*, 415–416. [CrossRef]
113. Meneghetti, F.; Ferraboschi, P.; Grisenti, P.; Reza, E.S.; Mori, M.; Ciceri, S. Crystallographic and NMR investigation of ergometrine and methylergometrine, two alkaloids from *Claviceps purpurea*. *Molecules* **2020**, *25*, 331. [CrossRef]
114. Amici, A.M.; Minghetti, A.; Scotti, T.; Spalla, C.; Tognoli, L. Production of peptide ergot alkaloids in submerged culture by three isolates of *Claviceps purpurea*. *Microbiol. Appl.* **1969**, *18*, 464–468. [CrossRef]
115. Kawasaki, Y. Mizoribine: A new approach in the treatment of renal disease. *Clin. Dev. Immunol.* **2009**, *2009*, 681482. [CrossRef] [PubMed]
116. Chen, J.; Arioka, M. Autophagy deficiency boosts the production of kojic acid in the filamentous fungus *Aspergillus oryzae*. *Biosci. Biotechnol. Biochem.* **2021**, *85*, 2429–2433. [CrossRef] [PubMed]
117. Krücken, J.; Harder, A.; Jeschke, P.; Holden-Dye, L.; O'Connor, V.; Welz, C.; Von Samson-Himmelstjerna, G. Anthelmintic cyclcooctadepsipeptides: Complex in structure and mode of action. *Trends Parasitol.* **2012**, *28*, 385–394. [CrossRef] [PubMed]
118. Kinkade, C.W.; Rivera-Núñez, Z.; Gorcyzca, L.; Aleksunes, L.M.; Barrett, E.S. Impact of fusarium-derived mycoestrogens on female reproduction: A systematic review. *Toxins* **2021**, *13*, 373. [CrossRef]
119. Bray, F.; Laversanne, M.; Weiderpass, E.; Soerjomataram, I. The ever-increasing importance of cancer as a leading cause of premature death worldwide. *Cancer* **2021**, *127*, 3029–3030. [CrossRef]
120. World Health Organization (WHO). *Global Health Estimates 2020: Deaths by Cause, Age, Sex, by Country and by Region, 2000–2019*; WHO: Geneva, Switzerland, 2020. Available online: https://www.who.int/health-topics/cancer#tab=tab_1 (accessed on 15 March 2022).
121. Sung, H.; Ferlay, J.; Siegel, R.L.; Laversanne, M.; Soerjomataram, I.; Jemal, A.; Bray, F. Global Cancer Statistics 2020: GLOBOCAN estimates of incidence and mortality worldwide for 36 cancers in 185 countries. *CA Cancer J. Clin.* **2021**, *71*, 209–249. [CrossRef]
122. INCA—Instituto Nacional do Câncer; Ministério da Saúde. Estatísticas de Câncer. Available online: https://www.inca.gov.br/numeros-de-cancer (accessed on 17 March 2022).
123. Somarelli, J.A.; Boddy, A.M.; Gardner, H.L.; DeWitt, S.B.; Tuohy, J.; Megquier, K.; Sheth, M.U.; Hsu, S.D.; Thorne, J.L.; London, C.A.; et al. Improving cancer drug discovery by studying cancer across the tree of life. *Mol. Biol. Evol.* **2020**, *37*, 11–17. [CrossRef]

24. Balde, E.S.; Andolfi, A.; Bruyère, C.; Cimmino, A.; Lamoral-Theys, D.; Vurro, M.; Damme, M.V.; Altomare, C.; Mathieu, V.; Kiss, R.; et al. Investigations of fungal secondary metabolites with potential anticancer activity. *J. Nat. Prod.* **2010**, *73*, 969–971. [CrossRef]
25. Strobel, G. The emergence of endophytic microbes and their biological promise. *J. Fungi* **2018**, *4*, 57. [CrossRef]
26. Chesnoff, S. The use of taxol as a trademark. *Nature* **1995**, *374*, 208. [CrossRef]
27. Sykłowska-Baranek, K.; Sygitowicz, G.; Maciejak-Jastrzębska, A.; Pietrosiuk, A.; Szakiel, A. Application of priming strategy for enhanced paclitaxel biosynthesis in *Taxus* x media hairy root cultures. *Cells* **2022**, *11*, 2062. [CrossRef]
28. Stierle, A.; Strobel, G.; Stierle, D.D. Taxol and taxane production by *Taxomyces andreanae*, an endophytic fungus of Pacific yew. *Science* **1993**, *260*, 214–216. [CrossRef]
29. Strobel, G.; Yang, X.; Sears, J.; Kramer, R.; Sidhu, R.S.; Hess, W.M.; Young, B. Taxol from *Pestalotiopsis microspora*, an endophytic fungus of *Taxus wallachiana*. *Microbiology* **1996**, *142*, 435–440. [CrossRef]
30. Gangadevi, V.; Muthumary, J. Taxol, an anticancer drug produced by an endophytic fungus *Bartalinia robillardoides* Tassi, isolated from a medicinal plant, *Aegle marmelos* Correa ex Roxb. *World J. Microbiol. Biotechnol.* **2008**, *24*, 717–724. [CrossRef]
31. Pandi, M.; Kumaran, R.S.; Choi, Y.; Kim, H.J.; Muthumary, J. Isolation and detection of taxol, an anticancer drug produced from *Lasiodiplodia theobromae*, an endophytic fungus of the medicinal plant *Morinda citrifolia*. *Afr. J. Biotechnol.* **2011**, *10*, 1428–1435. [CrossRef]
32. Kumaran, R.S.; Kim, H.J.; Hur, B.K. Taxol promising fungal endophyte, *Pestalotiopsis* species isolated from *Taxus cuspidata*. *J. Biosci. Bioeng.* **2010**, *110*, 541–546. [CrossRef]
33. Soliman, S.S.M.; Raizada, M.N. Interactions between co-habitating fungi elicit synthesis of Taxol from an endophytic fungus in host *Taxus* plants. *Front. Microbiol.* **2013**, *4*, 1–14. [CrossRef]
34. Palem, P.P.C.; Kuriakose, G.C.; Jayabaskaran, C. An Endophytic Fungus, *Talaromyces Radicus*, Isolated from *Catharanthus roseus*, Produces Vincristine and Vinblastine, Which Induce Apoptotic Cell Death. *PLoS ONE* **2015**, *10*, e0144476. [CrossRef]
35. Xiao, J.; Zhang, Q.; Gao, Y.Q.; Tang, J.J.; Zhang, A.L.; Gao, J.M. Secondary metabolites from the endophytic *Botryosphaeria dothidea* of *Melia azedarach* and their antifungal, antibacterial, antioxidant, and cytotoxic activities. *J. Agric. Food Chem.* **2014**, *62*, 3584–3590. [CrossRef]
36. Wang, Y.; Liu, H.X.; Chen, Y.C.; Sun, Z.H.; Li, H.H.; Li, S.N.; Yan, M.L.; Zhang, W.M. Two new metabolites from the endophytic fungus *Alternaria* sp. A744 derived from *Morinda officinalis*. *Molecules* **2017**, *22*, 765. [CrossRef] [PubMed]
37. Zhao, D.L.; Wang, D.; Tian, X.Y.; Cao, F.; Li, Y.Q.; Zhang, C.S. Anti-phytopathogenic and cytotoxic activities of crude extracts and secondary metabolites of marine-derived fungi. *Mar. Drugs* **2018**, *16*, 36. [CrossRef] [PubMed]
38. Bashyal, B.P.; Wijeratne, E.M.K.; Tillotson, J.; Arnold, A.E.; Chapman, E.; Gunatilaka, A.A.L. Chlorinated Dehydrocurvularins and Alterperylenepoxide A from *Alternaria* sp. AST0039, a Fungal Endophyte of *Astragalus lentiginosus*. *J. Nat. Prod.* **2017**, *80*, 427–433. [CrossRef] [PubMed]
39. Rönsberg, D.; Debbab, A.; Mándi, A.; Vasylyeva, V.; Böhler, P.; Stork, B.; Engelke, L.; Hamacher, A.; Sawadogo, R.; Diederich, M.; et al. Pro-apoptotic and immunostimulatory tetrahydroxanthone dimers from the endophytic fungus *Phomopsis longicolla*. *J. Org. Chem.* **2013**, *78*, 12409–12425. [CrossRef] [PubMed]
40. Ebrahim, W.; Aly, A.H.; Wray, V.; Mándi, A.; Teiten, M.H.; Gaascht, F.; Orlikova, B.; Kassack, M.U.; Lin, W.; Diederich, M.; et al. Embellicines A and B: Absolute configuration and NF-κB tran-scriptional inhibitory activity. *J. Med. Chem.* **2013**, *56*, 2991–2999. [CrossRef]
41. Uesugi, S.; Fujisawa, N.; Yoshida, J.; Watanabe, M.; Dan, S.; Yamori, T.; Shiono, Y.; Kimura, K. Pyrrocidine A, a metabolite of endophytic fungi, has a potent apoptosis-inducing activity against HL60 cells through caspase activation via the Michael addition. *J. Antibiot.* **2016**, *69*, 133–140. [CrossRef]
42. Gallo, M.B.C.; Falso, M.J.S.; Balem, F.; Menezes, D.; Rocha, N.; Balachandran, R.; Sturgeon, T.S.; Pupo, M.T.; Day, B.W. The anti-promyelocytic leukemia mode of action of two endophytic secondary metabolites unveiled by a proteomic approach. *Planta Med.* **2014**, *80*, 473–481. [CrossRef] [PubMed]
43. Lin, T.; Wang, G.; Zhou, Y.; Zeng, D.; Liu, X.; Ding, R.; Jiang, X.; Zhu, D.; Shan, W.; Chen, H. Structure elucidation and biological activity of two new trichothecenes from an endophyte, *Myrothecium roridum*. *J. Agric. Food Chem.* **2014**, *62*, 5993–6000. [CrossRef] [PubMed]
44. Liu, Y.; Stuhldreier, F.; Kurtán, T.; Mándi, A.; Arumugam, S.; Lin, W.; Stork, B.; Wesselborg, S.; Weber, H.; Henrich, B.; et al. Daldinone derivatives from the mangrove derived endophytic fungus *Annulohypoxylon* sp. *RSC Adv.* **2017**, *7*, 5381–5393. [CrossRef]
45. Koul, M.; Kumar, A.; Deshidi, R.; Sharma, V.; Sinqh, R.D.; Sinqh, J.; Sharma, P.R.; Shah, B.A.; Jaqlan, S.; Sinqh, S. Cladosporol. A triggers apoptosis sensitivity by ROS-mediated autophagic flux in human breast cancer cells. *BMC Cell Biol.* **2017**, *18*, 26. [CrossRef] [PubMed]
46. Zhou, L.; Qin, J.; Ma, L.; Li, H.; Li, L.; Ning, C.; Gao, W.; Yu, H.; Han, L. Rosoloactone: A natural diterpenoid inducing apoptosis in human cervical cancer cells through endoplasmic reticulum stress and mitochondrial damage. *Biomed. Pharmacother.* **2017**, *95*, 355–362. [CrossRef] [PubMed]
47. Li, C.S.; Li, X.M.; Gao, S.S.; Lu, Y.H.; Wang, B.G. Cytotoxic anthranilic acid derivatives from deep sea sediment-derived fungus *Penicillium paneum* SD-44. *Mar. Drugs* **2013**, *11*, 3068–3076. [CrossRef] [PubMed]

148. Zhan, J.; Burns, A.M.; Liu, M.X.; Faeth, S.H.; Gunatilaka, A.A.L. Search for cell motility and angiogenesis inhibitors with potential anticancer activity: Beauvericin and other constituents of two endophytic strains of *Fusarium oxysporum*. *J. Nat. Prod.* **2007**, *70*, 227–232. [CrossRef] [PubMed]
149. Wang, Q.X.; Li, S.F.; Zhao, F.; Dai, H.Q.; Bao, L.; Ding, R.; Gao, H.; Zhang, L.X.; Wen, H.A.; Liu, H.W. Chemical constituents from endophytic fungus *Fusarium oxysporum*. *Fitoterapia* **2011**, *82*, 777–781. [CrossRef] [PubMed]
150. Chen, J.; Huo, L.N.; Gao, Y.; Zhang, Y.L.; Chen, Y. Two new N-acetyl-glucosamine derivatives from the medical algae-derived endophytic fungus *Penicillium chrysogenum*. *Nat. Prod. Res.* **2022**, *36*, 3988–3991. [CrossRef]
151. Paek, S.M. Recent Synthesis and Discovery of brefeldin A analogs. *Mar. Drugs* **2018**, *16*, 133. [CrossRef]
152. Horiuchi, M.; Tokuda, H.; Ohnishi, K.; Yamashita, M.; Nishino, H.; Maoka, T. Porritoxins, metabolites of *Alternaria porri*, as anti-tumor-promoting active compounds. *Nat. Prod. Res.* **2006**, *20*, 161–166. [CrossRef]
153. Huang, Y.F.; Li, L.H.; Tian, L.; Qiao, L.; Hua, H.M.; Pei, Y.H. Sg17-1-4, a novel isocoumarin from a marine fungus *Alternaria tenuis* Sg17-1. *J. Antibiot.* **2006**, *59*, 355–357. [CrossRef]
154. Aly, A.H.; Edrada-Ebel, R.; Indriani, I.D.; Wray, V.; Muller, W.E.; Totzke, F.; Zirrgiebel, U.; Schachtele, C.; Kubbutat, M.H.; Lin, W.H.; et al. Cytotoxic metabolites from the fungal endophyte *Alternaria* sp. and their subsequent detection in its host plant *Polygonum senegalense*. *J. Nat. Prod.* **2008**, *71*, 972–980. [CrossRef]
155. Liu, Z.Z.; Zhu, J.; Sun, B.; Liu, S.; Geng, S.; Liu, X.; Li, C.L. Alternol inhibits proliferation and induces apoptosis in mouse lymphocyte leukemia (L1210) cells. *Mol. Cell Biochem.* **2007**, *306*, 115–122. [CrossRef]
156. Zhao, J.; Li, C.; Wang, W.; Zhao, C.; Luo, M.; Mu, F.; Fu, Y.; Zu, Y.; Yao, M. *Hypocrea lixii*, novel endophytic fungi producing anticancer agent cajanol, isolated from pigeon pea (*Cajanus cajan* [L.] Millsp.). *J Appl. Microbiol.* **2013**, *115*, 102–113. [CrossRef]
157. Du, L.; Zhu, T.J.; Li, L.Y.; Cai, S.X.; Zhao, B.Y.; Gu, Q.Q. Cytotoxic sorbicillinoids and bisorbicillinoids from a marine-derived fungus *Trichoderma* sp. *Chem. Pharm. Bull.* **2009**, *57*, 220–223. [CrossRef]
158. Lan, W.J.; Zhao, Y.; Xie, Z.L.; Liang, L.Z.; Shao, W.Y.; Zhu, L.P.; Yang, D.P.; Zhu, X.F.; Li, H.-J. Novel sorbicillin analogues from the marine fungus *Trichoderma sp.* associated with the seastar *Acanthaster planci*. *Nat. Prod. Commun.* **2012**, *7*, 1337–1340. [CrossRef]
159. Ma, L.; Liu, W.; Huang, Y.; Rong, X. Two acid sorbicillin analogues from saline lands-derived fungus *Trichoderma* sp. *J. Antibiot.* **2011**, *64*, 645–647. [CrossRef]
160. Liu, W.; Gu, Q.; Zhu, W.; Cui, C.; Fan, G. Two new benzoquinone derivatives and two new bisorbicillinoids were isolated from a marine-derived fungus *Penicillium terrestre*. *J. Antibiot.* **2005**, *58*, 441–446. [CrossRef]
161. Liu, W.; Gu, Q.; Zhu, W.; Cui, C.; Fan, G. Dihydrotrichodimerol and tetrahydrotrichodimerol, two new bisorbicillinoids, from a marine-derived *Penicillium terrestre*. *J. Antibiot.* **2005**, *58*, 621–624. [CrossRef]
162. Ying, Y.M.; Zhan, Z.J.; Ding, Z.S.; Shan, W.G. Bioactive metabolites from *Penicillium sp.* P-1, a fungal endophyte in *Huperzia serrata*. *Chem. Nat. Compd.* **2011**, *47*, 541–544. [CrossRef]
163. Li, D.; Cai, S.; Zhu, T.; Wang, F.; Xiao, X.; Gu, Q. New cytotoxic metabolites from a deep-sea-derived fungus, *Phialocephala* sp., strain FL30r. *Chem. Biodivers.* **2011**, *8*, 895–901. [CrossRef]
164. Li, D.; Cai, S.; Zhu, T.; Wang, F.; Xiao, X.; Gu, Q. Three new sorbicillin trimers, trisorbicillinones B, C, and D, from a deep ocean sediment derived fungus, *Phialocephala* sp. FL30r. *Tetrahedron* **2010**, *66*, 5101–5106. [CrossRef]
165. Li, D.; Wang, F.; Xiao, X.; Fang, Y.; Zhu, T.; Gu, Q.; Zhu, W. Trisorbicillinone A, a novel sorbicillin trimer, from a deep sea fungus, *Phialocephala sp.* FL30r. *Tetrahedron Lett.* **2007**, *48*, 5235–5238. [CrossRef]
166. Li, D.; Wang, F.; Cai, S.; Zeng, X.; Xiao, X.; Gu, Q.; Zhu, W. Two new bisorbicillinoids isolated from a deep-sea fungus, *Phialocephala* sp. FL30r. *J. Antibiot.* **2007**, *60*, 317–320. [CrossRef] [PubMed]
167. Yao, Y.; Li, J.; Jiang, C.S.; Zhao, X.X.; Miao, Z.H.; Liu, H.T.; Zheng, P.; Yao, W.X.; Li, W.Q. Trichodimerol and sorbicillin induced apoptosis of HL-60 cells is mediated by reactive oxygen species. *Pharmazie* **2015**, *70*, 394–398. [PubMed]
168. Bringmann, G.; Lang, G.; Gulder, T.A.M.; Tsuruta, H.; Mühlbacher, J.; Maksimenka, K.; Steffens, S.; Schaumann, K.; Stohr, R.; Wiese, J.; et al. The first sorbicillinoid alkaloids, the antileukemic sorbicillactones A and B, from a sponge-derived *Penicillium chrysogenum* strain. *Tetrahedron* **2005**, *61*, 7252–7265. [CrossRef]
169. Rehman, S.U.; Yang, L.J.; Zhang, Y.H.; Wu, J.S.; Shi, T.; Haider, W.; Shao, C.L.; Wang, C.Y. Sorbicillinoid derivatives from sponge-derived fungus *Trichoderma reesei* (HN-2016-018). *Front. Microbiol.* **2020**, *11*, 1334. [CrossRef] [PubMed]
170. Li, X.G.; Pan, W.D.; Lou, H.Y.; Liu, R.M.; Xiao, J.H.; Zhong, J.J. New cytochalasins from medicinal macrofungus *Crodyceps taii* and their inhibitory against human cancer cells activities. *Bioorg. Med. Chem. Lett.* **2015**, *25*, 1823–1826. [CrossRef]
171. Arora, D.; Sharma, N.; Singamaneni, V.; Sharma, V.; Kushwaha, M.; Abrol, V.; Guru, S.; Sharma, S.; Gupta, A.P.; Bhushan, S.; et al. Isolation and characterization of bioactive metabolites from *Xylaria psidii*, an endophytic fungus of the medicinal plant *Aegle marmelos* and their role in mitochondrial dependent apoptosis against pancreatic cancer cells. *Phytomedicine* **2016**, *23*, 1312–1320. [CrossRef]
172. Yedukondalu, N.; Arora, P.; Wadhwa, B.; Malik, F.A.; Vishwakarma, R.A.; Gupta, V.K.; Riyaz-Ul-Hassan, S.; Ali, A. Diapolic acid A–B from an endophytic fungus, *Diaporthe terebinthifolii* depicting antimicrobial and cytotoxic activity. *J. Antibiot.* **2017**, *70*, 212–215. [CrossRef]
173. Luo, D.Q.; Zhang, L.; Shi, B.Z.; Song, X.M. Two new oxysporone derivatives from the fermentation broth of the endophytic plant fungus *Pestalotiopsis karstenii* isolated from stems of *Camellia sasanqua*. *Molecules* **2012**, *17*, 8554–8560. [CrossRef]

174. Momesso, L.S.; Kawano, C.Y.; Ribeiro, P.H.; Nomizo, A.; Goldman, G.H.; Pupo, T.M. Chaetoglobosins produced by *Chaetomium globosum*, endophytic fungus found in association with *Viguiera robusta Gardn* (Asteraceae). *Quim. Nova* **2008**, *31*, 1680–1685. [CrossRef]
175. Li, H.; Xiao, J.; Gao, Y.; Tang, J.J.; Zhang, A.L.; Gao, J.M. Chaetoglobosins from *Chaetomium globosum*, an endophytic fungus in *Ginkgo biloba*, and their phytotoxic and cytotoxic activities. *J. Agric. Food Chem.* **2014**, *62*, 3734–3741. [CrossRef]
176. Hawas, U.W.; Abou El-Kassem, L.; Ahmed, E.F.; Alghamdi, R.A. Bioactive sulfonyl metabolites from the Red Sea endophytic fungus *Penicillium aculeatum*. *Nat. Prod. Res.* **2021**, *36*, 1–9. [CrossRef]
177. Cai, X.Y.; Wang, J.P.; Shu, Y.; Hu, J.T.; Sun, C.T.; Cai, L.; Ding, Z.T. A new cytotoxic indole alkaloid from the fungus *Penicillium polonicum* TY12. *Nat. Prod. Res.* **2020**, *36*, 2270–2276. [CrossRef]
178. Gubiani, J.R.; Oliveira, M.C.S.; Neponuceno, R.A.R.; Camargo, M.J.; Garcez, W.S.; Biz, A.R.; Soares, M.A.; Araujo, A.R.; Bolzani, V.D.S.; Lisboa, H.C.F.; et al. Cytotoxic prenylated indole alkaloid produced by the endophytic fungus *Aspergillus terreus* P63. *Phytochem. Lett.* **2019**, *32*, 162–167. [CrossRef]
179. Scopel, M.; Mothes, B.; Lerner, C.B.; Henriques, A.T.; Macedo, A.J.; Abraham, W.R. Arvoredol—An unusual chlorinated and biofilm inhibiting polyketide from a marine *Penicillium* sp. of the Brazilian coast. *Phytochem. Lett.* **2017**, *20*, 73–76. [CrossRef]
180. Adorisio, S.; Fierabracci, A.; Muscari, I.; Liberati, A.M.; Cannarile, L.; Thuy, T.T.; Sung, T.V.; Sohrab, H.; Hasan, C.M.; Delfino, D.V.; et al. Fusarubin and anhydrofusarubin isolated from a *Cladosporium* Species inhibit cell growth in human cancer cell lines. *Toxins* **2019**, *11*, 503. [CrossRef]
181. Gubiani, J.R.; Bernardi, D.I.; de Pala, C.C.P.; Seleghim, M.H.R.; Ferreira, A.G.; Batista, A.N.L.; Batista, J.M., Jr.; Oliveira, L.F.P.; Lira, S.P.; Burdette, J.E.; et al. Absolute configuration of cytotoxic anthraquinones from a Brazilian cave soil-derived fungus, *Aspergillus* sp. SDC28. *Arch. Pharm.* **2022**, *355*, e2100441. [CrossRef]
182. Kumar, A.; Patil, D.; Rajamohanan, P.R.; Ahmad, A. Isolation, purification and characterization of vinblastine and vincristine from endophytic fungus *Fusarium oxysporum* isolated from *Catharanthus roseus*. *PLoS ONE* **2013**, *8*, e71805. [CrossRef]
183. Yao, Q.; Wang, J.; Zhang, X.; Nong, X.; Xu, X.; Qi, S. Cytotoxic polyketides from the deep-sea-derived fungus *Engyodontium album* DFFSCS021. *Mar. Drugs* **2014**, *12*, 5902–5915. [CrossRef]
184. Rosa, L.H.; Gonçalves, V.N.; Caligiorne, R.B.; Alves, T.A.M.; Rabello, A.; Sales, P.A.; Romanha, A.J.; Sobral, M.E.G.; Rosa, C.A.; Zani, C.L. Leishmanicidal, trypanocidal, and cytotoxic activities of endophytic fungi associated with bioactive plants in Brazil. *Braz. J. Microbiol.* **2010**, *41*, 420–430. [CrossRef]
185. Alves, D.S.; Wildson, S.; Dayanne, F.; Francisco, V.; Fábio, M.S. Antioxidant, anticholinesterasic and cytotoxic activities of metabolites of endophytic fungi. *Braz. J. Dev.* **2020**, *6*, 73684–73691. [CrossRef]
186. Lou, J.; Fu, L.; Peng, Y.; Zhou, L. Metabolites from *Alternaria* fungi and their bioactivities. *Molecules* **2013**, *18*, 5891–5935. [CrossRef] [PubMed]
187. Shaaban, M.; Shaaban, K.A.; Abdel-Aziz, M.S. Seven naphtho-γ-pyrones from the marine-derived fungus *Alternaria alternata*: Structure elucidation and biological properties. *Org. Med. Chem. Lett.* **2012**, *2*, 6. [CrossRef] [PubMed]
188. Musetti, R.; Polizzotto, R.; Vecchione, A.; Borselli, S.; Zulini, L.; D'Ambrosio, M.; Di Toppi, L.S.; Pertot, I. Antifungal activity of diketopiperazines extracted from *Alternaria alternate* against *Plasmopara viticola*: An ultrastructural study. *Micron* **2007**, *38*, 643–650. [CrossRef] [PubMed]
189. Abbas, H.K.; Riley, R.T. The presence and phytotoxicity of fumonisins and AAL-toxin in *Alternaria alternata*. *Toxicon* **1996**, *34*, 133–136. [CrossRef]
190. Ostry, V. *Alternaria mycotoxins*: An overview of chemical characterization, producers, toxicity, analysis and occurrence in foodstuffs. *World Mycotoxin J.* **2008**, *1*, 175–188. [CrossRef]
191. Schreck, I.; Deigendesch, U.; Burkhardt, B.; Marko, D.; Weiss, C. The Alternaria mycotoxins alternariol and alternariol methyl ether induce cytochrome P450 1A1 and apoptosis in murine hepatoma cells dependent on the aryl hydrocarbon receptor. *Arch. Toxicol.* **2012**, *86*, 625–632. [CrossRef]
192. Mizushina, Y.; Maeda, N.; Kuriyama, I.; Yoshida, H. Dehydroalternusin is a specific inhibitor of mammalian DNA polymerase α. *Expert Opin. Inv. Drugs* **2011**, *20*, 1523–1534. [CrossRef]
193. Fernandes, M.; Dos, R.V.; Silva, T.A.C.; Pfenning, L.H.; Costa-Neto, C.M.; Heinrich, T.A.; Alencar, S.M.; Lima, M.A.; Ikegaki, M. Biological activities of the fermentation extract of the endophytic fungus *Alternaria alternata* isolated from *Coffea arabica* L. Braz. *J. Pharm. Sci.* **2009**, *45*, 677–685. [CrossRef]
194. Bhimba, B.V.; Franco, D.A.; Mathew, J.M.; Jose, G.M.; Joel, E.L.; Thangaraj, M. Anticancer and antimicrobial activity of mangrove derived fungi *Hypocrea lixii* VB1. *Chin. J. Nat. Med.* **2012**, *10*, 77–80. [CrossRef]
195. Duker-Eshun, G.; Jaroszewski, J.W.; Asomaning, W.A.; Oppong-Boachie, F.; Christensen, S.B. Antiplasmodial constituents of *Cajanus cajan*. *Phytother. Res.* **2004**, *18*, 128–130. [CrossRef]
196. Liu, X.L.; Zhang, X.J.; Fu, Y.J.; Zu, Y.G.; Wu, N.; Liang, L.; Efferth, T. Cajanol inhibits the growth of *Escherichia coli* and *Staphylococcus aureus* by acting on membrane and DNA damage. *Planta Med.* **2011**, *77*, 158–163. [CrossRef]
197. Cram, D.J. Mold metabolites. II. The structure of sorbicillin, a pigment produced by the mold *Penicillium notatum*. *J. Am. Chem. Soc.* **1948**, *70*, 4240–4243. [CrossRef]
198. Hou, X.; Zhang, X.; Xue, M.; Zhao, Z.; Zhang, H.; Xu, D.; Lai, D.; Zhou, L. Recent advances in sorbicillinoids from fungi and their bioactivities (covering 2016–2021). *J. Fungi* **2022**, *8*, 62. [CrossRef]

199. Meng, J.; Wang, X.; Xu, D.; Fu, X.; Zhang, X.; Lai, D.; Zhou, L.; Zhang, G. Sorbicillinoids from fungi and their bioactivities. *Molecules* **2016**, *21*, 715. [CrossRef]
200. Harned, A.M.; Volp, K.A. The sorbicillinoid family of natural products: Isolation, biosynthesis and synthetic studies. *Nat. Prod. Rep.* **2011**, *28*, 1790–1810. [CrossRef]
201. Ei-Elimat, T.; Raja, H.A.; Figueroa, M.; Swanson, S.M.; Falkinham, J.O.; Lucas, D.M.; Grever, M.R.; Wani, M.C.; Pearce, C.J.; Oberlies, N.H. Sorbicillinoid analogs with cytotoxic and selective anti *Aspergillus* activities from *Scytalidium album*. *J. Antibiot.* **2015**, *68*, 191–196. [CrossRef]
202. Lee, D.; Lee, J.H.; Cai, X.F.; Shin, J.C.; Lee, K.; Hong, Y.S.; Lee, J.J. Fungal metabolites, sorbicillinoid polyketides and their effects on the activation of peroxisome proliferator-activated receptor γ. *J. Antibiot.* **2005**, *58*, 615–620. [CrossRef]
203. Santos, S.N.; Ferraris, F.K.; de Souza, A.O.; Henriques, M.G.; Melo, I.S. Endophytic fungi from *Combretum leprosum* with potential anticancer and antifungal activity. *Symbiosis* **2012**, *58*, 109–117. [CrossRef]
204. Liu, S.; Guo, L.; Che, Y.; Liu, L. Pestaloficiols Q–S from the plant endophytic fungus *Pestalotiopsis fici*. *Phytotherapy* **2013**, *85*, 114–118. [CrossRef]
205. Mady, M.S.; Mohyeldin, M.M.; Ebrahim, H.Y.; Elsayed, H.E.; Houssen, W.E.; Haggag, E.G.; Soliman, R.F.; El Sayed, K.F. The indole *alkaloid meleagrin*, from the olive tree endophytic fungus *Penicillium chrysogenum*, as a novel lead for the control of c-Met-dependent breast cancer proliferation, migration and invasion. *Bioorg. Med. Chem.* **2016**, *24*, 113–122. [CrossRef]
206. Miethke, M.; Pieroni, M.; Weber, T.; Brönstrup, M.; Hammann, P.; Halby, L.; Arimondo, P.B.; Glaser, P.; Aigle, B.; Mülleet, R.; et al. Towards the sustainable discovery and development of new antibiotics. *Nat. Rev. Chem.* **2021**, *5*, 726–749. [CrossRef] [PubMed]
207. Yu, G.; Sun, Z.; Peng, J.; Zhu, M.; Che, Q.; Zhang, G.; Li, D. Secondary metabolites produced by combined culture of *Penicillium crustosum* and a *Xylaria* sp. *J. Nat. Prod.* **2019**, *82*, 2013–2017. [CrossRef] [PubMed]
208. Qin, Y.Y.; Huang, X.S.; Liu, X.B.; Mo, T.X.; Xu, Z.L.; Li, B.C.; Qin, X.Y.; Li, J.; Schaberle, T.F.; Yang, R.Y. Three new andrastin derivatives from the endophytic fungus *Penicillium vulpinum*. *Nat. Prod. Res.* **2020**, *36*, 3262–3270. [CrossRef] [PubMed]
209. Pinheiro, E.A.; Borges, F.C.; Pina, J.R.; Ferreira, L.R.; Cordeiro, J.S.; Carvalho, J.M.; Feitosa, A.O.; Campos, F.R.; Barison, A.; Souza, A.D.L.; et al. Annularins I and J: New metabolites isolated from endophytic fungus *Exserohilum rostratum*. *J. Braz. Chem. Soc.* **2016**, *27*, 1432–1436. [CrossRef]
210. Pina, J.R.S.; Silva-Silva, J.V.; Carvalho, J.M.; Bitencourt, H.R.; Watanabe, L.A.; Fernandes, J.M.P.; Souza, G.E.d.; Aguiar, A.C.C.; Guido, R.V.C.; Almeida-Souza, F.; et al. Antiprotozoal and antibacterial activity of ravenelin, a xanthone isolated from the endophytic fungus *Exserohilum rostratum*. *Molecules* **2021**, *26*, 3339. [CrossRef] [PubMed]
211. Pang, X.J.; Zhang, S.B.; Chen, H.L.; Zhao, W.T.; Yang, D.F.; Xian, P.J.; Xu, L.L.; Tao, Y.D.; Fu, H.Y.; Yang, X.L. Emericelactones A-D: Four novel polyketides produced by *Emericella* sp. XL 029, a fungus associated the leaves of *Panax notoginseng*. *Tetrahedron Lett.* **2018**, *59*, 4566–4570. [CrossRef]
212. Pang, X.J.; Zhang, S.B.; Xian, P.J.; Wu, X.; Yang, D.F.; Fu, H.Y.; Yang, X.L. Emericellins A and B: Two sesquiterpenoids with an unprecedented tricyclo[4,4,2,1]hendecane scaffold from the liquid cultures of endophytic fungus *Emericella* sp. XL 029. *Fitoterapia* **2018**, *131*, 55–58. [CrossRef]
213. Duan, X.X.; Qin, D.; Song, H.C.; Gao, T.C.; Zuo, S.H.; Yan, X.; Wang, J.Q.; Ding, X.; Di, Y.T.; Dong, J.Y. Irpexlacte AD, four new bioactive metabolites of endophytic fungus *Irpex lacteus* DR10-1 from the waterlogging tolerant plant *Distylium chinense*. *Phytochem. Lett.* **2019**, *32*, 151–156. [CrossRef]
214. Chen, J.; Bai, X.; Hua, Y.; Zhang, H.; Wang, H. Fusariumins C and D, two novel antimicrobial agents from *Fusarium oxysporum* ZZP-R1 symbiotic on *Rumex madaio* Makino. *Phytotherapy* **2019**, *134*, 1–4. [CrossRef]
215. Ibrahim, S.R.; Mohamed, G.A.; Al Haidari, R.A.; Zayed, M.F.; El-Kholy, A.A.; Elkhayat, E.S.; Ross, S.A. Fusarithioamide B, a new benzamide derivative from the endophytic fungus *Fusarium chlamydosporium* with potent cytotoxic and antimicrobial activities. *Bioorg. Med. Chem.* **2018**, *26*, 786–790. [CrossRef]
216. Khan, N.; Afroz, F.; Begum, M.N.; Rony, S.R.; Sharmin, S.; Moni, F.; Sohrab, M.H. Endophytic *Fusarium solani*: A rich source of cytotoxic and antimicrobial napthaquinone and aza-anthraquinone derivatives. *Toxicol. Rep.* **2018**, *5*, 970–976. [CrossRef]
217. Zhang, H.; Ruan, C.; Bai, X.; Zhang, M.; Zhu, S.; Jiang, Y. Isolation and identification of the antimicrobial agent beauvericin from the endophytic F5-19 with NMR and ESI-MS/MS. *BioMed. Res. Int.* **2016**, *2016*, 1084670. [CrossRef]
218. Deng, Z.; Li, C.; Luo, D.; Teng, P.; Guo, Z.; Tu, X.; Zou, K.; Gong, D. A new cinnamic acid derivative from plant-derived endophytic fungus *Pyronema* sp. *Nat. Prod. Res.* **2017**, *31*, 2413–2419. [CrossRef]
219. Shin, T.Y.; Bae, S.M.; Woo, S.D. Screening and characterization of antimicrobial substances originated from entomopathogenic fungi. *J. Asia-Pac. Entomol.* **2016**, *19*, 1053–1059. [CrossRef]
220. Mohamed, G.A.; Ibrahim, S.R.M.; Alhakamy, N.A.; Aljohani, O.S. Fusaroxazin, a novel cytotoxic and antimicrobial xanthone derivative from *Fusarium oxysporum*. *Nat. Prod. Res.* **2020**, *36*, 952–960. [CrossRef]
221. Jayasinghe, L.; Abbas, H.K.; Jacob, M.R.; Herath, W.H.; Nanayakkara, N.P. N-Methyl-4-hydroxy-2-pyridinone analogues from *Fusarium oxysporum*. *J. Nat. Prod.* **2006**, *69*, 439–442. [CrossRef]
222. Sudharshana, T.N.; Venkatesh, H.N.; Nayana, B.; Manjunath, K.; Mohana, D.C. Anti-microbial and anti-mycotoxigenic activities of endophytic *Alternaria alternata* isolated from *Catharanthus roseus* (L.) G. Don.: Molecular characterisation and bioactive compound isolation. *Mycology* **2018**, *10*, 40–48. [CrossRef]
223. Macelroy, R.D. Some comments on the evolution of extremophiles. *Biosystems* **1974**, *6*, 74–75. [CrossRef]

124. Dhakar, K.; Pandey, A. Wide pH range tolerance in extremophiles: Towards understanding an important phenomenon for future biotechnology. *Appl. Microbiol. Biotechnol.* **2016**, *100*, 2499–2510. [CrossRef]
125. Zhang, X.; Li, S.J.; Li, J.J.; Liang, Z.Z.; Zhao, C.Q. Novel natural products from extremophilic fungi. *Mar. Drugs* **2018**, *16*, 194. [CrossRef]
126. Coker, J.A. Extremophiles and biotechnology: Current uses and prospects. *F1000Research* **2016**, *5*, 396. [CrossRef] [PubMed]
127. Anwar, U.B.; Zwar, I.P.; de Souza, A.O. Biomolecules produced by extremophiles microorganisms and recent discoveries. In *New and Future Developments in Microbial Biotechnology and Bioengineering: Microbial Biomolecules: Properties, Relevance, and Their Translational Applications*; Rodrigues, A.G., Ed.; Elsevier: Amsterdam, The Netherlands, 2018; pp. 247–269. [CrossRef]
128. Baranova, A.A.; Alferova, V.A.; Korshuna, V.A.; Tyurina, A.P. Antibiotics from extremophilic micromycetes. *Russ. J. Bioorg. Chem.* **2020**, *46*, 903–971. [CrossRef] [PubMed]
129. Rogozhin, E.A.; Sadykova, V.S.; Baranova, A.A.; Vasilchenko, A.S.; Lushpa, V.A.; Mineev, K.S.; Georgieva, M.L.; Kul'ko, A.B.; Krasheninnikov, M.E.; Andreev, Y.A. A novel lipopeptaibol emericellipsin A with antimicrobial and antitumor activity produced by the extremophilic fungus *Emericellopsis alkalina*. *Molecules* **2018**, *23*, 2785. [CrossRef] [PubMed]
130. Selbmann, L.; Benkő, Z.; Coleine, C.; de Hoog, S.; Donati, C.; Druzhinina, I.; Emri, T.; Ettinger, C.L.; Gladfelter, A.S.; Gorbushina, A.A.; et al. Shed light in the DaRk LineagES of the fungal tree of life—Stres. *Life* **2020**, *10*, 362. [CrossRef]
131. Orwa, P.; Mugambi, G.; Wekesa, V.; Mwirichia, R. Isolation of haloalkaliphilic fungi from Lake Magadi in Kenya. *Heliyon* **2020**, *6*, e02823. [CrossRef]
132. Menezes, G.C.A.; Porto, B.A.; Amorim, S.S.; Zani, C.L.; Alves, T.M.A.; Sales, P.A., Jr.; Murta, S.M.F.; Simões, J.C.; Cota, B.B.; Rosa, C.A.; et al. Fungi in glacial ice of Antarctica: Diversity, distribution and bioprospecting of bioactive compounds. *Extremophiles* **2020**, *24*, 367–376. [CrossRef]
133. Khan, I.; Zhang, H.; Liu, W.; Zhang, L.; Peng, F.; Zhang, Q.; Zhang, G.; Zhang, W.; Zhang, C. Identification and bioactivity evaluation of secondary metabolites from Antarctic-derived *Penicillium chrysogenum* CCTCC M 2020019. *RSC Adv.* **2020**, *10*, 20738–20744. [CrossRef]
134. Wang, C.; Tang, S.; Cao, S. Antimicrobial compounds from marine fungi. *Phytochem. Rev.* **2021**, *20*, 85–117. [CrossRef]
135. Van Boeckel, T.P.; Brower, C.; Gilbert, M.; Grenfell, B.T.; Levin, S.A.; Robinson, T.P.; Teillant, A.; Laxminarayan, R. Global trends in antimicrobial use in food animals. *Proc. Natl. Acad. Sci. USA* **2015**, *112*, 5649–5654. [CrossRef]

MDPI AG
Grosspeteranlage 5
4052 Basel
Switzerland
Tel.: +41 61 683 77 34

Antibiotics Editorial Office
E-mail: antibiotics@mdpi.com
www.mdpi.com/journal/antibiotics

www.ingramcontent.com/pod-product-compliance
Lightning Source LLC
LaVergne TN
LVHW070452061126
851858LV00011B/220
* 9 7 8 3 7 2 5 8 8 2 1 9 9 *